Economic and Medicinal Plant Research

Volume 1

Economic and Medicinal Plant Research

Volume I

Edited by

H. WAGNER
Institut für Pharmazeutische Biologie
der Universität München
München, West Germany

HIROSHI HIKINO
Pharmaceutical Institute
Tohoku University
Sendai, Japan

NORMAN R. FARNSWORTH
Program for Collaborative Research in
the Pharmaceutical Sciences
College of Pharmacy, Health Sciences Center
University of Illinois at Chicago
Chicago, Illinois, U.S.A.

1985

ACADEMIC PRESS

(Harcourt Brace Jovanovich, Publishers)

London   Orlando   San Diego   New York
Toronto   Montreal   Sydney   Tokyo

ACADEMIC PRESS INC. (LONDON) LTD.
24–28 Oval Road
LONDON NW1 7DX

United States Edition published by
ACADEMIC PRESS, INC.
Orlando, Florida 32887

BRITISH LIBRARY CATALOGUING IN PUBLICATION DATA
Economic and medicinal plant research.
    1. Plants, Useful—-Research
    I. Wagner, H.   II. Hikino, Hiroshi
    III. Farnsworth, Norman R.
    581.6'1'072   QK98.4

LIBRARY OF CONGRESS CATALOGING IN PUBLICATION DATA

Main entry under title:

Economic and medicinal plant research.

    Includes index.
    1. Materia medica, Vegetable.   2. Medicinal plants.
3. Botany, Economic.   4. Plant products.   I. Wagner,
H. (Hildebert), Date   .   II. Hikino, Hiroshi,
Date   .   III. Farnsworth, Norman R.
RS164.E28   1985   581.6'1   85-1418
ISBN 0-12-730060-0 (alk. paper)

PRINTED IN THE UNITED STATES OF AMERICA

85 86 87 88   9 8 7 6 5 4 3 2 1

Contents

3
Gossypol: Pharmacology and Current Status as a Male Contraceptive

*Donald P. Waller, Lourens J. D. Zaneveld,
and Norman R. Farnsworth*

4
Immunostimulatory Drugs of Fungi and Higher Plants

H. Wagner and A. Proksch

5
Siberian Ginseng (*Eleutherococcus senticosus*): Current Status as an Adaptogen

*Norman R. Farnsworth, A. Douglas Kinghorn,
Djaja D. Soejarto, and Donald P. Waller*

6

Chemistry and Pharmacology of *Panax*

Shoji Shibata, Osamu Tanaka, Junzo Shoji, and Hiroshi Saito

Contributors

Numbers in parentheses indicate the pages on which the authors' contributions begin.

NORMAN R. FARNSWORTH (87, 155), Program for Collaborative Research in the Pharmaceutical Sciences, College of Pharmacy, Health Sciences Center, University of Illinois at Chicago, Chicago, Illinois 60680, U.S.A.

HIROSHI HIKINO (53), Pharmaceutical Institute, Tohoku University, Sendai 980, Japan

A. DOUGLAS KINGHORN (1, 155), Program for Collaborative Research in the Pharmaceutical Sciences, College of Pharmacy, Health Sciences Center, University of Illinois at Chicago, Chicago, Illinois 60680, U.S.A.

A. PROKSCH (113), Institut für Pharmazeutische Biologie der Universität München, D-8000 München 2, West Germany

HIROSHI SAITO (217), Faculty of Pharmaceutical Sciences, University of Tokyo, Tokyo 113, Japan

SHOJI SHIBATA (217), Meiji College of Pharmacy, Tokyo 154, Japan

JUNZO SHOJI (217), School of Pharmaceutical Sciences, Showa University, Tokyo 142, Japan

DJAJA D. SOEJARTO[1] (1, 155), Program for Collaborative Research in the Pharmaceutical Sciences, College of Pharmacy, Health Sciences Center, University of Illinois at Chicago, Chicago, Illinois 60612, U.S.A.

OSAMU TANAKA (217), Institute of Pharmaceutical Sciences, School of Medicine, Hiroshima University, Hiroshima 734, Japan

H. WAGNER (113), Institut für Pharmazeutische Biologie der Universität München, D-8000 München 2, West Germany

[1]Research Associate, Field Museum of Natural History—Botany, Chicago, Illinois 60605, U.S.A.

DONALD P. WALLER (87, 155), Program for Collaborative Research in the Pharmaceutical Sciences, College of Pharmacy, Health Sciences Center, University of Illinois at Chicago, Chicago, Illinois 60680, U.S.A.

LOURENS J. D. ZANEVELD[2] (87), Program for Collaborative Research in the Pharmaceutical Sciences, College of Pharmacy, Health Sciences Center, University of Illinois at Chicago, Chicago, Illinois 60680, U.S.A.

[2]Present address: Department of Obstetrics and Gynecology, Rush Presbyterian Hospital, Chicago, Illinois 60680, U.S.A.

Preface

The varied character of natural plant products, and indeed their very existence, pose fundamental questions to scientists. Many books have been published concerning the chemical aspects of these products; however, it is exceptional to find discussed within a single volume most aspects of particular genera or of particular pharmacological classes of natural substances, all having economic potential.

Thus, the intent of this volume is to identify areas of research in natural plant products that are of immediate or projected importance from a practical point of view and to review these areas in a concise and critical manner.

We feel that these topics will be of great interest to graduate students, research workers, and others interested in the discovery of natural products and in their further utilization as drugs, pharmacological tools, models for synthetic efforts, or other economic purposes. We hope decision makers in industry, government agencies, philanthropic foundations, and elsewhere will benefit from these timely reviews and consider these and related projects as worthwhile endeavors for further research.

Current Status of Stevioside as a Sweetening Agent for Human Use

A. DOUGLAS KINGHORN
DJAJA D. SOEJARTO

Program for Collaborative Research in the Pharmaceutical Sciences
College of Pharmacy, Health Sciences Center
University of Illinois at Chicago
Chicago, Illinois USA

1

ECONOMIC AND MEDICINAL PLANT RESEARCH
VOLUME 1

I. INTRODUCTION

Stevioside and extracts prepared from the leaves of the plant *Stevia rebaudiana* have been used widely in Japan since the mid-1970s as sweetening agents, taste modifiers, and sugar substitutes. They are currently accepted for general use as food additives in Japan, and to date there have been no adverse effects reported from the use of *Stevia* products by humans.

The sweet properties of *Stevia rebaudiana* have been known for centuries, being first realized in Paraguay, the natural habitat for this plant. Stevioside, the major sweet substance of this plant, was discovered in 1905 but was never seriously considered as a sugar substitute until the early 1970s, when a group of Japanese industrial firms specializing in the manufacture of food products decided to form a consortium in order to commercialize stevioside and *Stevia* extracts for the food industry in that country.

This chapter is concerned with *Stevia rebaudiana* and its sweet principles relative to their application in foods. The history, botany, chemistry, and pharmacology of extracts prepared from this species are discussed, as well as analytical methods for quantifying the stevioside content of such extracts. Additionally, all applicable pharmacologic data on stevioside and related sweet principles are discussed and evaluated in the context of human safety. Stability studies of stevioside under a variety of conditions are summarized, and the legitimate uses and applications of these materials in the food industry are presented.

Stevia rebaudiana and stevioside have periodically been the subject of reviews over the past 40 years. Topics such as the cultivation, constituents, and commercialization prospects for *S. rebaudiana,* as well as the chemistry, biological effects, safety, and utilization of stevioside as a sweetener have been addressed by Thomas (1937), Samaniego (1946), Klages (1951), Bell (1954), Fletcher (1955), Jacobs (1955), Nieman (1958), Sumida (1973), Matsumi (1974), Brucher (1974), Kato (1975), Yoshino (1975), Seidemann (1976a), Abe and Sonobe (1977), Chueh (1977), Felippe (1977), Akashi (1977), Okazaki *et al.* (1977), Morita (1977a), Kazuyama (1979), T. Fujita (1979), Ochi (1979), Hsin *et al.*

(1979), Tanaka (1979, 1980), Fujita and Edahiro (1979a,b), Tsuchiya (1979), H. Fujita (1980), Toffler and Orio (1981), Huang (1981), and Sakaguchi and Kan (1982). Most of the more recent papers are in Japanese, however. These papers are referred to when relevant, throughout this chapter. In addition, a number of general review articles on natural-product sweeteners including stevioside and the other *S. rebaudiana* sweet diterpene glycosides (Farnsworth, 1973; Seidemann, 1976b; Crosby, 1976; Morris, 1976; Inglett, 1976, 1978; Seidemann, 1977; Crammer and Ikan, 1977; Unterhalt, 1978; Bragg *et al.*, 1978; Lee, 1979; Crosby and Wingard, 1979; Kojima, 1980; Miyoshi, 1980a,b; Crosby and Furia, 1980; Tanaka, 1981; Heraud, 1981; DuBois, 1982) are mentioned when appropriate.

II. THE SWEET PRINCIPLES OF *STEVIA REBAUDIANA*

A. BOTANICAL ASPECTS

Stevioside is a sweet *ent*-kaurene glycoside constituent of *Stevia rebaudiana* (Bertoni) Bertoni, a plant native to elevated terrain at latitude ~25° S in the Amambay and Iguaçu districts on the borders of Brazil and Paraguay (Sumida, 1973; Soejarto *et al.*, 1983a). M. S. Bertoni originally described this plant, a herb of the Compositae (daisy family), which reaches a height of 80 cm when fully grown, as *Eupatorium rebaudianum* Bertoni, but later reassigned it to the genus *Stevia* (Bertoni, 1905). The correct name for this taxon therefore is *S. rebaudiana* (Bertoni) Bertoni, rather than *S. rebaudiana* (Bertoni) Hemsley (Hemsley, 1906), for reasons of priority (Soejarto *et al.*, 1983a). *Stevia rebaudiana* is known in the Guarani language in Paraguay as *Caá-êhé* (Gosling, 1901), *Kaá Hê-é* (Bertoni, 1905, 1918; Gosling, 1901; Hemsley, 1906), *Caá-êhé* and *Azucá-caá* (Cabrera, 1939), *Caá-hê-hê* or *Caá-enhem* (Mors and Rizzini, 1966), and *Ka-á Hê-e* (Soejarto *et al.*, 1983a), all of which mean "sweet herb."

Stevia is an entirely New World genus belonging to the tribe Eupatoriae within the Compositae (King and Robinson, 1967; Grashoff, 1972; Robinson and King, 1977). The distribution range of *Stevia* extends from the southwestern United States to northern Argentina, through Mexico, Central America, the South American Andes, and the Brazilian highlands. Records indicate that *Stevia* is not represented in the West Indies and Amazonia (King and Robinson, 1967; Grashoff, 1972). While more than 80 species of *Stevia* are known to occur in North America (Grashoff, 1974), the South American species do not appear to have been studied taxonomically in recent years. Estimates on the total number of species in the genus range from 150 to 300 (King and Robinson, 1967; Grashoff, 1972, 1974; Robinson and King, 1977; Soejarto *et al.*, 1983a).

B. CONSTITUENTS OF *STEVIA REBAUDIANA*

The dried leaves of *Stevia rebaudiana* contain ~42% (w/w) of water-soluble constituents (Bell, 1954). Chemical work to isolate the water-soluble sweet constituents of the plant, following preliminary observations by Bertoni (1905), began with experiments by Rasenack (1908) and Dieterich (1909). Dieterich (1909) isolated two sweet components, one of which, eupatorin, was renamed stevin by Bertoni (1918), in view of the nomenclatural change of the plant from *Eupatorium rebaudianum* to *S. rebaudiana* (Bertoni, 1918).

In a series of papers, Bridel and Lavieille reported several observations on the chemical nature of the sweet principles of *Stevia rebaudiana*, and stevin (= eupatorin) was renamed stevioside (Fig. 1, **1**) with preliminary

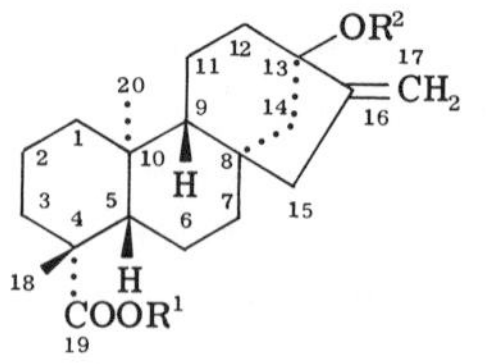

	R^1	R^2
1	β-Glu	β-Glu-β-Glu $(2 \to 1)$
2	—H	β-Glu-β-Glu $(2 \to 1)$
3	β-Glu	β-Glu-β-Glu $(2 \to 1)$ $\vert$ β-Glu $\quad (3 \to 1)$
4	—H	β-Glu-β-Glu $(2 \to 1)$ $\vert$ β-Glu $\quad (3 \to 1)$
5	β-Glu	β-Glu-α-Rha $(2 \to 1)$ $\vert$ β-Glu $\quad (3 \to 1)$
6	β-Glu-β-Glu $(2 \to 1)$	β-Glu-β-Glu $(2 \to 1)$ $\vert$ β-Glu $\quad (3 \to 1)$
7	β-Glu-β-Glu $(2 \to 1)$	β-Glu-β-Glu $(2 \to 1)$
8	β-Glu	β-Glu-α-Rha $(2 \to 1)$

(Glu = glucose; Rha = rhamnose)

FIG. I Structures of some *Stevia rebaudiana* constituents.

structural information on this glycoside and its aglycone being established (Bridel and Lavieille, 1931a–h). In 1952, a research group at the
National Institute of Arthritis and Metabolic Diseases, National Institutes of Health, Bethesda, Maryland, renewed the investigation into
the chemical structure of stevioside. This group was able to establish that
one sugar unit of stevioside occurs as a glucopyranose function attached
β to a carboxyl group, and a second sugar unit occurred as a sophorose
[2-O-(β-D-glucopyranosyl)-D-glucose] function attached β to an alcoholic
hydroxyl group of the aglycone (Wood *et al.*, 1955; Wood and Fletcher,
1956; Vis and Fletcher, 1956). The structures and stereochemistry of the
true aglycone moiety steviol (13-hydroxy-*ent*-kaurenoic acid) (Fig. 2, **17**)
and of isosteviol (Fig. 2, **18**), a product obtained on treatment of stevioside with boiling dilute sulfuric acid, were also established by this

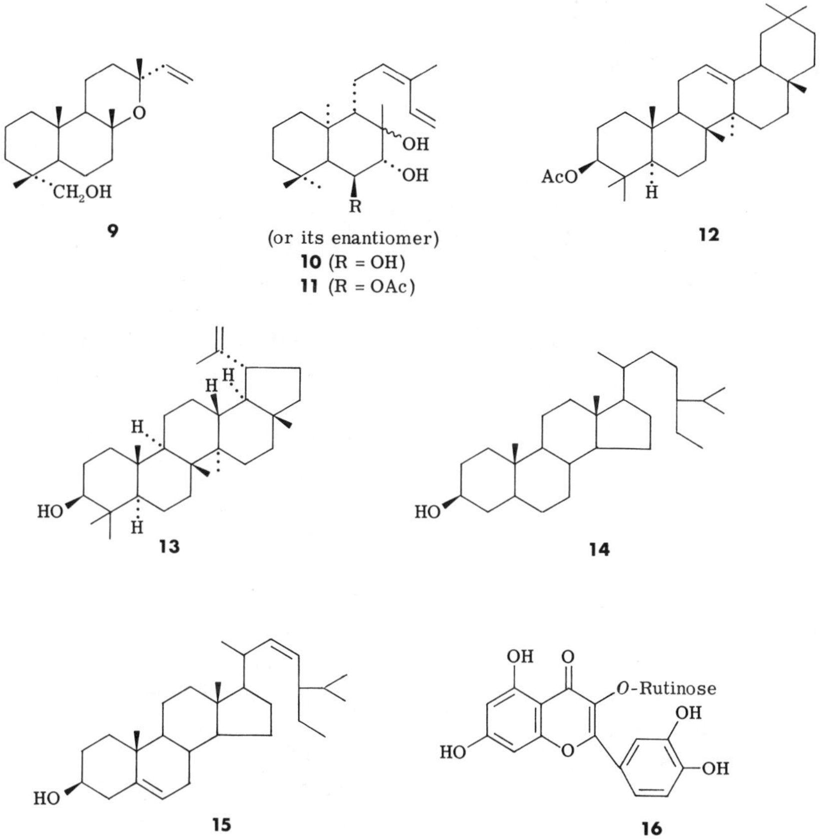

9

(or its enantiomer)
10 (R = OH)
11 (R = OAc)

12

13

14

15

16

FIG. I *Continued*

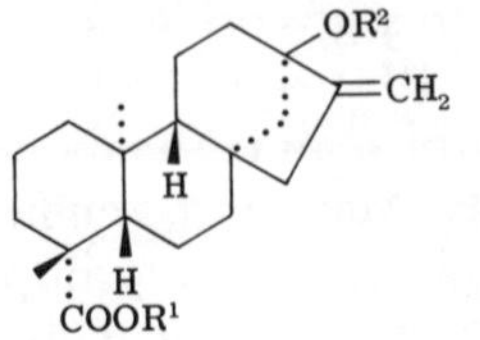

	R¹	R²
17	H	H
19	H	β-D-Glucopyranosyl
20	β-D-Glucopyranosyl	β-D-Glucopyranosyl

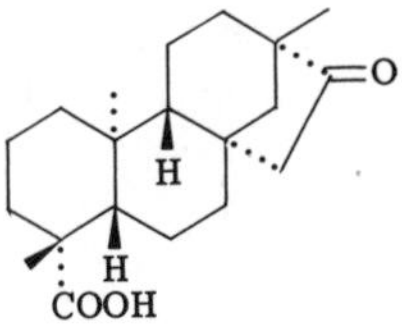

FIG. 2 Structures of steviol, isosteviol, and other hydrolytic products of stevioside.

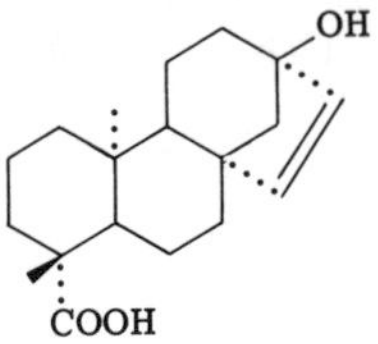

group (Mosettig and Nes, 1955; Dolder *et al.*, 1960; Mosettig *et al.*, 1961, 1963; Djerassi *et al.*, 1961). Steviol can be converted to isosteviol by a reaction involving a proton-catalyzed Wagner–Meerwein rearrangement (Mosettig *et al.*, 1963).

During the 1970s, further insight was gained into both the sweet and nonsweet constituents of *Stevia rebaudiana*. An additional seven sweet diterpene glycosides, based on the *ent*-kaurene skeleton (Fig. 1, **2–8**), were isolated from this plant by Japanese groups at Hiroshima University and Hokkaido University, including rebaudioside A (Fig. 1, **3**), which reportedly is sweeter than stevioside and has a more pleasant taste (Kohda *et al.*, 1976b; Yamasaki *et al.*, 1976; Kobayashi *et al.*, 1977; Sakamoto *et al.*, 1977a,b). Structural assignments of these new sweet principles relied heavily on carbon-13 nuclear magnetic resonance spectroscopy (Yamasaki *et al.*, 1976; Sakamoto *et al.*, 1977a,b). Nonsweet labdane diterpenes (Fig. 1, **9–11**), triterpenes (Fig. 1, **12, 13**), sterols (Fig. 1, **14, 15**), and a flavonoid (Fig. 1, **16**), have also been detected in the leaves or callus tissue of the plant (Nabeta *et al.*, 1976; Yamasaki *et al.*, 1976; Tanaka, 1979, 1980; Sholichin *et al.*, 1980). A total of 31 constituents of the volatile oil of *S. rebaudiana* herb have been identified by a combination of gas chromatography, infrared spectroscopy, and mass spectrometry (Fujita *et al.*, 1977). The chemical constituents of *S. rebaudiana* characterized to date are summarized in Table I. Table II provides

TABLE I

CHEMICAL CONSTITUENTS OF *STEVIA REBAUDIANA*

Compound class	Constituents	Plant part (Fig. 1)	Yield[a]	Reference
Diterpene glycoside	—	Leaves	—	Rasenack (1908)
Diterpene glycoside	Eupatorin (= stevioside, **1**)	Leaves/stems	—	Dieterich (1909); Bridel and Lavieille (1931f); Kobert (1915)
Diterpene glycoside	Stevioside (**1**)	Leaves	4.5	Bridel and Lavieille (1931a)
	Stevioside (**1**)	Leaves	7.0	Wood and Fletcher (1956)
	Stevioside (**1**)	Leaves	2.2	Kohda *et al.* (1976b)
	Stevioside (**1**)	Leaves	—	Kobayashi *et al.* (1977)
	Stevioside (**1**)	Callus	—	Komatsu *et al.* (1976); Kotani (1980)
Diterpene glycoside	Steviolbioside (**2**)	Leaves	0.04	Kohda *et al.* (1976b)
Diterpene glycoside	Rebaudioside A (**3**)	Leaves	1.43	Kohda *et al.* (1976b)
	Rebaudioside A (**3**)	Leaves	—	Kobayashi *et al.* (1977)
Diterpene glycoside	Rebaudioside B (**4**)	Leaves	0.44	Kohda *et al.* (1976b)
Diterpene glycoside	Rebaudioside C (**5**) (= dulcoside B)	Leaves	0.4	Sakamoto *et al.* (1977a)
		Leaves	0.013	Kohda *et al.* (1976b)
Diterpene glycoside	Rebaudioside D (**6**)	Leaves	0.03	Sakamoto *et al.* (1977a,b)
Diterpene glycoside	Rebaudioside E (**7**)	Leaves	0.03	Sakamoto *et al.* (1977a,b)
Diterpene glycoside	Dulcoside A (**8**)	Leaves	0.029	Kohda *et al.* (1976b)
Labdane diterpene	Jhanol (**9**)	Leaves	0.0063	Sholichin *et al.* (1980)
Labdane diterpene	Austroinulin (**10**)	Leaves	0.06	Sholichin *et al.* (1980)
Labdane diterpene	6-*O*-Acetylaustroinulin (**11**)	Leaves	0.15	Sholichin *et al.* (1980)
Triterpene	β-Amyrin acetate (**12**)	Leaves	—	Sholichin *et al.* (1980)
Triterpene	Lupeol (**13**)[b]	Leaves	—	Sholichin *et al.* (1980)
Sterol	β-Sitosterol (**14**)	Leaves	—	Sholichin *et al.* (1980)
Sterol	Stigmasterol (**15**)	Leaves	—	Sholichin *et al.* (1980)
Sterol	Stigmasterol (**15**)	Callus	0.00091	Nabeta *et al.* (1976)

(continued)

TABLE I (*Continued*)

Compound class	Constituents	Plant part (Fig. 1)	Yield[a]	Reference
Flavonoid glycoside	Rutin (**16**)	Callus	0.0073	Suzuki *et al.* (1976)
Tannins	Unidentified	Leaves	7.8	Chung and Lee (1979)
Volatile oil	See Table II	Herb[c]	0.12	Fujita *et al.* (1977)
Volatile oil	See Table II	Herb[d]	0.16	Fujita *et al.* (1977)
Volatile oil	See Table II	Inflorescence[d]	0.43	Fujita *et al.* (1977)

[a]Yields are expressed as the percentage (w/w) of a compound isolated from the dried plant part stated.
[b]Isolated in the form of three esters that were not identified.
[c]Cultivated in Paraguay.
[d]Cultivated in Japan.

TABLE II

VOLATILE OIL CONSTITUENTS FROM *STEVIA REBAUDIANA*
CULTIVATED IN PARAGUAY[a]

Compound class	Constituents (Fig. 1)	Yield of herb[b] (% w/w)
Alkanol	Octan-3-ol	0.00036
	Oct-1-en-3-ol	0.00084
Aldehyde	Hexan-1-ol	0.0011
Aromatic alcohol	Benzyl alcohol	0.0012
Monoterpene	Camphor	0.0017
	1,8-Cineole	0.00084
	p-Cymene	0.00084
	Geraniol	0.0016
	Limonene	0.0012
	Linalool	0.0067
	Linalool oxide[c]	0.0055
	α-Pinene	0.00048
	β-Pinene	0.0023
	λ-Terpinene	0.00024
	Terpinen-4-ol	0.0012
	α-Terpineol	0.0054
Sesquiterpene	λ-Cadinene	0.0036
	δ-Cadinene	0.0012
	α-Cadinol	0.0017
	tert-Cadinol	0.0028
	α-Calacorene	0.0012
	Calamenene	0.0018
	β-Caryophyllene	0.0013
	Caryophyllene oxide	0.019
	α-Copaene	0.00012
	α-Cubebene	0.00012
	β-Elemene	0.0006
	trans-β-Farnesene	0.00054
	α-Humulene	0.0029
	Nerolidol	0.031
	β-Selinene	0.0026

[a] A total of 22 compounds remained unidentified in the study by Fujita *et al.* (1977), including four sesquiterpene hydrocarbons, one sesquiterpene alcohol, and a ketone.
[b] The percentage (w/w) yield of volatile oil in the herb was 0.12.
[c] Both the cis and trans forms of this compound were detected.

a listing of the identified constituents of the volatile oil from the whole herb of *S. rebaudiana*, grown in Paraguay.

Abstracts of a number of patented procedures for the extraction and purification of stevioside have appeared in *Chemical Abstracts* since 1973. These methods may be classified into those involving solvent partition alone (Persinos, 1973; Asano *et al.*, 1975; Yamagami and Takato, 1976; Haga *et al.*, 1976a; Tanaka *et al.*, 1977a; Anonymous, 1980c, 1981e,f), and those involving solvent partition incorporating a decolorizing or a precipitating agent to remove impurities (Takamura *et al.*, 1977a,b, 1978; Mochida *et al.*, 1977b; Kikuchi and Sawaguchi, 1977; Kodaka, 1977; Yarita *et al.*, 1978; Gushiken, 1979; Sato *et al.*, 1980; Masuyama, 1980; Anonymous, 1980i,k,l,n, 1981d,h, 1982a; Wakabayashi, 1981), adsorption column chromatography (Mochida *et al.*, 1977a; Morita and Iwamura, 1977; Ogawa, 1979), the use of ion-exchange resins (Akashi *et al.*, 1975; Igoshi and Kato, 1976; Kubomura *et al.*, 1976; Haga *et al.*, 1976b; Uenishi *et al.*, 1977; Okane and Kamata, 1977; Sawaguchi and Kikuchi, 1977; Kiumi *et al.*, 1977; Kato *et al.*, 1977; Ohe *et al.*, 1977; Suzuki *et al.*, 1978; Ishizone, 1979; Kokai *et al.*, 1979; Itagaki and Kato, 1979; Itagaki and Ito, 1979a,b; Kuroda and Kamiyama, 1979; Miwa and Tsuji, 1979; Ise and Hirada, 1979; Miwa, 1979; Anonymous, 1980a,b,c,f,j, 1981g, 1982b; Yamada and Kajima, 1980), or electrolytic techniques (Miwa, 1978, 1979; Miwa *et al.*, 1979a,b). Isolation procedures have also been patented for rebaudioside A (K. Suzuki *et al.*, 1977; Tanaka *et al.*, 1977b), rebaudioside B (Fig. 1, **4**) (Tanaka *et al.*, 1977b), rebaudioside C (= dulcoside B) (Fig. 1, **5**) (Kobayashi *et al.*, 1978), and dulcoside A (Fig. 1, **8**) (Kobayashi *et al.*, 1978).

While the yields of stevioside presented in Table II are based on gravimetric determinations from standard phytochemical isolation procedures, a number of workers have estimated stevioside concentrations in *Stevia rebaudiana* leaves using modern analytical techniques (Table III). It may be seen from this table that high concentrations of stevioside have been determined in the plant based on variations in cultivation conditions, irrespective of the geographic origin of the sample analyzed. Concentration levels of rebaudioside A in *S. rebaudiana* leaves were found to range from 25 to 54% relative to stevioside, depending on the sample origin, as determined by high-pressure liquid chromatography (Hashimoto and Moriyasu, 1978). Our group estimated the levels of rebaudioside A and dulcoside A as 4.5% (w/w) and 0.9% (w/w), respectively, in a sample of *S. rebaudiana* leaves collected in Paraguay, using droplet countercurrent chromatography (Kinghorn *et al.*, 1982).

Considerable progress has been made in synthetic studies to produce stevioside and its sweet analogs. The initial step in such methodology is

TABLE III

RESULTS OF ANALYTICAL STUDIES TO DETERMINE THE STEVIOSIDE CONTENT OF
STEVIA REBAUDIANA LEAVES

Reference	Sample origin	Number of samples analyzed	Stevioside content (% w/w)
Mitsuhashi *et al.* (1975a)	Japan	5	5.13–7.27[a]
Mitsuhashi *et al.* (1975a)	South America	1	7.20[b]
Mitsuhashi *et al.* (1975b)	Japan	12	2.26–7.59[a]
Mitsuhashi *et al.* (1975b)	Japan	12	3.10–8.65[b]
Sakamoto *et al.* (1975)	Japan	5	3.8–18.5[a]
Sugisawa *et al.* (1977)	Japan (?)	4	6.7–11.0[c]
Sugisawa *et al.* (1977)	Japan (?)	4	8.6–15.4[b]
Hashimoto and Moriyasu (1978)	Japan	6	4.7–8.4[d]
Hashimoto and Moriyasu (1978)	Korea	1	7.8[d]
Hashimoto and Moriyasu (1978)	Paraguay	1	5.7[d]
Hashimoto and Moriyasu (1978)	Brazil	1	4.5[d]
Iwamura *et al.* (1979a)	Japan	3	4.8–12.4[b]
Iwamura *et al.* (1979a)	Korea	1	9.1[b]
Iwamura *et al.* (1979a)	Brazil	1	9.5[b]
Miyazaki *et al.* (1978)	Japan[e]	3	16.0–22.6[c]
Miyazaki *et al.* (1979)	Paraguay	1	12.8[c]
Mizukami *et al.* (1982)	Japan	6	8.6–11.6[f]

[a] Determined by gas-liquid chromatography.
[b] Determined by thin-layer chromatography–densitometry.
[c] Colorimetric determination.
[d] Determined by high-pressure liquid chromatography.
[e] Range of stevioside content (% w/w) of *S. rebaudiana* stems, cultivated in Japan (three samples) = 1.4–6.7.
[f] Determined enzymatically.

the synthesis of steviol. This is complicated by the need to produce a bicyclo[3,2,1]octane C/D ring system with a bridgehead tertiary hydroxyl group, but this has been achieved by several teams (Mori *et al.*, 1970, 1972; Cook and Knox, 1970; Zeigler and Kloek, 1977). Ogawa *et al.* (1980) were able to partially synthesize stevioside from steviolbioside (Fig. 1, **2**) (yield 61%) and to produce steviolbioside from the methyl ester of steviol (yield 29%) (Nozaki *et al.*, 1978; Ogawa *et al.*, 1980). Tanaka and co-workers have converted stevioside to rebaudioside A with a reaction yield of 70% (Kaneda *et al.*, 1977), and, in addition, have provided methods for the synthesis of rebaudiosides D and E (Kasai *et al.*, 1981).

C. CULTIVATION AND CELL CULTURE STUDIES ON
STEVIA REBAUDIANA SWEET CONSTITUENTS

It is because of the high concentration levels of stevioside occurring in *Stevia rebaudiana,* as well as the fact that acceptable crop yields of this compound can be obtained when the plant is cultivated in warm temperate to subtropical climates, that *S. rebaudiana* is now produced on a commercial scale in Japan and South Korea, as well as in Paraguay (Kato, 1975; Akashi, 1977; Soejarto *et al.,* 1983a). It has been found that *S. rebaudiana* will grow throughout Japan except in Hokkaido, the northernmost part of Japan (Kato, 1975). It is usual to transplant *S. rebaudiana* seedlings in Japan during April and May and to harvest the crop twice during the growing season (Akashi, 1977). The mean leaf yield is at a maximum in 3-year-old plants cultivated in Japan, although plants that mature earlier have a higher stevioside content (Sumida, 1980). *Stevia rebaudiana* grows preferentially on acid soils of pH 4–5 and requires high light intensities, with warm temperatures and minimal frost (Metivier and Viana, 1979b; Shock, 1982). The plant is affected adversely by water stress and saline conditions (Shock, 1982). Other investigators have also studied the effects of photoperiod response on the growth of *S. rebaudiana* (Valio and Rocha, 1977; Kudo and Koga, 1977; Zaidan *et al.,* 1980), as well as effects of growth regulators (Valio and Rocha, 1977), insect growth hormones (Felippe, 1980), and fertilizers (Lee *et al.,* 1980).

There have been several reports of experiments aimed at the growth of *Stevia rebaudiana* explants by plant tissue culture, with the ultimate aim of producing large quantities of stevioside, and thus to obviate the need to cultivate the plant (Nabeta *et al.,* 1976; Komatsu *et al.,* 1976; Kotani, 1980; Suzuki *et al.,* 1976; Handro *et al.,* 1977; Misawa, 1977; Wada *et al.,* 1981). Stevioside, however, appears to have been produced in only two such experiments (Komatsu *et al.,* 1976; Kotani, 1980).

D. DISTRIBUTION OF SWEET *ENT*-KAURENE
GLYCOSIDES IN THE GENUS *STEVIA*

While there is an abundance of ethnobotanical literature suggesting that many *Stevia* species have found use as medicinal plants (Soejarto *et al.,* 1983a,b) phytochemical information is available to date on only about 20 members of this large genus. Investigators have reported several types of sesquiterpenes in plants of the genus *Stevia* (exclusive of *S. rebaudiana*), including those based on bisabolane (Bohlmann *et al.,* 1976, 1977, 1982), germacrane (Bohlmann *et al.,* 1976, 1977, 1982; Salmon *et al.,* 1975), guaiane (Bohlmann *et al.,* 1979, 1982; Salmon *et al.,* 1973, 1977), humulane (Bohlmann *et al.*, 1977, 1979), α-longipinane (Bohlmann *et al.*, 1976, 1977, 1979; Roman *et al.*, 1981), and pseudoguaiane (Rios *et al.*, 1967) skeletons, as well as clerodane (Bohlmann *et al.*, 1982; Angeles *et al.*, 1982), kaurane (Bohlmann *et al.*, 1979, 1982; Quijano *et al.*, 1982), and labdane-type (Bohlmann *et al.*, 1976, 1982; Quijano *et al.*, 1982; Ortega *et al.*, 1980) diterpenes, and sterols (Dominguez *et al.*, 1974; Bohlmann *et al.*, 1979; Quijano *et al.*, 1982), triterpenes (Dominguez *et al.*, 1974; Bohlmann *et al.*, 1982), flavonoids (Dominguez *et al.*, 1974; Bohlmann *et al.*, 1976; Ortega *et al.*, 1980), chromenes (Bohlmann *et al.*, 1976; Kohda *et al.*, 1976a; Quijano *et al.*, 1982), and miscellaneous volatile derivatives (Montes, 1969; Dominguez *et al.*, 1974; Bohlmann *et al.*, 1977, 1979, 1982; Ghisalberti *et al.*, 1979). In addition, a group of nonsweet *ent*-kaurene glycosides, paniculosides I–V, has been reported in the species *S. paniculata* Lag. (Kohda *et al.*, 1976c; Yamasaki *et al.*, 1977) and *S. ovata* Lag. (Kaneda *et al.*, 1978).

In studies directed toward the elucidation of the distribution of sweet *ent*-kaurene glycosides within the genus *Stevia*, we have tested for sweet effects produced by *Stevia* leaves obtained from an herbarium collection (Soejarto *et al.*, 1982) and obtained fresh in the field (Soejarto *et al.*, 1983a). Intense sweetness was exhibited only by leaves of *S. rebaudiana* among those tested, and such an effect persisted even in an herbarium specimen over 60 years old (Soejarto *et al.*, 1982). Stevioside was detected in a herbarium specimen of a second species in this genus, *S. phlebophylla* A. Gray, after analysis of crude leaf extracts using a combination of chromatographic techniques (Hovanec-Brown *et al.*, 1982). Interestingly, this *S. phlebophylla* sample, although collected in Mexico in 1889, still exhibited a slightly sweet taste (Soejarto *et al.*, 1982). We found no evidence, however, of the presence of steviol glycosides in 108 of the 110 species of *Stevia* leaf herbarium samples examined (Hovanec-Brown *et al.*, 1982).

E. ANALYTICAL METHODS FOR STEVIOSIDE AND RELATED *ENT*-KAURENE GLYCOSIDES

It may be seen from the publications cited in Table III that stevioside in *Stevia rebaudiana* has been qualitatively and quantitatively analyzed using five major methods: gas–liquid chromatography (Mitsuhashi *et al.*, 1975a,b; Sakamoto *et al.*, 1975), thin-layer chromatography–densitometry (Mitsuhashi *et al.*, 1975a,b; Sugisawa *et al.*, 1977; Iwamura *et al.*, 1979a), colorimetric determination (Sugisawa *et al.*, 1977; Miyazaki *et al.*, 1978), high-pressure liquid chromatography (Hashimoto and Moriyasu, 1978), and enzymatically (Mizukami *et al.*, 1982). The use of gas–liquid chromatography for this purpose required the conversion of stevioside to the methyl ester of either isosteviol (Mitsuhashi *et al.*, 1975a) or steviol (Sakamoto *et al.*, 1975). The colorimetric methods referred to in Table III utilized either the Carr–Price reaction (Sugisawa *et al.*, 1977) or the anthrone reagent (Miyazaki *et al.*, 1978). It was found that glucose occurring as a result of the hydrolysis of stevioside with crude hesperidinase was produced in a quantitative fashion and that stevioside estimations by this enzymic method correlate favorably with those obtained by existing methods of analysis (Mizukami *et al.*, 1982). Other methods for the quantitation of stevioside and the other sweet *S. rebaudiana* glycosides have also involved gas–liquid chromatography (Minamisono and Azuno, 1978; Shirakawa and Onishi, 1979; Nakajima *et al.*, 1979; Tezuka *et al.*, 1980), thin-layer chromatography–densitometry (Iwamura *et al.*, 1980, 1982), colorimetric estimation (Hayashi and Noda, 1975; Metivier and Viana, 1979a; Angelucci, 1981a), and high-pressure liquid chromatography (Chen and Yeh, 1978; Hashimoto *et al.*, 1978; Hashimoto, 1979; Hirokado *et al.*, 1980; Ahmed *et al.*, 1980; Nakajima *et al.*, 1980; Ahmed and Dobberstein, 1982a,b; Chang and Cook, 1983). Several of these methods have been specifically designed for the analysis of these compounds in foods (Minamisono and Azuno, 1978; Shirakawa and Onishi, 1979; Nakajima *et al.*, 1979, 1980; Tezuka *et al.*, 1980; Hirokado *et al.*, 1980).

F. STABILITY STUDIES ON STEVIOSIDE AND ITS SWEET ANALOGS

Stevioside is a stable molecule over the pH range 3–9, when heated at 100°C for 1 hr (Abe and Sonobe, 1977; Ochi, 1979; Fujita and Edahiro, 1979a), but rapid decomposition occurs at pH levels greater than 9 under these conditions (Fujita and Edahiro, 1979a) (Table IV). Steviolbioside produced from stevioside by alkaline hydrolysis (Wood and Fletcher, 1956), would presumably be the major decomposition product obtained at pH 10, although this fact was not categorically stated by Fujita and Edahiro (1979a) (Table IV). Much interest has been expressed in the characterization of the products of reaction of stevioside under acidic hydrolytic conditions, particularly from the viewpoint of the design of analytical methods for this compound. The results of these experiments are summarized in Table V. It may be seen from this table that steviol (Fig. 2, **17**), the aglycone of stevioside, may be produced from the latter compounds by the use of several enzymes (Bridel and Lavieille, 1931d,h; Ruddat *et al.*, 1965; Sakamoto *et al.*, 1975), by an oxidation–elimination reaction sequence with sodium periodate and potassium hydroxide (Matsui *et al.*, 1978; Nozaki *et al.*, 1978; Ogawa *et al.*, 1980), and

TABLE IV

STABILITY PROFILE OF
STEVIOSIDE AT VARIOUS pH
FOLLOWING HEATING IN
AQUEOUS SOLUTIONS AT
100°C FOR 1 HR

pH	Stevioside remaining[a] (%)
3	98
5	99
7	100
8	97
9	99
10	47

[a]These data were obtained from a review article (Fujita and Edahiro, 1979a) in which no mention was made of the analytical method(s) used.

by reaction with 0.4% hydrochloric acid in aqueous methanol (Nabeta *et al.,* 1977; Iwamura *et al.,* 1979b). However, isosteviol (Fig. 2, **18**), which has a different ring C/D stereochemistry than steviol is a more typical reaction product of stevioside when in acid media (Table V). The report of the production of compound **21** (Fig. 2) following reflux of stevioside with weak aqueous methanolic solutions of hydrochloric and sulfuric acids (Iwamura *et al.,* 1979b) is questionable, since the elimination of the C-17 methyl group would not be expected under these conditions. It may be pointed out that no spectral evidence for the structure of **21** was provided by Iwamura *et al.* (1979b).

Decomposition profiles for stevioside (**1**) and rebaudioside A (**3**) have recently been reported when they were individually formulated at a level of 0.1% (w/v) in carbonated phosphoric and citric acid beverages (Chang and Cook, 1983). In the case of stevioside, 36 and 17% losses of original concentrations were reported after storage at 37°C for 4 weeks in phosphoric and citric acid beverages, respectively. Steviolbioside (**2**), glucose, and one unknown degradation product were identified from **1** by analytical high-pressure liquid chromatography and thin-layer chromatography. Rebaudioside A showed no significant decomposition following storage at 37°C for 1 month in either phosphoric acid or citric acid containing beverages. However, exposure of beverages containing rebaudioside A to 3000 langleys of sunlight for 1 week resulted in losses of

TABLE V

ENZYMIC AND ACID HYDROLYTIC PRODUCTS OF STEVIOSIDE (FIG. 1, I)

Diterpenoid hydrolytic product (Figs. 1, 2)	Reagent	Reaction/time and/or temperature	Yield (% w/w)	Reference
Steviol (17)	*Helix pomata* hepatic pancreatic juice	5 days	90.5	Bridel and Lavieille (1931d,h)
	Pectinase, pH 4	140 hr, 37°C	90	Ruddat et al. (1965)
	Hesperidinase (crude, pH 4)	48 hr, 40°C	90	Sakamoto et al. (1975)
	Sodium iodate; potassium hydroxide	16 hr, 25°C, 1 hr, reflux	75	Nozaki et al. (1978); Ogawa et al. (1980); Matsui et al. (1978)
	0.4% Hydrochloric acid in aqueous methanol	5 hr, reflux 5 hr, reflux	49 78.4	Nabeta et al. (1977) Iwamura et al. (1979b)
Isosteviol (18)	5% Sulfuric acid	100°C	NS[a]	Bridel and Lavieille (1931e)
	30% Sulfuric acid	30 min, 100°C	96.2	Sugisawa et al. (1977)
		30 min, 100°C	96.5	Iwamura et al. (1979a)

16

Compound	Reagent	Conditions	Yield (%)	Reference
	25% Sulfuric acid	1 hr, 100°C	91.2	Iwamura et al. (1979a)
	20% Sulfuric acid	2 hr, reflux[b]	97.0	Iwamura et al. (1979a)
	6 N Hydrochloric acid	1 hr, 100°C	88.8	Iwamura et al. (1979a)
	48% Hydrobromic acid	19 hr, 25°C	87	Mosettig et al. (1963)
	2.5 N Trifluoroacetic acid	10 min, 100°C	90.0	Iwamura et al. (1979a)
	2.5 N Trifluoroacetic acid	1 hr, 100°C	96.0	Iwamura et al. (1979a, 1982)
	2.5 N Trifluoroacetic acid	5 hr, reflux	99.0	Iwamura et al. (1979b)
	0.4% Hydrochloric acid in aqueous methanol	5 hr, reflux	7.1	Iwamura et al. (1979b)
	1.5% Hydrochloric acid in aqueous methanol	5 hr, reflux	93.5	Iwamura et al. (1979b)
	2.5 N Monochloroacetic acid	5 hr, reflux	97.4	Iwamura et al. (1979b)
	2.5 N Dichloroacetic acid	5 hr, reflux	96.9	Iwamura et al. (1979b)
Steviol-13-O-β-D-glucopyranoside (19)	0.4% Hydrochloric acid in aqueous methanol	5 hr, reflux	4	Nebeta et al. (1977)
Steviol-13-O-β-D-glucopyranoside, 19-O-β-D-glucopyranosyl ester (20)	0.4% Hydrochloric acid in aqueous methanol	5 hr, reflux	19	Nebeta et al. (1977)

(continued)

17

TABLE V (*Continued*)

Diterpenoid hydrolytic product (Figs. 1, 2)	Reagent	Reaction/time and/or temperature	Yield (% w/w)	Reference
Steviolbioside (**2**)	0.4% Hydrochloric acid in aqueous methanol	5 hr, reflux	29	Nebeta et al. (1977)
	0.22% Citric acid	4 hr, 100°C[c]	NS	Chang and Cook (1983)
	0.04% Phosphoric acid	4 hr, 100°C[c]	NS	Chang and Cook (1983)
Compound **21**	0.4% Hydrochloric acid in aqueous methanol	5 hr, reflux	10.7	Nebeta et al. (1977)
	1.5% Sulfuric acid in aqueous methanol	5 hr, reflux	5.6	Iwamura et al. (1979b)

[a]NS, not stated.
[b]We have found that rebaudiosides A–E, steviolbioside, and dulcoside A all produce isosteviol under these conditions (Hovanec-Brown et al., 1982).
[c]Rebaudioside A (**3**) under the same conditions yielded rebaudioside B (**4**) (Chang and Cook, 1983).

TABLE VI

STABILITY OF STEVIOSIDE WHEN FORMULATED IN SOY SAUCE AND OTHER MEDIA[a]

Formulation[b]	Treatment	Recovery of stevioside[c] (%)
Water	None	100
Soy sauce	None	91
Vinegar	None	84
Vegetable protein hydrolysate	None	84
Water	80°C, 1 hr	92
Water	80°C, 2 hr	96
Water	80°C, 3 hr	96
Water	80°C, 5 hr	92
Soy sauce	80°C, 1 hr	92
Soy sauce	80°C, 2 hr	77
Soy sauce	80°C, 3 hr	72
Vegetable protein hydrolysate	80°C, 1 hr	92
Vegetable protein hydrolysate	80°C, 2 hr	62
Vegetable protein hydrolysate	80°C, 3 hr	26
Water	30°C, 10 days	101
Water	30°C, 20 days	99
Water	40°C, 30 days	99
Soy sauce	30°C, 10 days	98
Soy sauce	30°C, 20 days	99
Soy sauce	30°C, 30 days	90
Vegetable protein hydrolysate	30°C, 10 days	98
Vegetable protein hydrolysate	30°C, 20 days	98
Vegetable protein hydrolysate	30°C, 30 days	97
Glucose 3%[d]	80°C, 5 hr	77.5
Acetic acid 0.16%[d]	80°C, 5 hr	103
Succinic acid 0.03%[d]	80°C, 5 hr	101
Lactic acid 1.8%[d]	80°C, 5 hr	104
Sodium chloride 18%[d]	80°C, 5 hr	98

[a] From Shirakawa and Onishi (1979).
[b] Stevioside originally added to all formulations = 0.1% (1 mg/ml).
[c] % Recoveries of stevioside were obtained by gas–liquid chromatography on a 10% SE-30 column at 270°C, after the hydrolysis of stevioside-containing mixtures with 0.5 N HCl, extraction of the resultant isosteviol with n-hexane, and trimethylsilylation in pyridine.
[d] Aqueous solution.

about 20% of this sweetening agent. Both stevioside and rebaudioside A were stable when formulated in acidulated beverages that were stored at 60°C for over 5 days (Chang and Cook, 1983).

The percentage recoveries of stevioside, following addition to soy sauce, vegetable protein hydrolysate, and other media, at 0.1% concentration and normal and elevated temperatures, are summarized in Table VI. While stevioside seems to be relatively stable in food substances at slightly elevated temperatures, considerable decomposition occurs at higher temperatures of storage (Table VI). Shirakawa and Onishi (1979), who conducted this work, did not attempt to characterize any of the stevioside breakdown products encountered. Studies have not been reported involving the stability of stevioside *in vivo* or in the presence of digestive fluids *in vitro*.

III. SENSORY PROPERTIES OF THE *STEVIA REBAUDIANA* SWEET GLYCOSIDES

Stevioside (purity 93–95%) exhibits a persistent aftertaste, with bitterness and astringency, although when only 50% pure a less desirable aftertaste is experienced (Isima and Kakayama, 1976). However, the quality of sweetness of stevioside is considered to be preferable to that of glycyrrhizin or sodium saccharin and equal to or better than that of aspartame or sodium cyclamate (Abe and Sonobe, 1977). Using an incomplete-paired comparison organoleptic human taste panel, the sweetness of stevioside (pure) was found to be about 300 times that of sucrose at 0.4% sucrose concentration, 150 times sweeter at 4% sucrose concentration, and 100 times sweeter at 10% sucrose concentration (Isima and Kakayama, 1976). The functional groups of stevioside (Fig. 1, 1) thought to be essential for its sweet properties are the saccharide groups at C-13 and C-19, and the exomethylene group at C-17 (Isima and Kakayama, 1976; Kamiya *et al.*, 1979; Tanaka, 1980). DuBois and colleagues (1981) have partially synthesized a sweet sulfopropyl ester of steviol (Fig. 3, **22**) and have concluded that the carbohydrate portions of the stevioside molecule are not necessary for the exhibition of a sweet taste.

The relative sweet intensities of the *Stevia rebaudiana* sweet *ent*-kaurene glycosides have been reported by Tanaka (1980). Stevioside (Fig. 1, **1**) and rebaudioside E (Fig. 1, **7**) are about 100 to 150 times sweeter than sucrose (concentration not stated), while rebaudioside A (Fig. 1, **3**) and rebaudioside D (Fig. 1, **6**), which have a branched sugar unit at C-13, are perhaps 30% sweeter than **1** and **7** (Tanaka, 1980). It has been found

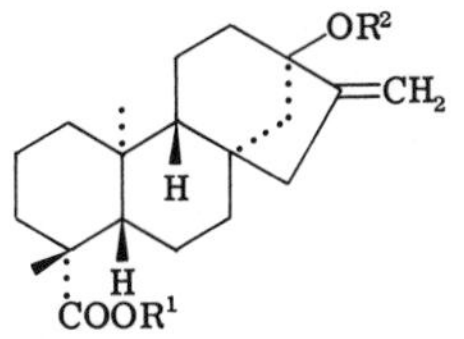

	R¹	R²
22	$(CH_2)_3SO_3Na$	β-D-Sophorosyl

FIG. 3 Sweet sulfopropyl ester analog of stevioside, nondegradable to steviol.

that replacement of a glucose residue with rhamnose decreases the sweetness of the resultant compound. Thus, rebaudioside C (Fig. 1, 5) and dulcoside A (Fig. 1, 8) are considerably less sweet than rebaudioside A and stevioside, respectively.

Several attempts have been made to improve the unpleasant aftertaste of stevioside. For example, it has been found that these effects are not removed by mixing it with acetic acid, citric acid, or sodium chloride, although some improvement in taste quality was achieved by the addition of sucrose, fructose, or glucose (Isima and Kakayama, 1976). According to Kim and Lee (1979), sucrose, glucose, fructose, invert sugar, and sodium saccharin, in decreasing order of effectiveness, improve the organoleptic quality of stevioside, when used in combination with the latter compound. A number of patented procedures for the taste improvement of stevioside have suggested advantageous combinations of this substance with other sweet *Stevia rebaudiana* diterpene glycosides (T. Morita *et al.*, 1977b,c; Morita, 1977c), amino acids (Morita, 1977d; Shimizu and Ochi, 1978; Anonymous, 1981a), peptides (Yamamota and Ishida, 1979), alkali salts of inorganic and organic acids (Ochi and Shimizu, 1978, 1979), alginic acid and gums (Anonymous, 1981b), somatin and D-xylose (Anonymous, 1980c,l,m), and maltitol and glucono-δ-lactone (Matsuoka, 1978; Anonymous, 1980g). The bitter afterflavor of *Stevia* extract may also be removed by column chromatography followed by calcium hydroxide treatment (Anonymous, 1980d). In addition, several Japanese patents have described the use of stevioside or an aqueous extract of *S. rebaudiana* to improve the taste of artificial sweeteners (Sasaki and Murakami, 1977), sucrose (Kukuchi and Suguri, 1977), and fructose (Anonymous, 1980e), to improve the cool mouth feel of soft drinks (Fujita and Edahiro, 1980), and to improve tobacco flavor and aroma (Morita, 1977f) and dough (when used in combination with propylene glycol) (Shidehara, 1980). Similar patents have been issued for *S. rebaudiana*-derived sweet compounds other than stevioside (Morita, 1977b,e; E. Morita *et al.*, 1977; T. Morita *et al.*, 1977a; Miyake, 1979, 1980; Anonymous, 1981c), including rebaudioside A (Morita, 1977b,e; E. Morita *et al.*, 1977; T. Morita *et al.*, 1977a).

IV. SAFETY AND TOXICITY ASSESSMENT OF EXTRACTS AND PURE COMPOUNDS FROM *STEVIA REBAUDIANA*

A. ACUTE TOXICITY

In early work, stevioside as purified by Bridel and Lavielle (1931a) was reported to be nontoxic to the rabbit, guinea pig, and fowl and to be excreted without structural modification (Pomaret and Lavieille, 1931). The LD_{50} of an extract of *Stevia rebaudiana* leaves containing 50% stevioside, when administered i.p. to rats, was reported to be 3.4 g/kg (Lee *et al.*, 1979). In general, two types of extracts of *S. rebaudiana* leaves have been tested for acute toxicity. Type A (crude) extracts contain about 20% and type B extracts contain about 40% stevioside (Akashi and Yokoyama, 1975). Type A extracts have an LD_{50} when administered orally to mice of ~17 g/kg, and type B extracts, more than 42 g/kg (Fujita and Edahiro, 1979a; Akashi and Yokoyama, 1975; Anonymous, 1981i). Without giving details, Kim and co-workers (Lee *et al.*, 1979) have reported that stevioside has an LD_{50} of 8.2 g/kg when administered orally to rats.

In a similar type of study, our group was unable to show acute toxicity when separate 2.0 g/kg doses of stevioside, rebaudiosides A–C, steviolbioside, or dulcoside A were administered to mice by oral intubation (Medon *et al.*, 1982). Also, no significant differences in body or organ weights were evident at sacrifice 2 weeks after compound administration (Medon *et al.*, 1982).

B. SUBACUTE TOXICITY

Three-month subacute toxicity studies have been carried out using 5-week-old male and female SLC–Wistar strain rats. A water extract of *Stevia rebaudiana* leaves, having a known stevioside concentration, was mixed with standard laboratory feed to give final stevioside concentrations in the feed of 0.28% (group A), 1.4% (group B), and 7.0% (group C). Fifteen males and 15 females were used in each group, e.g., groups A, B, and C, as well as a control group. Body weight and food consumption were measured, and a number of hematological and biochemical tests were carried out, as well as urinalysis; each animal was sacrificed and autopsied at the end of the 50-day study, and gross observations of the major organs were made. Tissues were taken from 10 male and 10 female animals in each group for histopathological examination. Statistical analyses were carried out on all data (Akashi and Yokoyama, 1975). No deaths were observed during the entire study, nor were there any particular symptoms noted in the animals. Response to light and sound were normal, and the appearance of the feces, the luster of the hair, and animal movements were all normal (Akashi and Yokoyama, 1975). There was a significant decrease in body weight in the males of group C (high dose, 7.0% stevioside) beginning on the fifth week of the experiment; however, there was a significant increase in the weight gains of the females in group A (low dose), similar to the control groups (Akashi and Yokoyama, 1975). No marked differences in food consumption were observed relative to controls in any of the groups (Akashi and Yokoyama, 1975). Some differences were noted in hematological parameters studied, but they were randomly arranged and were within normal physiological limits (Akashi and Yokoyama, 1975). Any changes in the biochemical parameters measured during the study were not dose-related and were considered to be normal (Akashi and Yokoyama, 1975). There were no significant changes in the urinanalyses relative to controls (Akashi and Yokoyama, 1975). Organ weights and gross characteristics on autopsy were considered to be normal following correlation of any differences with the histopathological examinations of the lung, heart, liver, kidney and bone marrow, but these were not dose-related (Akashi and Yokoyama, 1975).

In these rather extensive studies it was concluded that up to 7.0% stevioside, when fed to male and female rats for 3 months, produced no remarkable toxic effects.

A similar study was carried out in South Korea with an aqueous extract of *Stevia rebaudiana* leaves containing about 50% stevioside. Two levels of extract were mixed with laboratory chow for feeding studies. One was prepared so that each rat would receive 0.25 g stevioside daily in 15 g feed, and the other so that each rat would receive 0.5 g stevioside daily in 15 g feed. Animals were fed the experimental diets daily for 56 days, after which time they were sacrificed (Lee *et al.*, 1979). The usual parameters were examined during the course of the experiment, such as measurement of body weight, hematology (RBC, WBC, Hb, hematocrit), blood chemistry (total protein, albumin, α_1-, α_2-, β-, and γ-globulin, glucose, triglycerides, total cholesterol, creatinine, urea, inorganic phosphate, calcium, and GOT and LDH), and histological examination of liver tissue (Lee *et al.*, 1979).

There were no abnormalities reported from this study relative to controls that were dose-related, except for a significant decrease in serum LDH levels (Lee *et al.*, 1979). These results correlate well with those reported in the similar Japanese study indicated previously.

C. EFFECTS ON REPRODUCTION

It is claimed that Paraguayan Matto Grosso Indians use the leaves and stems of *Stevia rebaudiana* in the form of a tea as a contraceptive (Planas and Kuć, 1968). Planas and Kuć (1968) reported an antifertility effect for aqueous extracts of *S. rebaudiana* administered for periods up to 2 months following the addition of 10 ml of a 5% *S. rebaudiana* extract to drinking water of both male and female rats. This would seem to give some credence to the alleged folkloric use. However, workers in several other laboratories have been unable to confirm this antifertility effect in rats (Farnsworth, 1973; Akashi and Yokoyama, 1975; Fujita and Edahiro, 1979a; Anonymous, 1981i). Later inquiries made in several locations in northeastern Paraguay did not confirm the use of *S. rebaudiana* extracts for contraceptive purposes (Soejarto *et al.*, 1983a).

In a detailed study, an extract of *Stevia rebaudiana* leaves containing 53.1% stevioside was mixed with standard laboratory ration to give three dose levels of stevioside for feeding to three different groups of rats, for example, 0.28, 1.40, and 7.0%. After receiving normal ration for several days, 13-week-old male and female SLC–Wistar strain rats were fed the stevioside-containing ration for 21 days. After this period, one male was placed with two females for mating. Female rats were examined daily for evidence of a vaginal plug, which was considered as day 0 of pregnancy. On day 20, half of the pregnant animals were laparotomized and the remaining half were allowed to have spontaneous deliveries. At the time of mating, the stevioside-containing ration was removed and substituted with standard laboratory ration. The results are shown in Table VII. These data show a lack of antifertility effect for the extracts tested (Akashi and Yokoyama, 1975). In this study (Akashi and Yokoyama, 1975), there was no evidence of any teratogenic effects or other abnormalities in the fetuses or offspring.

In a later study it was reported that male and female rats, maintained with up to 3.0% stevioside in the diet, exhibited no abnormal signs in mating performance and fertility. Furthermore, no skeletal or other teratogenic effects attributable to stevioside were observed in fetuses after birth (Mori *et al.*, 1981).

D. MUTAGENICITY POTENTIAL

Stevioside and crude extracts derived from *Stevia rebaudiana* have been thoroughly evaluated for mutagenic activity. Utilizing *Salmonella ty-*

TABLE VII

PREGNANCY RATE OF RATS FED STEVIOSIDE-CONTAINING RATION[a]

Group[b]	Number of females	Number of pregnancies (Confirmed day of pregnancy)							Total	Pregnancy rate (%)	Pregnancy rate as compared to control group (%)
		1	2	3	4	5	6	7			
Control	10	1	0	2	3	2	0	0	8	80	100
A	10	0	2	2	2	0	0	0	6	60	75
B	10	1	1	3	2	1	0	0	8	80	100
C	10	0	2	3	2	0	0	0	7	70	88

[a]From Akashi and Yokoyama (1975).
[b]Group A, 0.28% stevioside in ration; B, 1.4%; C, 7.0%.

phimurium strains TA98, TA100, TA1535, TA1538, and TM677, or *Escherichia coli* strain WP2, either in the presence or absence of a 9000 *g* supernatant metabolic activating system derived from rat liver (S-9), these substances have been reported as nonmutagenic by five independent laboratories (Akashi and Yokoyama, 1975; Okumura *et al.*, 1978; Fujita and Edahiro, 1979a; Anonymous, 1981i; Medon *et al.*, 1982). One laboratory reported a slight positive effect only with the crude product (not stevioside), in the presence of S-9, with strain TA100 (not TA98) (Anonymous, 1981i). Mutagenicity tests performed with recombination-deficient (rec⁻) strains of *Bacillus subtilis* (H17 and M45) also showed the substances to be innocuous, both in the presence or absence of S-9, as did host-mediated tests with mice bearing *S. typhimurium* strain G46 (Okumura *et al.*, 1978).

No activity was demonstrated in the silkworm oocyte test system, and the substances neither enhanced sister chromatid exchange with human fetal fibroblasts, nor did they induce grossly visible chromosomal aberrations with these cells or rat medullary cells. A slight positive result was reported for tests examining the cellular growth and chromosomal structure of fibroblasts derived from Chinese hamster lung with *S. rebaudiana* crude extract (Anonymous, 1981i).

We have found that steviol (Fig. 2, **17**) is highly mutagenic in a forward mutation assay utilizing *S. typhimurium* strain TM677, in the presence of a metabolic activating system derived from a 9000 g supernatant fraction from the liver of Aroclor 1254-pretreated rats. Unmetabolized steviol was not active in this system, and neither was isosteviol (Fig. 2, **18**), regardless of activation (Pezzuto *et al.*, 1983).

Thus, a reasonably broad range of tests have been performed with stevioside and crude *S. rebaudiana* extracts and no significant mutagenic or genotoxic activity has been reported. However, our recent demonstration of the mutagenic effects of metabolite(s) of steviol offers cause for concern about the safety for human consumption of stevioside, since many bacterial mutagens are also carcinogens. Tests which may be of future value include the evaluation of mutagenic potential in cultured mammalian cells and the examination of the karyotype of human subjects who routinely ingest stevioside.

E. EFFECTS ON CARBOHYDRATE METABOLISM

Based on recent reports on the hypoglycemic activity of extracts of *Stevia rebaudiana* (Oviedo *et al.*, 1970; H. Suzuki *et al.*, 1977; Miguel, 1977; von Schmelling *et al.*, 1977), the leaves, including flowers and stems, of the plant are now used in popular medicine in Paraguay (in the form of a tea) as a remedy for diabetes (Soejarto *et al.*, 1983a). However, data in support of *S. rebaudiana* leaves having an effect on carbohydrate metabolism are equivocal. Virtually all data in support of extracts of this plant having a hypoglycemic effect have been published only in the form of abstracts, with little evidence, other than claims, that such an effect was produced.

The earliest study of this type was in Paraguay by Oviédo and co-workers (1970), who administered a dried aqueous extract of *Stevia rebaudiana* leaves orally to each of 25 healthy adults, without specifying the dose employed, in a "double-blind" study. They claimed an average 35.2% fall in normal blood sugar levels from initial values, between 6 and 8 hr after giving the extracts (Oveido *et al.*, 1970).

A further study carried out in Brazil as an abstract in 1981 involved the daily administration of an aqueous extract (equivalent to 1 g stevioside) to each of 15 normal human subjects ranging in age from 19 and 25, including both men and women (Alvares *et al.*, 1981). The extract was given in divided doses every 6 hr, so that each person received 250 mg stevioside equivalent daily. The first dose of the extract was given 12 hr after a glucose tolerance test, and the last dose was given 2 hr before a second glucose tolerance test 4 days later. It was reported only that an "accentuated hypoglycemic response" was observed in the patients studied, but no mention was made of the results from a control group, or even if controls were included in the study (Alvares *et al.*, 1981).

H. Suzuki and co-workers (1977) have studied the effect of feeding rats with a high-carbohydrate diet containing 10% dried *Stevia rebaudiana* leaves (corresponding to about 0.5% stevioside in the diet). After 2 weeks of feeding, there was a significant decrease in liver glycogen ($p < .01$), which further decreased at 4 weeks of feeding ($p < .001$). On the other hand, blood glucose was unaffected after 2 weeks of feeding, but was significantly decreased ($p < 0.01$) at 4 weeks, from a control value of 119 ± 3.1 mg/100 ml to 100 ± 8.0 mg/100 ml.

Lee and co-workers (1979) reported no change in blood glucose levels when crude extracts of *Stevia rebaudiana* leaves were fed to rats for 56 days, each rat consuming from 0.5 to 1.0 g extract/day. Akashi and Yokoyama (1975) also reported no dose-related effects on blood glucose after feeding studies in rats for 56 days with extracts of *S. rebaudiana* in which the final concentration of stevioside in the ration was 7.0%.

Addition of stevioside (0.1%) to a high-carbohydrate diet in rats is reported to cause a decrease of liver glycogen ($p < 0.05$), but not of blood glucose levels. Rats fed a high-fat diet containing 0.1% stevioside had no significant changes in blood glucose or liver glycogen levels, in comparison with those of rats fed a high-fat diet alone (H. Suzuki *et al.*, 1977).

Using the isolated rat pancreas, perfusion with 15 μM stevioside did not influence the induction by L-arginine HCl (19.2 mM) of insulin or glucagon secretion, which was indicative of a lack of a direct action on pancreatic α and β cells, and hence these data would not be supportive of studies claiming hypoglycemic effects *in vivo* for stevioside (Usami *et al.*, 1980).

In a study by Brazilian investigators, stevioside was administered i.v. to alloxan-treated rabbits at a dose of 7.0 mg/kg. These authors claim that their data support those of others who have reported hypoglycemic and/or antihyperglycemic effects for *Stevia rebaudiana* leaf extracts (Pinheiro and Gasparini, 1981). Clearly, however, the graphic data presented in this report show only a transient antihyperglycemic effect in the treated rabbits, but the variability of results in this type of assay, even under ideal experimental conditions, requires statistical treatment to ascertain validity. The data in this report did not receive such an analysis (Pinheiro and Gasparini, 1981).

Thus, even though some equivocal data exist suggesting that both *Stevia rebaudiana* leaf extracts and stevioside lower blood sugar levels in animals and/or humans, other more reliable animal data clearly are not in support of an effect on blood glucose levels.

F. ANTIMITOTIC ACTIVITY

In antimitotic tests used to evaluate the effects of hot water extracts of *Stevia rebaudiana* leaves on the cell cycle of *Allium cepa* meristems (onion root), no mitotic alterations or alterations of the distribution of cells in G_1, S, and G_2 stages of the cell cycle were seen. Although cell-cycle durations were prolonged by the infusions of *S. rebaudiana,* no specific toxicological effects on the cell cycle were observed (Schvartzman *et al.*, 1975, 1977).

G. ANTIMICROBIAL ACTIVITY

Water-soluble and lower-alcohol-soluble extracts prepared from *Stevia rebaudiana* leaves, have been reported to be separated by chromatography into eight unidentified compounds. Two of the components were shown to be active against several species of test bacteria, especially against *Pseudomonas aeruginosa* and *Proteus vulgaris,* while the remaining six components had sweet properties and little antibacterial activity (Anonymous, 1980h).

Stevioside was found to be a less favorable substrate for *Streptococcus mutans* serotypes with regard to acid production than either sucrose, glucose, or fructose (Berry and Henry, 1981), which suggests that it could be used in oral preparations as a sweetener in lieu of common sugars, to lower the incidence of dental caries.

Attempts have been made to study the effect of stevioside on the growth and metabolic activity of microorganisms commonly associated with dental caries. In one such study, stevioside markedly suppressed the growth of *Streptococcus mutans*, as did glucose, both being evaluated at a concentration of 0.5%. *Lactobacillus plantarum* behaved similarly. When stevioside, sorbitol, and xylitol were compared for their inhibitory effects against *L. plantarum* and *L. casei*, stevioside and xylitol completely inhibited microbial growth; sorbitol was only slightly effective. Combinations of stevioside and glucose, as well as glucose and sorbitol, showed moderate and marked inhibitory effects against *L. plantarum,* and *L. casei* and *S. mutans*, respectively. A combination of glucose and xylitol against the two *Lactobacillus* species showed moderate inhibitory activity, but against *S. mutans* the effect was marked. Against *S. mutans* strain HS-1, combinations of sucrose with stevioside, sorbitol, or xylitol were identical in their lack of inhibitory effects. Similar results were obtained with *S. mutans* strains IB and GS-5. However, the latter two strains showed differences in susceptibility to the effects of combined sucrose and xylitol. This combination of sugars almost completely inhibited the growth of *S. mutans* strain GS-5, but only slightly inhibited strain IB. Dextran sucrase from *S. mutans* was inhibited about 18% by $10^{-2} M$ stevioside, 10% by the same concentration of glucose, 5% by xylitol, and there was no inhibition by sorbitol. The inhibitory effects of dextran sucrase on *S. mutans* strains IB, HS-1, and GS-5 differed only slightly. Stevioside inhibited invertase about 20%, whereas sorbitol and xylitol had no effect (Yabu *et al.*, 1977).

While there are some differences in the action of stevioside, sucrose, xylitol, sorbitol, and glucose against common oral bacteria or their derived enzymes, the differences do not appear to be sufficient to differentiate these sweeteners as one being preferential to another.

H. CARDIOVASCULAR EFFECTS

In a brief report based on work carried out in Brazil, a tea prepared from *Stevia rebaudiana* leaves was administered daily for 30 days to 18 human subjects ranging in age from 20 to 40 years and having normal blood pressure. Weight, vitality, arterial pressure, and cardiac and electrocardiogram (ECG) data were recorded. It was concluded that the only effect brought about by ingestion of the tea was a discrete lowering of the systolic and diastolic arterial blood pressure of $\sim$9.5%. There was also a discrete prolongation noted in the electric systole in the ECG (Boeckh, 1981).

In the same report, a single oral dose of *Stevia rebaudiana* tea was prepared from 3.0 g leaves and given to each of 10 normal adult subjects. The same parameters were monitored as previously noted. A lowering of systolic blood pressure by 9.5% was noted, without alteration of the diastolic pressure. Bradycardia was found, with a discrete shortening of the duration of electric systole (QT_c) (Boeckh, 1981). Insufficient details were available in this report to determine whether the lowering of blood pressure reported would be of any consequence in the human situation.

I. EFFECTS ON SMOOTH MUSCLE

Stevioside (10^{-4} M) (83% pure) is reported to markedly enhance the contractile response of guinea pig ileum *in vitro* stimulated with K^+ (165 mM), Ca^{2+} (0.3 mM), or acetylcholine. Stevioside, on the other hand, was observed to inhibit acetylcholine-induced bradycardia in the rat right atrium *in vitro* (Turrin, 1981). On the basis of these and related data, it was postulated that stevioside has an effect on membrane transport, possibly facilitating the transport of cations (Turrin, 1981).

J. METABOLISM OF STEVIOSIDE AND REBAUDIOSIDE A

In spite of the fact that acute oral administration of large doses of stevioside and/or *Stevia rebaudiana* extracts and long-term studies with feeding either of these materials to laboratory animals have shown them to be virtually devoid of toxic effects, one must consider the limited data available on metabolites of the major sweet principles of this plant.

Wingard and co-workers (1980) have reported that stevioside and rebaudioside A are both degraded to the aglycone steviol by rat intestinal microflora *in vitro*. In this study, bacterial suspensions were prepared by removing the cecal contents from a freshly sacrificed Sprague–Dawley rat and suspending 0.1 g (fresh weight)/ml in anaerobic Krebs–Ringer 0.25 M phosphate buffer (pH 7.4) containing 0.25 mg dithiothreitol/ml. The suspensions were gassed with an oxygen-free mixture of N_2 and CO_2 (9:1) and passed through Pyrex wool. Aliquots of 1.0 ml substrate solution were prepared, gassed, tightly stoppered, and incubated at 37°C for periods up to 6 days. The bacterial cells were sedimented by centrifugation and the supernatant and pellet fractions were analyzed sepa-

TABLE VIII

DEGRADATION OF DITERPENE GLYCOSIDE SWEETENERS BY WHOLE CELLS OF RAT CECAL BACTERIA IN VITRO[a]

Substrate	Incubation period (days)	Fraction[b]	Percentage recovered as steviol[c]	Percentage of theoretical recovered
Stevioside (2.5 g/ml)	2	Supernatant	71	107
		Pellet	36	
Rebaudioside A (3.0 mg/ml)	2	Supernatant	60	65
		Pellet	5	
	4	Supernatant	74	83
		Pellet	9	
	6	Supernatant	51	108
		Pellet	57	
Steviol (0.2 mg/ml)	2	Supernatant	80	100
		Pellet	20	
	4	Supernatant	80	87
		Pellet	7	
	6	Supernatant	80	105
		Pellet	25	

[a] From Wingard et al. (1980), with permission of Birkhäuser Boston, Inc.
[b] Cecal bacterial suspensions were prepared by removing the cecal contents from a freshly sacrificed Sprague–Dawley rat and suspending 0.1 g fresh weight/ml in anaerobic Krebs–Ringer 0.25 M phosphate buffer (pH 7.4) containing 0.25 mg dithiothreitol/ml. The suspensions were gassed with oxygen-free N_2/CO_2 (9/1) and passed through Pyrex wool. One milliliter aliquots of the anaerobic cecal bacterial cell suspensions containing 0.4 mg/ml α-D-glucose were mixed with 1.0 ml of the substrate solution (final concentrations given above), gassed, tightly stoppered, and incubated at 37°C for periods up to 6 days. The bacterial cells were sedimented by centrifugation and the supernatant and pellet fractions analyzed separately.
[c] Steviol was determined by isocratic high-pressure liquid chromatography (HPLC) with an apparatus equipped with a monochromator operating at 200 nm. Quantitation was based on peak height using a 30 cm × 4 mm inner diameter reversed-phase μBondapak C-18 column (Waters Associates) eluted with 50% acetonitrile/50% 0.03 M KH_2PO_4 aqueous buffer at pH 4.8 (retention time 7 min).

rately. Steviol was determined by isocratic high-pressure liquid chromatography in a quantitative manner.

As can be seen from Table VIII, incubation of stevioside for 2 days under these conditions resulted in 100% conversion of the compound to the aglycone steviol. Under the same conditions, rebaudioside A was converted to steviol with about 65% conversion after 2 days of incubation, 83% conversion after 4 days incubation, and 100% conversion after 6 days incubation. Steviol itself was recovered unchanged in 87–100% yield following 2–6 days incubation. Cell-free extracts converted stevioside and rebaudioside A more slowly, with 50% conversion of stevioside to steviol after 7 days incubation and only 2% conversion of rebaudioside A to steviol being achieved after 7 days. Wingard and co-workers (1980) have extrapolated these data to humans and suggest that the human bowel could convert over 0.4 g rebaudioside A to steviol each hour.

These same workers (Wingard *et al.*, 1980) extended their *in vitro* studies to determine the extent of absorption-distribution of [17-^{14}C]steviol in normal intact and bile duct cannulated rats following oral and/or intracecal administration. The data are presented in Table IX and suggest that [17-^{14}C]steviol is almost completely absorbed from the lower bowel of the rat after oral or intracecal administration. Radioactivity was largely excreted in the urine of bile-duct ligated animals and in the bile of bile-duct-cannulated and intact animals. Very little $^{14}CO_2$ (0.02% of dose) was observed, indicating the stability of the exocyclic methylene label. Certain facts appeared to be presumed that may not be true in suggesting that humans would metabolize stevioside and/or rebaudioside A to steviol, which would then be absorbed, following oral administration. Stevioside and rebaudioside A were metabolized in the Wingard studies to steviol *in vitro* by rat cecal flora. Humans do not have a cecum, and it may be anticipated that the microbial flora of the human intestinal tract contains different microorganisms than does the rat cecum. Thus, it remains to be shown that the microbial flora of other species, or that of humans, will metabolize the *Stevia* sweet principles in a manner analogous to that described by Wingard and co-workers (1980).

In view of the mutagenic activity of metabolized steviol mentioned earlier (Pezzuto *et al.*, 1983), there would seem to be an urgent need to determine whether steviol is metabolically obtained on ingestion of the *Stevia rebaudiana* sweet glycoside in humans. If the products of metabolism from stevioside and rebaudioside A following oral ingestion by humans do not include steviol, such products must be shown to be absorbed and must ultimately be identified.

TABLE IX

RECOVERY OF RADIOACTIVITY AFTER ADMINISTRATION OF [17-^{14}C] STEVIOL TO INTACT BILE-DUCT-LIGATED OR BILE-DUCT-CANNULATED RATS[a]

Group	Radioactivity (% administered dose)							
	Urine	Feces	Organs	Gut content	CO_2	Bile	Cage	Total
Intact oral[b]	1.48 ± 0.36	96.36 ± 7.07	0.03 ± 0.01	0.22 ± 0.17	0.10 ± 0.01[c]	—	0.08 ± 0.04	98.26 ± 6.98
Ligated oral[d]	96.00 ± 0.28	3.31 ± 1.66	0.05 ± 0.01	0.17 ± 0.17	—	—	0.83 ± 0.09	100.4 ± 1.3
Ligated intracecal[e]	94.3 ± 3.88	6.03 ± 4.28	0.01 ± 0.003	0.03 ± 0.02	0.02 ± 0.001	—	1.16 ± 0.76	101.9 ± 4.4
Cannulated intracecal[f]	—	—	—	—	—	105.5 ± 8.45	—	—

[a]From Wingard et al. (1980), with permission of Birkhäuser Boston, Inc.

[b]X ± SD, n = 3, normal intact rats dosed orally with 1 ml (1.7 μCi, 0.7 mg) of [17-^{14}C] steviol in a 0.5% klucel suspension.

[c]X ± SD, n = 2, normal intact rats dosed with [17-^{14}C] steviol, as above.

[d]X ± SD, n = 2, bile-duct-ligated rats dosed orally with 1 ml (0.92 μCi, 0.37 mg) of [17-^{14}C] steviol in a 0.5% klucel suspension.

[e]X ± SD, n = 5, bile-duct-ligated rats dosed intracecally with 1 ml (2.63 μCi, 0.92 mg) of [17-^{14}C] steviol in a 0.5% klucel suspension.

[f]X ± SD, n = 2, bile-duct-cannulated rats dosed intracecally with 1 ml (1.7 μCi, 0.7 mg) of [17-^{14}C] steviol in a 0.5% klucel suspension.

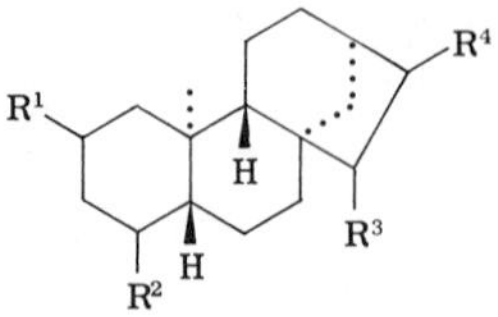

	R¹	R²	R³	R⁴
23 Atractyligenin	OH	α-COOH	OH	=CH$_2$
25 Atractylitriol	OH	CH$_2$OH	OH	=CH$_2$
26 Dihydroatractylitriol	OH	CH$_2$OH	OH	CH$_3$
27 Diketoatractyligenin methyl ester	=O	α-COOCH$_3$	=O	=CH$_2$

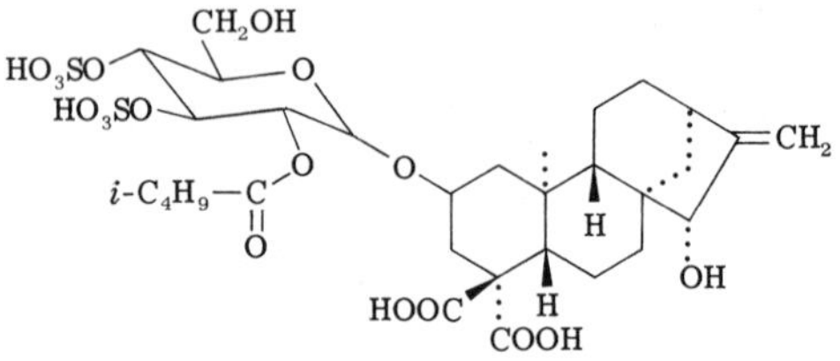

24 Carboxyatractyloside

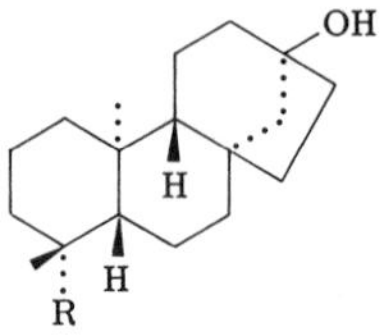

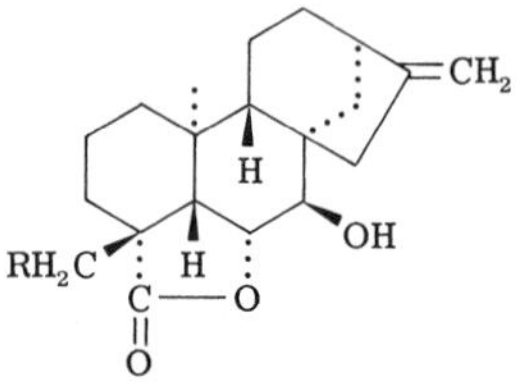

28 Dihydrosteviol (R = COOH)
29 13-Hydroxystevane (R = CH$_3$)

30 7-Hydroxykaurenolide (R = H)
31 7,18-Dihydroxykaurenolide (R = OH)

FIG. 4 Structures of some analogs of steviol that exhibit inhibitory effects on oxidative phosphorylation.

K. EFFECTS OF STEVIOSIDE, STEVIOL, AND RELATED *ENT*-KAURENE DITERPENOIDS ON OXIDATIVE PHOSPHORYLATION

A further reason for discussing the metabolic products of stevioside and rebaudioside A is that a report has been published that shows that steviol, the *in vitro* metabolite of both of these compounds, inhibits the

TABLE X

COMPARATIVE ACTIVITIES ON OXIDATIVE PHOSPHORYLATION OF STEVIOL, STEVIOSIDE, AND SOME STRUCTURAL ANALOGS[a]

Inhibitor (Figs. 1–3)	Concentration required for 50% inhibition of ATP Synthesis $(M \times 10^4)$[b,c]
Steviol (**17**)	0.4
Stevioside (**1**)	12.0
Atractyligenin (**23**)[d]	2.1
Atractylitriol (**25**)	2.7
Dihydroatractylitriol (**26**)	5.0
Diketoatractyligenin methyl ester (**27**)	3.0
Dihydrosteviol (**28**)	1.0
13-Hydroxystevane (**29**)	1.2
7-Hydroxykaurenolide (**30**)	0.8
7,18-Dihydroxykaurenolide (**31**)	7.5

[a]From Vignais *et al.* (1966).
[b]Inhibitors in 9.5% (v/v) ethanol (10–20 μl); rat liver mitochondria (2.5–3.5 mg protein) in 0.1 ml of 0.25 M sucrose were added. Final volume 2 ml, incubation at 28°C for 5 min.
[c]Incubation medium: 100 mM KCl; 16 mM phosphate buffer (pH 7.3), labeled with ^{32}P; 6 mM MgCl$_2$; 0.5 mM ADP; 1 mg hexokinase (type IV, Sigma); 20 mM glucose and 10 mM β-hydroxybutyrate.
[d]The atractyligenin concentration for ID$_{50}$ depends on the concentration of ADP.

mitochondrial translocation of adenine nucleotides (Vignais *et al.*, 1966). The report indicates that steviol behaves similarly to atractyligenin (Fig. 4, **23**). ADP, however, does not competitively antagonize the effects of steviol, but it does antagonize the effects of atractyligenin (Vignais *et al.*, 1966). The toxic effects of carboxyatractyloside (Fig. 4, **24**), a derivative of atractyligenin, are well known (Luciani *et al.*, 1978). Comparative activities of several structurally related compounds to steviol on oxidative phosphorylation in rat mitochondria are listed in Table X.

It should be noted that the molecular mechanisms of action of many respiratory poisons have been determined precisely, and several others have been categorized in a general group, for example, uncoupling agents. This suggests a specific bioaffinity of such agents for mitochondrial components required for normal oxidative phosphorylation. Although steviol can inhibit oxidative phosphorylation (Table X), no specificity was demonstrated (Vignais *et al.*, 1966). All of the reported partial reactions related to oxidative phosphorylation were inhibited by steviol. Since key respiratory components are membrane-bound, it is not unlikely that steviol (which is not very water-soluble) simply binds to the lipophilic components of the isolated mitochondria utilized in *in vitro* experimental systems (Vignais *et al.*, 1966).

Additionally, since many other steviol derivatives [e.g., *ent*-kaurene derivatives, such as 7-hydroxykaurenolide (Fig. 4, **30**), 13-hydroxystevane (Fig. 4, **29**), atractylitriol (Fig. 4, **25**), dihydroatractylitriol (Fig. 4, **26**) and diketoatractyligenin methyl ester (Fig. 4, **27**)] do not affect the translocation of adenine nucleotides, but they do inhibit electron transfer between NADH and cytochrome b in the respiratory chain, it is difficult to make structure–activity correlations with this group of compounds relative to their effects on oxidative phosphorylation (Vignais *et al.*, 1966). It should be pointed out that steviol is a *gem*-4,4-dimethyl *ent*-kaurene derivative, whereas atractyligenin, atractylitriol, dihydroatractylitriol, and diketoatractyligenin methyl ester are all 4-demethyl *ent*-kaurene derivatives and thus would be expected to behave differently in biological systems.

The effects by steviol on oxidative phosphorylation and the translocation of adenine nucleotides in mitochondria most likely are of academic interest only, since replicative feeding studies, including different levels of stevioside and *Stevia rebaudiana* extracts in the diets of laboratory animals fed over extended periods of time, have shown no deleterious effects (Lee *et al.*, 1979; Akashi and Yokoyama, 1975; Anonymous, 1981i). If indeed the *Stevia ent*-kaurene derivatives are metabolized to steviol, and the metabolites are absorbed from the gut, evidence is lacking from all of the feeding studies conducted to date that would suggest any type of toxicity for these metabolites.

L. ANDROGENIC AND ANTIANDROGENIC EFFECTS OF STEVIOL AND "DIHYDROISOSTEVIOL"

Steviol has been studied in the chick-comb antiandrogen assay, which involved the inunction of the compound to the comb of 2-day-old white leghorn male chicks stimulated by a single injection of testosterone (Dorfman and Nes, 1960). Steviol was inuncted once daily for 7 days, and then the combs were removed and weighed. Doses of steviol utilized in three separate experiments were 0.5, 2.0, and 3.0 mg/comb (total dose given to each chick over the seven day treatment period). The results showed that at the 2.0- and 3.0-mg/comb dose levels, a "tendency toward anti-androgen activity was observed." Two trials at the 0.5-mg/comb level were negative (Dorfman and Nes, 1960).

A compound of apparently unknown structure, but which has been referred to as "dihydroisosteviol," has also been tested in the chick-comb antiandrogen assay as just described (Dorfman and Nes, 1960). "Dihydroisosteviol" has not been described as a metabolic product of any of the *Stevia ent*-kaurene sweeteners, nor has it been shown to be a degradative product of this class of compounds. It failed to produce an antiandrogen effect at dose levels of 0.5 and 1.5 mg/comb, but did show a statistically significant effect at 3.0 mg/comb (Dorfman and Nes, 1960).

"Dihydroisosteviol" did not stimulate the seminal vesicles, prostate, or *levator ani,* nor was this compound effective in inhibiting the action of testosterone at total dose levels of 5.0 and 20 mg per animal when injected subcutaneously in 28-day-old Charles River male castrated rats. Injections of the test compound were given in divided doses daily for 7 days (Dorfman and Nes, 1960).

M. EFFECTS OF STEVIOL ON PLANT GERMINATION

Steviol has also been reported to show gibberellinlike activity in the lettuce and cucumber elongation tests and on the growth of bean plants (Valio and Rocha, 1976). This type of plant growth-regulating activity is not unique to *ent*-kaurene or gibberellin derivatives, but has been shown for a large number of naturally occurring compounds of diverse chemical structure.

N. BIOLOGICAL ACTIVITIES OF TRACE CONSTITUENTS OF STEVIA REBAUDIANA

As shown in Tables I and II, *Stevia rebaudiana* contains, in addition to the major *ent*-kaurene glycoside sweet principles, three labdane diterpenes, two triterpenes, two sterols, three alkanes, 12 monoterpenes, 15 sesquiterpenes, and other constituents. Most of these are trace constituents of the essential oil of *S. rebaudiana,* and none of these would be anticipated to contribute to adverse effects in humans if ingested.

V. USE OF STEVIOSIDE AND STEVIA REBAUDIANA EXTRACTS AS SWEETENING AGENTS

The use of stevioside and *Stevia rebaudiana* extracts for sweetening and for medicinal purposes appears to be restricted, at present, to two geographic locations, Paraguay/Brazil and Japan. Discussion of the current use of stevioside/*S. rebaudiana* extracts in these two regions follows.

A. USE IN PARAGUAY AND BRAZIL

In Paraguay, *Stevia rebaudiana* ("the sweet herb of Paraguay") has been quoted in the literature as having been used to sweeten beverages such as *maté* (*Ilex paraguayensis* St.-Hil.) by various native tribes since before colonization by the Spaniards in the sixteenth century (Gosling, 1901; Bertoni, 1905, 1918; Cabrera, 1939; Mors and Rizzini, 1966; Sumida, 1973; Abe and Sonobe, 1977; Felippe, 1977; Soejarto *et al.*, 1983a,b). As mentioned in Section IV,C, aqueous decoctions of *S. rebaudiana* leaves reputedly have been employed as a contraceptive agent (Planas and Kuć, 1968; Medon *et al.*, 1982) and for the treatment of hyperglycemia (L. E. de Gasperi, personal communication, 1981; Soejarto *et al.*, 1983a). The directions for use of one proprietary *S. rebaudiana* product (El Dulce Te del Paraguay, Luis E. de Gasperi, Asunción, Paraguay) are as follows.

This product (25 g) consists of the leaves, flowers, and fine branchlets of milled *Stevia rebaudiana*. It is recommended for the treatment of hyperglycemia and is claimed to have a wide margin of safety, although users are advised to seek the advice of a physician when using this product. A dose of up to 5 g/day, dissolved in the form of a tea, is recommended for patients with high initial blood sugar levels, and maintenance doses of 1 g/day are suggested for stabilized patients.

The basic directions for use require that tea be prepared by boiling briefly the stipulated amount of powder in water, and the decoction obtained is drunk either hot or cold. The tea so produced may be mixed with juices of lemon, orange, or pomelo, or can be consumed with *maté*. A variant of this, namely, the production of a concentrate or "honey," is obtained by boiling 50 g of the product with 1 liter of water until the consistency of a syrup is produced. This is then added dropwise to sweeten coffee and fruit juices. Both preparations described above may be refrigerated.

It is also claimed that the product has tonic, stomachic, and hypotensive actions, and can also help regulate obesity (L. E. de Gasperi, personal communication, 1981).

In Brazil, there has been a recent upsurge in interest in the use of *Stevia rebaudiana* as a sweetening agent, as evidenced by the organization of a symposium in June 1981 dealing with the history, cultivation, phytochemistry, analysis, and pharmacology of this plant and its constituents (Angelucci, 1981b). Two products, *S. rebaudiana* tea and *S. rebaudiana* capsules, manufactured by Eyra Distribuidora de Productos Alimenticios e Vegetais Ltda., were officially approved for sale in Brazil in 1980 (Anonymous, 1980o).

B. USE IN JAPAN

The most widespread use of stevioside and purified *Stevia rebaudiana* extracts is currently taking place in Japan (Kato, 1975; Abe and Sonobe, 1977; Akashi, 1977; Kazuyama, 1979; Anonymous, 1977, 1981c). Their use for sweetening has intensified since 1980, after a temporary ban was placed on sodium saccharin as a sweetener in that country (Anonymous, 1977, 1981j). Not only is *S. rebaudiana* grown in Japan but substantial amounts of this crop are grown in South Korea, Taiwan, and other countries in Southeast Asia for export to Japan (Shock, 1982). This rapid development of interest in stevioside is remarkable when it is considered that the plant was first successfully introduced by Sumida from Paraguay as recently as 1971 (Sumida, 1973; Kato, 1975).

Current estimates for the use of *Stevia rebaudiana* leaves in Japan for the production of stevioside or *S. rebaudiana* extracts range from a high of 700–1000 tons (dry weight) in 1979 (Fujita and Edahiro, 1979a) to 650–750 tons in 1981 (M. Ogura, personal communication, 1982). In the latter case, 450–500 tons would be produced in Japan and 200–250 tons imported from South Korea, the People's Republic of China, Taiwan, and other southeast Asian countries (M. Ogura, personal communication, 1982).

Most of the experimental studies on stevioside and *Stevia rebaudiana* extracts reported in the scientific literature have been carried out in Japan by a consortium of 11 firms collectively known as the "*Stevia* Konwakai" (*Stevia* Consortium). A list of these firms is presented in Table XI. In addition, a considerable number of other firms are active either in the production, formulation, or distribution of *Stevia* sweetening products, and have been awarded patents in this area.

TABLE XI

JAPANESE FIRMS CONSTITUTING THE "*STEVIA* KONWAKAI"

Dainippon Ink Kagaku Kogyo Co., Ltd., Tokyo
Fuji Kagaku Co., Ltd, Tokyo
Ikeda Toka Kogyo Co., Ltd., Hiroshima
Maruzen Kasei Co., Ltd., Tokyo
Morita Kagaku Kogyo Co., Ltd., Osaka
Nikken Kagaku Co., Ltd., Tokyo
Sanyo Kokusaku Pulp Co., Ltd., Tokyo
Sekisui Kagaku Kogyo Co., Ltd., Osaka
Tama Seikagaku Co., Ltd., Tokyo
Tokiwa Shokubutsu Kagaku Laboratories Co., Ltd., Chiba
Yoto Foods Co., Ltd., Tokyo

According to Akashi (1977), stevioside is appropriate for use as a sweetening agent in all foods, since it is highly stable to heat and acids; its sweetness characteristics are similar to those of sucrose; it is mild and free of aftertaste; it can be considered a noncaloric sweetening agent; it is nonfermentive; and it does not become yellowish when heated. In regard to the latter point, it has been shown that when stevioside and glutamic acid (50 mg of each) were heated at 100°C at pH 3 for 5 hr, no browning of the mixture was observed (Abe and Sonobe, 1977). Stevioside, however, suffers from several disadvantages when used as a sweetening agent in foods, being low in hygroscopicity, solubility, and permeability (Akashi, 1977). In addition, *Stevia rebaudiana* extracts possess a grassy taste (Ochi, 1979).

The *Stevia rebaudiana*-derived sweetening products sold by Maruzen Kasei K.K., Tokyo, Japan, in 1978, are used for the following types of food products, in descending order of sales volume: Japanese-style pickles; dried seafoods; flavoring materials; fish meat products; vegetables and seafoods boiled down with soy sauce; miscellaneous categories of use; and confectionery products (Kazuyama, 1979). Japanese-style vegetables, dried seafoods, soy sauce, and miso (bean paste) products are generally formulated with high levels of sodium chloride, which is added

TABLE XII

JAPANESE COMMERCIAL FORMULATION
OF TANGLES BOILED DOWN WITH SOY
SAUCE, CONTAINING STEVIOSIDE[a]

Ingredient[b]	Weight used (kg)
Tangles	18.0
Monosodium glutamate	12.0
Sodium chloride	8.0
Malmiron C[c]	0.508
Nucleic acids	0.265
Glycine	0.25
Seasoning sauce	0.04

[a] From Kazuyama (1979).
[b] Representative of both primary and secondary seasoning procedures.
[c] Product of Maruzen Kasei K.K., Tokyo, Japan. Formulation: Total stevioside 5%; sodium citrate 20%; contains glycyrrhizin (Kazuyama, 1979). Total stevioside content of formulation = 0.065% (w/w).

TABLE XIII

JAPANESE COMMERCIAL FORMULATION OF A
CARBONATED DRINK CONTAINING
STEVIOSIDE[a]

Ingredient	Weight/volume used
Granule sugar	17.8 kg
Histevia-100[b]	147 g
Citric acid	280 g
Sodium citrate	22 g
Cider essence	220 ml
Carbonated water	to 200 liters

[a] From Fujita and Edahiro (1979a).
[b] Formulation includes 10% stevioside. Total stevioside content of formulation = 0.007% (w/v).

for preservation and flavoring. It has been found that stevioside possesses a mellow taste that suppresses the pungent effects of the sodium
chloride (Kazuyama, 1979). In soft drinks, stevioside does not form a
precipitate under acidic conditions (pH ~3). It thus offers a significant
advantage over glycyrrhizin in this respect (Abe and Sonobe, 1977).

In Tables XII to XV, the formulations of a number of food products
commercially available in Japan, containing stevioside as a sweetening
agent, are presented. This information is taken from two review articles
(Kazuyama, 1979; Fujita and Edahiro, 1979a), and from the examples
cited it may be calculated that the concentration levels of stevioside in the
formulations, both savory and dessert items, are not greater than 0.1%
(w/w or w/v).

TABLE XIV

JAPANESE COMMERCIAL FORMULATION OF ORANGE
JUICE DRINK CONTAINING STEVIOSIDE[a]

Ingredient	Weight/volume used
Granule sugar	18 kg
Histevia-100[b]	150 g
Citric acid	440 g
Malic acid	60 g
Orange essence	220 ml
1/5 Orange juice concentrate	4.4 g
Water	to 200 liters

[a] From Fujita and Edahiro (1979a).
[b] Formulation includes 10% stevioside. Total stevioside content of formulation = 0.0075% (w/v).

TABLE XV
JAPANESE COMMERCIAL FORMULATION OF ICE CREAM
CONTAINING STEVIOSIDE[a]

Ingredient	Percentage
Butter	6.5
Whole milk	8.0
Nonfat dry milk	6.4
Dry dextrose syrup	6.34
Granule sugar	6.66
Licostevia S-1[b]	0.05
Emulsifier	0.3
Stabilizer	0.2
Water	65.55

[a] From Fujita and Edahiro (1979b).
[b] Formulation includes 10% stevioside and 6% glycyrrhizin.
Total stevioside content of formulation = 0.005% (w/w).

VI. CONCLUSIONS AND SUMMARY

1. *Stevia rebaudiana* extracts and/or stevioside, the most abundant *ent*-kaurene glycoside constituent of this plant, are useful sweetening agents, currently approved for food use in Japan, Paraguay, and Brazil.

2. *Stevia rebaudiana* extracts have been used for centuries to sweeten beverages in Paraguay. Within the last decade, *S. rebaudiana* extracts and stevioside have been widely used in Japan to sweeten a variety of food products, including Japanese-style pickles, dried sea foods, fish meat pastes, soy sauce and bean paste products, fruit-flavored drinks and other beverages, and dessert items such as ice cream and chewing gum.

3. Stevioside is considered to be appropriate for use as a sweetening agent for all foods, since it is highly stable to acids and heat, it has similar sweetness characteristics to sucrose, it can be regarded as noncaloric, it is nonfermentive, and it does not discolor when heated. These are highly desirable properties not associated with other sucrose substitutes.

4. Crude and purified *Stevia rebaudiana* extracts, as well as crystalline stevioside, have been tested for acute and subacute toxicity in rodents, and are considered to be safe for human consumption as a result of this work.

5. While it has been claimed in a single experimental paper that an aqueous extract of *Stevia rebaudiana* produced antifertility effects when incorporated into the drinking water of male and female rats, workers in

several other laboratories have been unable to confirm these antifertility effects. There has been no evidence that *S. rebaudiana* extracts or constituents produce teratogenic symptoms in rodents.

6. A variety of mutagenicity tests performed on *Stevia rebaudiana* extracts and stevioside has yielded almost entirely negative data. However, the recent demonstration of mutagenic activity of metabolized steviol offers cause for concern about the safety for human consumption of the *S. rebaudiana* sweet glycosides, in view of the demonstrated correlation of mutagenicity with carcinogenic activity for many compounds. Clearly, further studies dealing with these aspects are needed before the *S. rebaudiana ent*-kaurene glycosides can be pronounced safe for human ingestion as sugar substitutes.

7. A number of meeting abstract reports, from laboratories in certain South American countries, have claimed that stevioside has activity in several physiological systems. Full papers expanding on these preliminary reports have not so far appeared in the literature.

8. There are limited data on the *in vivo* and *in vitro* metabolism of the sweet principles of *Stevia rebaudiana,* including stevioside. Both stevioside and rebaudioside A have been shown to be degraded to the aglycone steviol by rat cecal microflora *in vitro*. Steviol is almost completely absorbed from the lower bowel of the rat after oral or intracecal administration. No products of metabolism of the sweet principles of *S. rebaudiana* have as yet been determined in humans; it remains to be shown if the microbial flora of human will metabolize stevioside and rebaudioside A to steviol.

9. Steviol inhibits mitochondrial translocation of adenine nucleotides. If, in the future, this compound is shown to be a metabolite of the *Stevia rebaudiana* sweet principles when administered to humans, further investigations of its toxicological effects may be necessary.

10. A compound of uncertain structure and source, "dihydroisosteviol," has weak to equivocal antiandrogenic activity in a chick-comb assay, but is devoid of such antiandrogenic activity when evaluated in castrated rats. "Dihydroisosteviol" is not known to occur naturally, nor has it been identified as a chemical degradative product of any *Stevia rebaudiana*-derived constituent.

11. Since *Stevia rebaudiana* extracts and/or stevioside have been widely employed as sweetening agents in Japan and other countries during the past decade, as evidenced by the need for up to 750 tons of *S. rebaudiana* leaves per year (estimated amount of stevioside would be 60 tons) in Japan alone, and no adverse reactions have been reported during this period, one might conclude, on the basis of those observations, that these materials offer no potential toxicity risk to humans.

ACKNOWLEDGMENTS

The contributions of Dr. N. R. Farnsworth, Dr. J. M. Pezzuto, and Dr. P. J. Medon are acknowledged for certain aspects of this manuscript. Our laboratory investigations referred to in this chapter were supported by contract NO1-NIH-DE-02425 with the National Institute of Dental Research, National Institutes of Health, Bethesda, Maryland. Dr. A. I. Bakal, ABIC International Consultants, Inc., Pine Brook, New Jersey, is to be thanked for making valuable suggestions towards the organization of this chapter. We are also very grateful to Dorothy Guilty for typing the manuscript.

REFERENCES

Abe, K., and Sonobe, M. (1977). *New Food Ind.* **19** (1), 67.
Ahmed, M. S., and Dobberstein, R. H. (1982a). *J. Chromatogr.* **236**, 523.
Ahmed, M. S., and Dobberstein, R. H. (1982b). *J. Chromatogr.* **245**, 373.
Ahmed, M. S., Dobberstein, R. H., and Farnsworth, N. R. (1980). *J. Chromatogr.* **192**, 387.
Akashi, H. (1977). *Shokuhin Kogyo* **20** (24), 20.
Akashi, H., and Yokoyama, Y. (1975). *Shokuhin Kogyo* **18** (20), 34.
Akashi, H., Yokoyama, Y., and Osada, M. (1975). *Jpn. Kokai Tokkyo Koho* **75 24,300;** *Chem. Abstr.* **83**, 5377b (1975).
Alvares, M., Bazzone, R. B., Godoy, G. L., Cury, R., and Botion, L. M. (1981). *Abstract Pap., Semin. Bras.* Stevia rebaudiana *Bertoni, 1st, 1981* p. XIII.I.
Angeles, E., Folting, K., Grieco, P. A., Huffmann, J. C., Miranda, R., and Salmón, M. (1982). *Phytochemistry* **21**, 1804.
Angelucci, E. (1981a). *Abstr., Semin. Brasil.* Stevia rebaudiana *Bertoni, 1st, 1981* pp. X.1–X.3. (1981a).
Angelucci, E. (1981b). "I Seminari Brasileiro sobre *Stevia rebaudiana.* Campinas 25 e 1981" (Abstr. Pap.). Governo do Estado de São Paulo, Secretaria de Agricultura e Abastecimento, Instituto de Technologia de Alimentos.
Anonymous (1977). *Confect. Prod.,* October issue, p. 408.
Anonymous (1980a). *Jpn. Kokai Tokkyo Koho* **80 28,080;** *Chem. Abstr.* **94**, 103787g (1981).
Anonymous (1980b). *Jpn. Kokai Tokkyo Koho* **80 40,596;** *Chem. Abstr.* **95**, 151116e (1981).
Anonymous (1980c). *Jpn. Kokai Tokkyo Koho* **80 46,695;** *Chem. Abstr.* **94**, 155288q (1981).
Anonymous (1980d). *Jpn. Kokai Tokkyo Koho* **80 47,871;** *Chem. Abstr.* **94**, 155261a (1981).
Anonymous (1980e). *Jpn. Kokai Tokkyo Koho* **80 54,871;** *Chem. Abstr.* **93**, 93830d (1980).
Anonymous (1980f). *Jpn. Kokai Tokkyo Koho* **80 81,567;** *Chem. Abstr.* **93**, 148477b (1980).
Anonymous (1980g). *Jpn. Kokai Tokkyo Koho* **80 88,675;** *Chem. Abstr.* **93**, 166396d (1980).
Anonymous (1980h). *Jpn. Kokai Tokkyo Koho* **80 92,323;** *Chem. Abstr.* **94**, 7721k (1981).
Anonymous (1980i). *Jpn. Kokai Tokkyo Koho* **80 92,400;** *Chem. Abstr.* **94**, 84457f (1981).
Anonymous (1980j). *Jpn. Kokai Tokkyo Koho* **80 111,768;** *Chem. Abstr.* **93**, 237303j (1980).
Anonymous (1980k). *Jpn. Kokai Tokkyo Koho* **80 120,770;** *Chem. Abstr.* **93**, 237280z (1980).
Anonymous (1980l). *Jpn. Kokai Tokkyo Koho* **80 138,372;** *Chem. Abstr.* **94**, 82556p (1981).
Anonymous (1980m). *Jpn. Kokai Tokkyo Koho* **80 156,562;** *Chem. Abstr.* **94**, 138134u (1981).
Anonymous (1980n). *Jpn. Kokai Tokkyo Koho* **80 159,770;** *Chem. Abstr.* **94**, 101655p (1981).
Anonymous (1980o) Concessão de Registro e Medicamento, Nos. 3,875/80 and 3,876/80. Diaro Oficial, Brazil.
Anonymous (1981a). *Jpn. Kokai Tokkyo Koho* **81 11,772;** *Chem. Abstr.* **94**, 173143t (1981).
Anonymous (1981b). *Jpn. Kokai Tokkyo Koho* **81 55,174;** *Chem. Abstr.* **95**, 60213t (1981).
Anonymous (1981c). *Jpn. Kokai Tokkyo Koho* **81 99,498;** *Chem. Abstr.* **95**, 202297z (1981).

Anonymous (1981d). *Jpn. Kokai Tokkyo Koho* **81** 99,768; *Chem. Abstr.* **95**, 167391b (1981).
Anonymous (1981e). *Jpn. Kokai Tokkyo Koho* **81**, 121,454; *Chem. Abstr.* **96**, 5171y (1982).
Anonymous (1981f). *Jpn. Kokai Tokkyo Koho* **81** 121,455; *Chem. Abstr.* **96**, 5172z (1982).
Anonymous (1981g). *Jpn. Kokai Tokkyo Koho* **81** 137,866; *Chem. Abstr.* **96**, 50966x (1982).
Anonymous (1981h). *Jpn. Kokai Tokkyo Koho* **81** 160,962; *Chem. Abstr.* **96**, 141434p (1982).
Anonymous (1981i). "Safety of *Stevia.*" Tama Biochemical Co., Ltd., Tokyo.
Anonymous (1981j). *Jpn. Chem. Week,* March 26, p. 6.
Anonymous (1982a). *Jpn. Kokai Tokkyo Koho* **82** 02,656; *Chem. Abstr.* **96**, 141436r (1982).
Anonymous (1982b). *Jpn. Kokai Tokkyo Koho* **82** 05,663; *Chem. Abstr.* **96**, 141440n (1982).
Asano, K., Tomomatsu, S., and Kawasaki, M. (1975). *Jpn. Kokai Tokkyo Koho* **75** 88,100; *Chem. Abstr.* **84**, 5323b (1976).
Bell, F. (1954). *Chem. Ind. (London)* pp. 897–898.
Berry, C. W., and Henry, C. A. (1981). *J. Dent. Res.* **60**, 430.
Bertoni, M. S. (1905). *An. Cient. Paraguay. Ser. I* **5**, 1.
Bertoni, M. S. (1918). *An. Cient. Paraguay. Ser. II* **6**, 129.
Boeckh, E. M. A. (1981). *Abstr. Pap., Semin. Bras.* Stevia rebaudiana *Bertoni, 1st, 1981* pp. XI.I–XI.II.
Bohlmann, F., Zdero, C., and Schöneweiss, S. (1976). *Chem. Ber.* **109**, 3366.
Bohlmann, F., Suwita, A., Natu, A. A., Czerson, H., and Suwita, A. (1977). *Chem. Ber.* **110**, 3572.
Bohlmann, F., Dutta, L. N., Dorner, W., King, R. M., and Robinson, H. (1979). *Phytochemistry* **18**, 673.
Bohlmann, F., Zdero, C., King, R. M., and Robinson, H. (1982). *Phytochemistry* **21**, 2021.
Bragg, R. W., Chow, Y., Dennis, L., Ferguson, L. N., Howell, S., Morga, G., Ogino, G., Pugh, H., and Winters, M. (1978). *J. Chem. Educ.* **55**, 281.
Bridel, M., and Lavieille, R. (1931a). *J. Pharm. Chim.* **14**, 99.
Bridel, M., and Lavieille, R. (1931b). *J. Pharm. Chim.* **14**, 154.
Bridel, M., and Lavieille, R. (1931c). *J. Pharm. Chim.* **14**, 161.
Bridel, M., and Lavieille, R. (1931d). *J. Pharm. Chim.* **14**, 321.
Bridel, M., and Lavieille, R. (1931e). *J. Pharm. Chim.* **14**, 369.
Bridel, M., and Lavieille, R. (1931f). *Bull. Soc. Chim. Biol.* **13**, 656; *Biol. Abstr.* **6**, 21498 (1932).
Bridel, M., and Lavieille, R. (1931g). *C. R. Hebd. Seances, Acad. Sci.* **192**, 1123.
Bridel, M., and Lavieille, R. (1931h). *C. R. Hebd. Seances Acad. Sci.* **193**, 72.
Brucher, H. (1974). *Naturwiss. Rundsch.* **27**, 231.
Cabrera, A. L. (1939). "Las Compuestas Utiles Cultivadas en la Republica Argentina, 1939," p. 14. Ministerio de Obras Publicas de la Provincia de Buenos Aires, Buenos Aires, Argentina.
Chang, S. S., and Cook, J. M. (1983). *J. Agric. Food Chem.* **31**, 409.
Chen, W.-S., and Yeh, C.-S. (1978). *T'ai-wan T'ang Yeh Yen Chiu So Yen Chiu Hui Pao* **79**, 43; *Chem. Abstr.* **89**, 211995d (1978).
Chueh, W.-J. (1977). *Tsu Pin Kung Yeh (Taiwan)* **9**, 34.
Chung, M. H., and Lee, M. H. (1979). *Saengyak Hakhoe Chi (Hanguk Saengyak Hakhoe)* **9**, 149.
Cook, I. F., and Knox, J. R. (1970). *Tetrahedron Lett.* p. 4091.
Crammer, B., and Ikan, R. (1977). *Chem. Soc. Rev.* **6**, 431.
Crosby, G. A. (1976). *CRC Crit. Rev. Food Sci. Nutr.* **7**, 297.
Crosby, G. A., and Furia, T. E. (1980). *In* "CRC Handbook of Food Additives" (T. E. Furia, ed.), 2nd ed., Vol. 2, pp. 187–227. CRC Press, Boca Raton, Florida.
Crosby, G. A., and Wingard, R. E., Jr. (1979). *In* "Developments in Sweeteners" (C. A. M.

Hough, K. J. Parker, and A. J. Vlitos, eds.), Vol. 1, pp. 135–164. Applied Science Publishers, London.

Dieterich, K. (1909). *Pharm. Zentralbl.* **50,** 435; *Chem. Abstr.* **3,** 2485-1 (1909).

Djerassi, C., Quitt, P., Mosettig, E., Cambie, R. C., Rutledge, P. S., and Briggs, L. H. (1961). *J. Am. Chem. Soc.* **83,** 3720.

Dolder, F., Lichti, H., Mosettig, E., and Quitt, P. (1960). *J. Am. Chem. Soc.* **82,** 246.

Domínguez, X. A., González, A., Angeles-Zamudio, M., and Garza, A. (1974). *Phytochemistry* **13,** 2001.

Dorfman, R. I., and Nes, W. R. (1960). *Endocrinology* **67,** 282.

DuBois, G. E. (1982). *Annu. Rep. Med. Chem.* **17,** 323.

DuBois, G. E., Dietrich, P. S., Lee, J. F., McGarraugh, G. V., and Stephenson, R. A. (1981). *J. Med. Chem.* **24,** 1269.

Farnsworth, N. R. (1973). *Am. Perfum. Cosmet.* **88,** 27.

Felippe, G. M. (1977). *Cienc. Cult. (Sao Paulo)* **29,** 1240.

Felippe, G. M. (1980). *Cienc. Cult. (Sao Paulo)* **32,** 1384.

Fletcher, H. G., Jr. (1955). *Chemurg. Dig.* **14** (7/8), 7.

Fujita, H. (1980). *Jpn. Fudo Saiensu* **19** (4), 56.

Fujita, H., and Edahiro, T. (1979a). *Shokuhin Kogyo* **22** (20), 66.

Fujita, H., and Edahiro, T. (1979b). *Shokuhin Kogyo* **22** (22), 65.

Fujita, H., and Edahiro, T. (1980). *Jpn. Kokai Tokkyo Koho* **80** 13,017; *Chem. Abstr.* **93,** 6373r (1980).

Fujita, S.-I., Taka, K., and Fujita, Y. (1977). *Yakugaku Zasshi* **97,** 692.

Fujita, T. (1979). *New Food Ind.* **21** (9), 16.

Ghisalberti, E. L., Jefferies, P. R., and Stuart, A. D. (1979). *Aust. J. Chem.* **32,** 1627.

Gosling, C. (1901). *Kew Bull.* p. 173.

Grashoff, J. L. (1972). Ph.D. Dissertation, University of Texas at Austin.

Grashoff, J. L. (1974). *Brittonia* **26,** 347.

Gushiken, T. (1979). *Jpn. Kokai Tokkyo Koho* **79** 122,300; *Chem. Abstr.* **92,** 74646m (1980).

Haga, T., Ise, R., and Kobayashi, A. (1976a). *Jpn. Kokai Tokkyo Koho* **76 131,900;** *Chem. Abstr.* **86,** 121718z (1977).

Haga, T., Ise, R., and Kobayashi, A. (1976b). *Jpn. Kokai Tokkyo Koho* **76 149,300;***Chem. Abstr.* **87,** 136338v (1977).

Handro, W., Hell, K. G., and Kerbauy, G. B. (1977). *Planta Med.* **32,** 115.

Hashimoto, Y. (1979). *Jpn. Kokai Tokkyo Koho* **79 09,991;** *Chem. Abstr.* **91,** 4132w (1980).

Hashimoto, Y., and Moriyasu, M. (1978). *Shoyakugaku Zasshi* **32,** 209.

Hashimoto, Y., Moriyasu, M., Nakamura, S., Ishiguro, S., and Komuro, M. (1978). *J. Chromatogr.* **161,** 403.

Hayashi, T., and Noda, M. (1975). *Jpn. Kokai Tokkyo Koho* **75 67,192;** *Chem. Abstr.* **83,** 110854c (1975).

Hemsley, W. B. *In* "Hooker's Icones Plantarum" (D. Prain, ed.), 4th Ser., Vol. 9. Dulau & Co., London.

Heraud, G. (1981). *Ann. Falsif. Expert. Chim. Toxicol.* **74,** 605.

Hirokado, M., Nakajima, I., Nakajima, K., Mizoiri, S., and Endo, F. (1980). *Shokuhin Eiseigaku Zasshi* **21,** 451 (1980).

Hovanec-Brown, J., Makapugay, H. C., Nanayakkara, N. P. D., Soejarto, D. D., Medon, P. J., Pezzuto, J. M., Kinghorn, A. D., and Kamath, S. K. (1982). *Abstr. Pap., 23rd Annu. Meet. Am. Soc. Pharmacogn.* Abstract No. 56.

Hsin, Y.-Y., Yang, Y.-W., and Chang, W.-C. (1979). *K'o Hsueh Fa Chan Yueh K'an* **7,** 1049.

Huang, B. (1981). *Shipin Kexue (Beijing)* **24,** 1.

Igoshi, M., and Kato, H. (1976). *Jpn. Kokai Tokkyo Koho* **76 52,200;** *Chem. Abstr.* **85,** 193045e (1976).

Inglett, G. E. (1976). *J. Toxicol. Environ. Health* **2**, 207.

Inglett, G. E. (1978). *In* "Health and Sugar Substitutes" (B. Guggenheim, ed.), pp. 184–190. Karger, Basel.

Ise, R., and Hirada, I. (1979). *Jpn. Kokai Tokkyo Koho* **79** 132,599; *Chem. Abstr.* **92**, 109391d (1980).

Ishizone, H. (1979). *Jpn. Kokai Tokkyo Koho* **79** 12,400; *Chem. Abstr.* **90**, 148704m (1979).

Isima, N., and Kakayama, O. (1976) *Shokuhin Sogo Kenkyusho Kenkyu Hokoku* **31**, 80.

Itagaki, K., and Ito, T. (1979a). *Jpn. Kokai Tokkyo Koho* **79** 78,386; *Chem. Abstr.* **91**, 156321v (1979).

Itagaki, K., and Ito, T. (1979b). *Jpn. Kokai Tokkyo Koho* **79** 78,388; *Chem. Abstr.* **91**, 156322w (1979).

Itagaki, K., and Kato, T. (1979). *Jpn. Kokai Tokkyo Koho* **79** 41,899; *Chem. Abstr.* **91**, 74856j (1979).

Iwamura, J.-I., Kinoshita, R., and Hirao, N. (1979a). *Koen Yoshishu Koryo, Terupen Oyobi Seiyu Kagaku Ni Kansuru Toronkai, 23rd* p. 249.

Iwamura, J.-I., Kinoshita, R., and Hirao, N. (1979b). *Koen Yoshishu Koryo, Terupen Oyobi Seiyu Kagaku Ni Kansura Toronkai, 23rd*, p. 253.

Iwamura, J.-I., Kinoshita, R., and Hirao, N. (1980). *Nippon Nogei Kagaku Kaishi* **54**, 195.

Iwamura, J., Kinoshita, R., Ishima, T., Katayama, O., Morita, T., and Hirao, N. (1982). *Nippon Nogei Kagaku Kaishi* **56**, 87.

Jacobs, M. B. (1955). *Am. Perfum. Essent. Oil Rev.* **66** (6), 44; *Chem. Abstr.* **50**, 2883i (1956).

Kamiya, S., Konishi, F., and Esaki, S. (1979). *Agric. Biol. Chem.* **43**, 1863.

Kaneda, N., Kasai, R., Yamasaki, K., and Tanaka, O. (1977). *Chem. Pharm. Bull.* **25**, 2466.

Kaneda, N., Kohda, H., Yamasaki, K., Tanaka, O., and Nishi, K. (1978). *Chem. Pharm. Bull.* **26**, 2266.

Kasai, R., Kaneda, N., Tanaka, O., Yamasaki, K., Sakamoto, I., Morimoto, K., Okada, S., Kitahata, S., and Furukawa, H. (1981). *Nippon Kagaku Kaishi* **5**, 726; *Chem. Abstr.* **95**, 169682w (1981).

Kato, I. (1975). *Shokuhin Kogyo* **18** (20), 44.

Kato, R., Sakaguchi, Y., and Motoi, N. (1977). *Jpn. Kokai Tokkyo Koho* **77** 136,200; *Chem. Abstr.* **88**, 110519a (1978).

Kazuyama, S. (1979). *Shokuhin to Kagaku* **21** (4), 90.

Kikuchi, H., and Sawaguchi, Y. (1977). *Jpn. Kokai Tokkyo Koho* **77** 57,199; *Chem. Abstr.* **87**, 116646r (1977).

Kim, H.-S., and Lee, H.-J. (1979). *Hanguk Sikp'um Kwahakhoe Chi* **11**, 56; *Chem. Abstr.* **91**, 209305w (1979).

King, R. M., and Robinson, H. (1967). *Sida* **3**, 165.

Kinghorn, A. D., Nanayakkara, N. P. D., Soejarto, D. D., Medon, P. J., and Kamath, S. (1982). *J. Chromatogr.* **237**, 478.

Kiumi, M., Nakazawa, T., Sasaki, S., and Fukumura, T. (1977). *Jpn. Kokai Tokkyo Koho* **77** 100,500, 1–6; *Chem. Abstr.* **88**, 20811k (1978).

Klages, A. (1951). *Pharm. Zentralhalle* **90**, 253.

Kobayashi, M., Horikawa, S., Degrandi, I. H., Ueno, J., and Misuhashi, H. (1977). *Phytochemistry* **16**, 1405.

Kobayashi, M., Horikawa, H., Ueno, J., Mitsuhashi, H., Kubomura, S., Miyazaki, K., and Chida, S. (1978). *Jpn. Kokai Tokkyo Koho* **78** 113,065; *Chem. Abstr.* **90**, 53396y (1979).

Kobert, R. (1915). *Ber. Dtsch. Pharm. Ges.* **25**, 162.

Kodaka, K. (1977). *Jpn. Kokai Tokkyo Koho* **77** 110,872; *Chem. Abstr.* **88**, 91353y (1978).

Kohda, H., Yamazaki, K., Tanaka, O., and Nishi, K. (1976a). *Phytochemistry* **15**, 846.

Kohda, H., Kasai, R., Yamasaki, K., Murakami, K., and Tanaka, O. (1976b). *Phytochemistry* **15**, 981.

Kohda, H., Tanaka, O., and Nishi, K. (1976c). *Chem. Pharm. Bull.* **24**, 1040.

Kojima, K. (1980). *Shoku no Kagaku* **56**, 40; *Chem. Abstr.* **95**, 22970e (1981).

Kokai, M., Miyamori, S., Sasaki, S., and Fukumura, T. (1979). *Jpn. Kokai Tokkyo Koho* **79** 20,000; *Chem. Abstr.* **90**, 182809j (1979).

Komatsu, K., Nozaki, W., Takamura, M., and Nakaminami, M. (1976). *Jpn. Kokai Tokkyo Koho* **76** 19,169; *Chem. Abstr.* **84**, 16317h (1976).

Kotani, C. (1980). *Jpn. Kokai Tokkyo Koho* **80** 19,009; *Chem. Abstr.* **93**, 4133p (1980).

Kubomura, S., Ueno, J., Chida, S., and Kanaeda, J. (1976). *Jpn. Kokai Tokkyo Koho* **76** 91,300; *Chem. Abstr.* **86**, 5777u (1977).

Kudo, M., and Koga, Y. (1977). *Nettai Nogyo* **20**, 211.

Kukuchi, H., and Suguri, N. (1977). *Jpn. Kokai Tokkyo Koho* **77** 47,959; *Chem. Abstr.* **87**, 83372t (1977).

Kuroda, A., and Kamiyama, S. (1979). *Jpn. Kokai Tokkyo Koho* **79** 76,600; *Chem. Abstr.* **91**, 156325z (1979).

Lee, C. K. (1979). *World Rev. Nutr. Diet.* **33**, 142.

Lee, J. L., Kang, K. H., Park, H. W., Ham, Y. S., and Park, C. H. (1980). *Nongsa Sihom Yongu Pogo* **22** (Crop), 138; *Chem. Abstr.* **94**, 207746j (1981).

Lee, S. J., Lee, K. R., Park, J. R., Kim, K. S., and Tchai, B. S. (1979). *Hanguk Sikp'um Kwahakhoe Chi* **11**, 224.

Luciani, S., Carpenedo, F., and Tarjan, E. M. (1978). *In* "Atractyloside, Chemistry, Biochemistry and Toxicology" (R. Santi and S. Luciani, eds.), pp. 109–124. Piccin Medical Book, Pandora, Italy.

Masuyama, F. (1980). *Jpn. Kokai Tokkyo Koho* **80** 07,039; *Chem. Abstr.* **93**, 44387f (1980).

Matsui, M., Ogawa, T., Mori, K., and Nozaki, M. (1978). *Jpn. Kokai Tokkyo Koho* **78** 63,364; *Chem. Abstr.* **89**, 163799u (1978).

Matsumi, S. (1974). *Shokuhin Kogyo* **17** (10), 60–64; *Chem. Abstr.* **81**, 150346d (1974).

Matsuoka, S. (1978). *Jpn. Kokai Tokkyo Koho* **78** 44,666; *Chem. Abstr.* **89**, 89117b (1978).

Medon, P. J., Pezzuto, J. M., Hovanec-Brown, J. M., Nanayakkara, N. P. D., Soejarto, D. D., Kamath, S. K., and Kinghorn, A. D. (1982). *Fed. Proc., Fed. Am. Soc. Exp. Biol.* **41**, 1568.

Metivier, J., and Viana, A. M. (1979a). *J. Exp. Bot.* **30**, 805.

Metivier, J., and Viana, A. M. (1979b). *J. Exp. Bot.* **30**, 1211.

Miguel, O. (1977). *Rev. med. Paraguay* **7**, 200.

Minamisono, H., and Azuno, K. (1978). *Nenpo—Kagoshima-ken Kogyo Shikenjo* **24**, 66.

Misawa, M. (1977). *Plant Tissue Cult., Biotechnol. Appl., Proc. Int. Congr., 1st, 1976* pp. 17–26.

Mitsuhashi, H., Ueno, J., and Sumida, T. (1975a). *Yakugaku Zasshi* **95**, 127.

Mitsuhashi, H., Ueno, J., and Sumida, T. (1975b). *Yakugaku Zasshi* **95**, 1501.

Miwa, K. (1978). *Jpn. Kokai Tokkyo Koho* **78** 105,500; *Chem. Abstr.* **90**, 3392y (1979).

Miwa, K. (1979). *Jpn. Kokai Tokkyo Koho* **79** 103,900; *Chem. Abstr.* **92**, 82420y (1980).

Miwa, K., and Tsuji, H. (1979). *Jpn. Kokai Tokkyo Koho* **79** 90,200; *Chem. Abstr.* **91**, 191698v (1979).

Miwa, K., Maeda, S., and Murata, Y. (1979a). *Jpn. Kokai Tokkyo Koho* **79**, 89,066; *Chem. Abstr.* **91**, 173681c (1979).

Miwa, K., Maeda, S., and Murata, Y. (1979b). *Jpn. Kokai Tokkyo Koho* **79** 90,199; *Chem. Abstr.* **91**, 206904k (1979).

Miyake, T. (1979). *Jpn. Kokai Tokkyo Koho* **79** 05,070; *Chem. Abstr.* **90**, 150478j (1979).

Miyake, T. (1980). *U.S. Patent* 4,219,571; *Chem. Abstr.* **94**, 29062y (1981).

Miyazaki, Y., Watanabe, H., and Watanabe, T. (1978). *Eisei Shikensho Hokoku* **96**, 86.

Miyoshi, H. (1980a). *New Food Ind.* **22** (1), 61.

Miyoshi, Y. (1980b). *Gekkan Shokuhin* **24** (6), 53.

Mizukami, H., Shiiba, K., and Ohasi, H. (1982). *Phytochemistry* **21**, 1927.

Mochida, K., Ikura, K., Kimura, R., and Ikehara, K. (1977a). *Jpn. Kokai Tokkyo Koho* **77 53,899;** *Chem. Abstr.* **87,** 100862x (1977).
Mochida, K., Ikura, K., Kimura, R., and Ikehara, K. (1977b). *Jpn. Kokai Tokkyo Koho* **77 53,900;** *Chem. Abstr.* **87,** 116645q (1977).
Montes, A. L. (1969). *An. Soc. Cient. Argent., ser II* **187,** 21.
Mori, K., Nakahara, Y., and Matsui, M. (1970). *Tetrahedron Lett.* p. 2411.
Mori, K., Nakahara, Y., and Matsui, M. (1972). *Tetrahedron* **28,** 3217.
Mori, N., Sakanoue, M., Takeuchi, M., Shimpo, K., and Tanabe, T. (1981). *Shokuhin Eiseigaku Zasshi* **22,** 409.
Morita, E. (1977a). *Shokuhin to Kagaku* **19** (4), 83.
Morita, E. (1977b). *Jpn. Kokai Tokkyo Koho* **77 102,469;** *Chem. Abstr.* **88,** 20806n (1978).
Morita, E. (1977c). *Jpn. Kokai Tokkyo Koho* **77 117,474;** *Chem. Abstr.* **88,** 49257v (1978).
Morita, E. (1977d). *Jpn. Kokai Tokkyo Koho* **77 122,676;** *Chem. Abstr.* **88,** 73503t (1978).
Morita, E. (1977e). *Jpn. Kokai Tokkyo Koho* **77 125,640;** *Chem. Abstr.* **88,** 41568b (1977).
Morita, E. (1977f). *Jpn. Kokai Tokkyo Koho* **77 148,700;** *Chem. Abstr.* **88,** 118021c (1978).
Morita, E. (1978). *Jpn. Kokai Tokkyo Koho* **78 59,074;** *Chem. Abstr.* **89,** 145372v (1978).
Morita, E., Morita, T., and Fujita, M. (1977). *Jpn. Kokai Tokkyo Koho* **77 117,473;** *Chem. Abstr.* **88,** 48258w (1978).
Morita, T., and Iwamura, J. (1977). *Jpn. Kokai Tokkyo Koho* **77 83,980;** *Chem. Abstr.* **88,** 168688w (1978).
Morita, T., Fujita, I., and Iwamura, J. (1977a). *Jpn. Kokai Tokkyo Koho* **77 27,226;** *Chem. Abstr.* **88,** 5026q (1978).
Morita, T., Morita, E., and Fujita, I. (1977b). *Jpn. Kokai Tokkyo Koho* **77 57,366;** *Chem. Abstr.* **87,** 132564t (1977).
Morita, T., Fujita, M., and Morita, E. (1977c). *Jpn. Kokai Tokkyo Koho* **77 105,260;** *Chem. Abstr.* **88,** 49255t (1978).
Morris, J. A. (1976). *Lloydia* **36,** 25.
Mors, W. B., and Rizzini, C. T. (1966). "Useful Plants of Brazil," p. 93. Holden-Day, San Francisco, California.
Mosettig, E., and Nes, W. R. (1955). *J. Org. Chem.* **20,** 884.
Mosettig, E., Quitt, P., Beglinger, U., Waters, J. A., Vorbrueggin, H., and Djerassi, C. (1961). *J. Am. Chem. Soc.* **83,** 3163.
Mosettig, E., Beglinger, U., Dolder, F., Lichti, H., Quitt, P., and Waters, J. A. (1963). *J. Am. Chem. Soc.* **85,** 2305.
Nabeta, K., Kasai, T., and Sugisawa, H. (1976). *Agric. Biol. Chem.* **40,** 2103.
Nabeta, K., Ito, K., and Sugisawa, H. (1977). *Nippon Nogei Kagaku Kaishi* **51,** 179.
Nakajima, I., Hirokado, M., Usami, H., Mizoiri, S., and Endo, F. (1979). *Tokyo-toritsu Eisei Kenkyusho Kenkyu Nempo* **30–31,** 153.
Nakajima, I., Hirokado, M., Nakajima, K., Mizoiri, S., and Endo, F. (1980). *Tokyo-toritsu Eisei Kenkyusho Kenkyu Nempo* **31–1,** 180.
Nieman, C. (1958). *Zucker Suesswaren Wirtsch.* **11,** 124.
Nozaki, N., Ogawa, T., and Matsui, M. (1978). *Tennen Yuki Kagobutsu Toronkai Koen Yoshishu, 21st* p. 213.
Ochi, T. (1979). *New Food Ind.* **21** (9), 28.
Ochi, T., and Shimizu, T. (1978). *Jpn. Kokai Tokkyo Koho* **78 148,575;** *Chem. Abstr.* **90,** 150472c (1979).
Ochi, T., and Shimizu, T. (1979). *Jpn. Kokai Tokkyo Koho* **79 02,381;** *Chem. Abstr.* **90,** 150474e (1979).
Ogawa, S. (1979). *Jpn. Kokai Tokkyo Koho* **79 73,158;** *Chem. Abstr.* **91,** 191697u (1979).
Ogawa, T., Nozaki, N., and Matsui, M. (1980). *Tetrahedron* **36,** 2641.

50   A. D. KINGHORN AND D.D. SOEJARTO

Ohe, Y., Okane, H., Watanabe, M., Shibasato, M., and Kamata, Z. (1977). *Jpn. Kokai Tokkyo Koho* 77 120,170; *Chem. Abstr.* 88, 122894g (1978).

Okane, H., and Kamata, Z. (1977). *Jpn. Kokai Tokkyo Koho* 77 57,198; *Chem. Abstr.* 87, 99049d (1977).

Okazaki, K., Nakayama, Y., and Owada, K. (1977). *Seikatsu Eisei* 21, 185.

Okumura, M., Fujita, Y., Imamura, M., and Aikawa, K. (1978). *Shokuhin Eiseigaku Zasshi* 19, 486.

Ortega, A., Martínez, R., and García, C. L. (1980). *Rev. Latinoam. Quim.* 11, 45.

Oviédo, C. A., Fronciani, G., Moreno, R., and Maas, L. C. (1970). *Excerpta Med.* 208, 92.

Persinos, G. J. (1973). U.S. Patent 3,723,410; *Chem. Abstr.* 79, 103868z (1973).

Pezzuto, J. M., Nanayakkara, N. P. D., and Kinghorn, A. D. (1983). *Proc. Am. Assoc. Cancer Res.* 24, 88.

Pinheiro, C. E., and Gasparini, O. T. (1981). *Abstr. Pap., Semin. Bras.* Stevia rebaudiana Bertoni, 1st, 1981 pp. XV.I–XV.IV.

Planas, G. M., and Kuć, J. (1968). *Science* 162, 1007.

Pomaret, M., and Lavieille, R. (1931). *Bull. Soc. Chim. Biol.* 13, 1248; *Chem. Abstr.* 26, 3619 (1932).

Quijano, L., Calderón, J. S., Gómez, F., Vega, J. L., and Riós, T. (1982). *Phytochemistry* 21, 1369.

Rasenack, P. (1908). *Arb. Biol. Abt. Land- Forstwirtsch. Kais. Gesundheitsamte* 28, 420; *Chem. Abstr.* 3, 688-2 (1909).

Ríos, T., Romo de Vivar, A., and Romo, J. (1967). *Tetrahedron* 23, 4265.

Robinson, H., and King, R. M. (1977). *In* "The Biology and Chemistry of the Compositae" (V. H. Heywood, J. B. Harborne, and B. L. Turner, eds.), pp. 437–485. Academic Press, London.

Román, L. U., del Río, R. E., Hernández, J. D., Joseph-Nathan, P., Zabel, V., and Watson, W. H. (1981). *Tetrahedron* 37, 2769.

Ruddat, M., Heftmann, E., and Lang, A. (1965). *Arch. Biochem. Biophys.* 110, 496.

Sakaguchi, M., and Kan, T. (1982). *Cienc. Cult. (Sao Paulo)* 34, 235.

Sakamoto, I., Kohda, H., Murakami, K., and Tanaka, O. (1975). *Yakugaku Zasshi* 95, 1507.

Sakamoto, I., Yamasaki, K., and Tanaka, O. (1977a). *Chem. Pharm. Bull.* 25, 844.

Sakamoto, I., Yamasaki, K., and Tanaka, O. (1977b). *Chem. Pharm. Bull.* 25, 3437.

Salmón, M., Díaz, E., and Ortega, A. (1973). *J. Org. Chem.* 38, 1759.

Salmón, M., Ortega, A., and Diaz, E. (1975). *Rev. Latinoam. Quim.* 6, 45.

Salmón, M., Diaz, E., and Ortega, A. (1977). *Rev. Latinoam. Quim.* 8, 172.

Samaniego, C. C. (1946). *Rev. Pharm. (Buenos Aires)* 88, 199; *Chem.Abstr.* 44, 501c (1947).

Sasaki, K., and Murakami, K. (1977). *Jpn. Kokai Tokkyo Koho* 77 47,960; *Chem. Abstr.* 87, 83371s (1977).

Sato, T., Kuroda, J., and Mihara, K. (1980). *Jpn. Kokai Tokkyo Koho* 80 39,731; *Chem. Abstr.* 93, 44371w (1980).

Sawaguchi, Y., and Kikuchi, H. (1977). *Jpn. Kokai Tokkyo Koho* 77 05,800; *Chem. Abstr.* 87, 136337u (1977).

Schvartzman, J. B., Krimer, D. B., and Moreno Azorero, R. (1975). *Rev. Soc. Cient. Paraguay* 15, 52.

Schvartzman, J. B., Krimer, D. B., and Moreno Azorero, R. (1977). *Experientia* 33, 663.

Seidemann, J. (1976a). *Lebensmittelindustrie* 23, 553.

Seidemann, J. (1976b). *Nahrung* 20, 655.

Seidemann, J. (1977). *Edesipar* 28, 141.

Shidehara, N. (1980). *Jpn. Kokai Tokkyo Koho* 80 50,866; *Chem. Abstr.* 93, 69072n (1980).

Shimizu, T., and Ochi, T. (1978). *Jpn. Kokai Tokkyo Koho* **78** 148,573; *Chem. Abstr.* **90**, 150471b (1979).

Shirakawa, T., and Onishi, T. (1979). *Kagawa-ken Hakko Shokuhin Shikenjo Hokoku* **71**, 35.

Shock, C. C. (1982). *Calif. Agric.* Sept.–Oct., p. 4.

Sholichin, M., Yamasaki, K., Miyami, R., Yahara, S., and Tanaka, O. (1980). *Phytochemistry* **19**, 326.

Soejarto, D. D., Kinghorn, A. D., and Farnsworth, N. R. (1982). *J. Nat. Prod.* **45**, 590.

Soejarto, D. D., Compadre, C. M., Medon, P. J., Kamath, S. K., and Kinghorn, A. D. (1983a). *Econ. Bot.* **37**, 74.

Soejarto, D. D., Compadre, C. M., and Kinghorn, A. D. (1983b). *Bot. Mus. Leafl., Harv. Univ.* **29**, 1.

Sugisawa, H., Kasai, T., and Suzuki, H. (1977). *Nippon Nogei Kagaku Kaishi* **51**, 175.

Sumida, T. (1973). *Misc. Publ., Hokkaido Natl. Agric. Exp. Stn.* **2**, 69.

Sumida, T. (1980). *J. Cent. Agric. Exp. Stn.* **31**, 1; *Biol Abstr.* **70**, 68862 (1980).

Suzuki, H., Ikeda, T., Matsumoto, T., and Noguchi, M. (1976). *Agric. Biol. Chem.* **40**, 819.

Suzuki, H., Kasai, T., Sumihara, M., and Sugisawa, H. (1977). *Nogyo Kagaku Zasshi* **51**, 171.

Suzuki, K., Shimoda, Z., and Sasaki, H. (1977). *Jpn. Kokai Tokkyo Koho* **77** 62,300; *Chem. Abstr.* **87**, 184915c (1977).

Suzuki, K., Shimoda, Z., and Sasaki, H. (1978). *Jpn. Kokai Tokkyo Koho* **78** 13,700; *Chem. Abstr.* **89**, 161879w (1978).

Takamura, K., Kawaguchi, M., and Isono, C. (1977a). *Jpn. Kokai Tokkyo Koho* **77** 47,961; *Chem. Abstr.* **87**, 132568x (1977).

Takamura, K., Kawaguchi, M., and Isono, C. (1977b). *Jpn. Kokai Tokkyo Koho* **77** 51,069; *Chem. Abstr.* **87**, 116644p (1977).

Takamura, K., Kawaguchi, M., and Isono, C. (1978). *Jpn. Kokai Tokkyo Koho* **78** 91,173; *Chem. Abstr.* **89**, 213864j (1978).

Tanaka, O. (1979). *Annu. Rep. Nat. Prod. Inst., Seoul Natl. Univ.* **18**, 146.

Tanaka, O. (1980). *Saengyak Hakhoe Chi (Hanguk Saengyak Hakhoe)* **11**, 219.

Tanaka, O. (1981). *Kagaku no Ryoiki* **35**, 590; *Chem. Abstr.* **95**, 138425k (1981).

Tanaka, O., Yamazaki, K., Kasai, R., and Koda, H. (1977a). *Jpn. Kokai Tokkyo Koho* **77** 41,275; *Chem. Abstr.* **87**, 116643n (1977).

Tanaka, O., Yamazaki, K., Kasai, R., and Kanda, H. (1977b). *Jpn. Tokkyo Koho* **77** 83,731; *Chem. Abstr.* **87**, 199466f (1977).

Tezuka, S., Yamano, T., Shitou, T., and Tadauchi, N. (1980). *Shokuhin Kogyo* **23** (18), 43.

Thomas, E. (1937). *Bull. Assoc. Chim.* **54**, 844; *Chem. Abstr.* **32**, 944-9 (1938).

Toffler, F., and Orio, O. A. (1981). *Riv. Soc. Ital. Sci. Aliment.* **10**, 225.

Tsuchiya, S. (1979). *New Food Ind.* **21** (9), 12.

Turrin, M. Q. A. (1981). *Abstr. Pap., Semin. Bras.* Stevia rebaudiana *Bertoni, 1st, 1981* pp. XII.I–XII.IV.

Uenishi, H., Haga, T., and Kobayashi, A. (1977). *Jpn. Kokai Tokkyo Koho* **77** 23,100; *Chem. Abstr.* **87**, 4314z (1977).

Unterhalt, B. (1978). *Pharm. Heute* **2**, 111.

Usami, M., Seino, Y., Takai, J., Nakahara, H., Seino, S., Ikeda, M., and Imura, H. (1980). *Horm. Metab. Res.* **12**, 705; *Chem. Abstr.* **94**, 77652n (1981).

Valio, I. F. M., and Rocha, R. F. (1976). *Z. Pflanzenphysiol.* **78**, 90.

Valio, I. F. M., and Rocha, R. F. (1977). *Nippon Sakumotsu Gakkai Kiji* **46**, 243; *Chem. Abstr.* **91**, 169834q (1979).

Vignais, P. V., Duee, E. D., Vignais, P. M., and Huet, J. (1966). *Biochim. Biophys. Acta* **118**, 465.

Vis, E., and Fletcher, H. G., Jr. (1956). *J. Am. Chem. Soc.* **78**, 4709.
von Schmelling, G. A., de Carvalho, F. V., and Domingos Espinosa, A. (1977). *Cienc. Cult. (Sao Paulo)* **29**, 599.
Wada, Y., Tamura, T., Kodama, T., Yamaki, T., and Uchida, Y. (1981). *Yakugaku* **30**, 215; *Chem. Abstr.* **94**, 203745k (1981).
Wakabayashi, T. (1981). *Braz. Pedido PI BR* **81 03,228**; *Chem. Abstr.* **96**, 141428q (1982).
Wingard, R. E., Jr., Brown, J. P., Enderlin, F. E., Dale, J. A., Hale, R. L., and Seitz, C. T. (1980). *Experientia* **36**, 519.
Wood, H. B., Jr., and Fletcher, H. G., Jr. (1956). *J. Am. Chem. Soc.* **78**, 207.
Wood, H. B., Jr., Allerton, R., Diehl, H. W., and Fletcher, H. G., Jr. (1955). *J. Org. Chem.* **20**, 875.
Yabu, M., Takase, M., Toda, K., Tanimoto, K., Yasutake, A., and Iwamoto, Y. (1977). *Hiroshima Daigaku Shigaku Zasshi* **9**, 12.
Yamada, S., and Kajima, S. (1980). *Jpn. Kokai Tokkyo Koho* **80 162,953**; *Chem. Abstr.* **94**, 101656q (1981).
Yamagami, K., and Takato, M. (1976). *Jpn. Kokai Tokkyo Koho* **76 56,500**; *Chem. Abstr.* **85**, 107672c (1976).
Yamamoto, A., and Ishida, K. (1979). *Jpn. Kokai Tokkyo Koho* **79 13,498**; *Chem. Abstr.* **91**, 73365m (1979).
Yamasaki, K., Kohda, H., Kobayashi, T., Kasai, R., and Tanaka, O. (1976). *Tetrahedron Lett.* p. 1005.
Yamasaki, K., Kohda, H., Kobayashi, T., Kaneda, N., Kasai, R., Tanaka, O., and Nishi, K. (1977). *Chem. Pharm. Bull.* **25**, 2895.
Yarita, K., Asai, S., Matsushita, K., and Maruyama, H. (1978). *Jpn. Kokai Tokkyo Koho* **78 113,066**; *Chem. Abstr.* **90**, 152556g (1979).
Yoshino, H. (1975). *Nippon Shoyu Kenkyusho Zasshi* **1**, 104.
Zaidan, L. B. P., Dietrich, S. M. C., and Felippe, C. M. (1980). *Nippon Sakumotsu Gakkai Kiji* **49**, 569.
Zeigler, F. E., and Kloek, J. A. (1977). *Tetrahedron* **33**, 373.

2

Recent Research on Oriental Medicinal Plants

HIROSHI HIKINO

Pharmaceutical Institute
Tohoku University
Sendai, Japan

I. INTRODUCTION

Oriental medicine (*zhong-yi* in Chinese and *kampō* in Japanese) originated in China a few thousand years ago and is undoubtedly the most widely employed traditional medicine in the world. An important part of this medicine consists of therapy using crude drugs (*zhong-yao* in Chinese and *wakan-yaku* in Japanese). The knowledge of materia medica for these therapies is the summation of thousands of years of human experience in the selection of natural resources for preventative and curative purposes, and has been accumulated in a variety of herbals (*pents'ao*'s) and medical books.

In the past five hundred years, traditional medicine has evolved separately in China and Japan, so that the ideas, methodologies, and opinions on the effectiveness of these materia medica are now somewhat different in the two countries.

In China, traditional medicine, having a proud and long history of its own, encountered a period of severe constraint at the beginning of the twentieth century, but this attitude has been reassessed since about 1950. Regular medical education is now conducted either in Oriental medicine or in Western medicine. The *Pharmacopoeia of the People's Republic of*

ECONOMIC AND MEDICINAL PLANT RESEARCH
VOLUME 1

China, issued in 1978, describes 882 crude drugs consisting of 637 plant drugs, 158 plant extracts or isolated ingredients, 46 animal drugs, 8 animal extracts, and 33 mineral drugs. This pharmacopoeia also includes accounts of 270 multiitem medical preparations. Because new crude drugs are still being unearthed from folkloric usages in China, about 115 items of the 637 plant drugs have not been recorded in the previous *pents'ao*'s (But *et al.*, 1980).

In Japan, Oriental medicine was widely employed until the nineteenth century, but has not been authorized since about 1900. Meanwhile, the knowledge and technology of this medicine has been kept alive by a minority of physicians who initially learned Western medicine but who subsequently became interested in Oriental medicine. Because of the limitations of Western medicine, including the adverse effects of synthetic drugs, and the significant therapeutic results of Oriental medicine, Oriental medicine has now achieved a good reputation, and the number of medical facilities adopting Oriental medicine as part of their medical treatment is increasing. *Japanese Pharmacopoeia X*, issued in 1981, contains 119 crude drugs, composed of 102 plant drugs, 11 plant extracts or isolated ingredients, five animal drugs, and one mineral drug. Among these, only 29 are crude drugs used in Western medicine or its equivalent. The remaining 90 are those employed in Oriental medicine or in folkloric remedies.

In Oriental medicine, a crude drug is rarely used alone but rather is utilized as an ingredient in a composite medical preparation. Consequently, the most important problem is the therapeutic efficacy of multiitem preparations. Since the 1970s investigations of the substances used in Oriental medicine have amassed much knowledge of the active principles and pharmacological actions of these crude drugs, but we are still as confused as the five blind men with the elephant. Hence the clinical use of such crude drugs is often empirical, and is based on observations from clinical trials without experimental support. Furthermore, pharmacological data from laboratory animals have contributed little to the clinical application of crude drugs because the models used to clarify pharmacological actions are too unlike the natural disease states. Despite some efforts, therefore, the therapeutic efficacy of many multiitem preparations has not yet been well substantiated on a scientific basis.

This chapter is an attempt to summarize recent chemical and pharmacological data that have been accumulated in an attempt to corroborate the therapeutic effectiveness of materia medica employed in Oriental medicine. Because of the tremendous amount of work conducted in this field, this chapter must be limited to a few topics. The results of recent investigations on only the eight most frequently used crude drugs are outlined here and consideration of purely phytochemical work is mainly excluded. Because the purpose of this chapter is to describe the achievements of more recent research, previous results are mentioned only briefly and citation of the literature published before 1975 is omitted. Information published previously may be obtained from Akamatsu (1970), Institute of Materia Medica (1959–1977), Jiang-su New Medical School (1977), Liu *et al.* (1975, 1979), and Takagi *et al.* (1982). Structural formulas are given only for uncommon natural products that are connected with physiological actions.

II. GLYCYRRHIZAE RADIX (LICORICE)

Glycyrrhizae Radix originates from *Glycyrrhiza glabra* Linné and its varieties, and from *G. uralensis* Fischer, of the family Leguminosae. In Oriental medicine, its main use has been to sweeten decoctions, to mitigate the actions of drastic drugs, and to relieve pain caused by muscle contraction. In Europe it has been utilized as a corrective, an expectorant, an antitussive, and an antiulcer agent. Although licorice contains a variety of constituents, its clinical effectiveness may be exerted mainly by three groups of substances: the saponins, the flavonoids, and the polysaccharides.

The licorice saponins consist of one major component, glycyrrhizin (glycyrrhizic acid), and 13 minor components whose proportions vary depending on the species and collecting location. Glycyrrhizin is responsible for the licoricelike sweetness and in fact is 50 times sweeter than sucrose. Undoubtedly this property accounts for the widespread use of this crude drug in many composite prescriptions. It has been thought that licorice is effective on a sore throat due to its sweetness and the acceleration of tracheal mucous secretion mediated by the saponin glycyrrhizin. This also explains its use as an expectorant and an antitussive. On the other hand, it is claimed that a mechanism involving the central nervous system participates in the antitussive action of glycyrrhizin.

Glycyrrhizin: R = D-Glc A-β (1→2)-D-Glc A
Glycyrrhetinic acid: R = H

Glycyrrhizin was first developed in Japan as an antidote in the early 1940s; its active portion was considered to be glucuronic acid. From this usage, interest was then directed towards a utility for allergy. The antiallergic and antiinflammatory actions of glycyrrhizin were first observed during its clinical usage and were substantiated later with experimental models. Thus glycyrrhizin and glycyrrhetinic acid also show a broad range of antiinflammatory activity in all stages of inflammation. It is considered that the antiinflammatory activity of glycyrrhizin is elicited through steroidlike action. Glycyrrhizae Radix was recently shown to mediate antiallergic activity by the passive cutaneous anaphylaxis method in rats (Hata and Sankawa, 1982).

Licorice has been used as a folk medicine for the treatment of gastric ulcers in Europe, and on this basis its antiulcerous effect was recognized clinically and long-term administration of a large amount of licorice was widely carried out as a remedy for gastric ulcer. In parallel with the chemical investigations of the active principles, pharmacological tests on extracts and their fractions using experimental models were performed. An aqueous extract reduced secretion of gastric juice in rats and significantly inhibited formation of gastric ulcers in pylorus-ligated rats through intraperitoneal or intraduodenal administration. A methanol extract elicited similar actions. Glycyrrhizin produced no effect on gastric juice secretion, but did prevent gastric ulcer formation. Meanwhile, the 3-O-hemisuccinate of glycyrrhetinic acid, carbenoxolone, was developed as a remedy for stomach ulcers.

It has been demonstrated that experimental gastric ulcers in rats were improved by carbenoxolone through p.o. administration, but not through s.c. or i.v. injection (Dajani et al., 1979; Derelanko and Long, 1981). It seems likely that the mechanisms of the antiulcerous effects in the stomach involve acceleration of mutin excretion due to an increase in synthesis of glycoprotein at the gastric mucosa, prolongation of lifespan of the epithelial cells, and antipepsin activity (Parke, 1976; Bickel and Kauffmann, 1981).

As a result of the wide clinical use of large doses of licorice, an adverse effect called licorice-induced pseudoaldosteronism, with effects such as hypertension, edema, and hypopotassemia, became significant (Epstein et al., 1977), and continued use has been discouraged. It was subsequently revealed that this mineralocorticoidlike activity was mediated by glycyrrhizin, in which the essential portion was demonstrated to be its aglycone glycyrrhetinic acid. In order to avoid the pseudoaldosteronism, deglycyrrhizinated licorice (DGL) was prepared. It was found that DGL protected rats against experimental ulcers (Rees et al., 1979), and a mechanism of its protective action was proposed (Van Marle et al., 1981).

However, its clinical value has not been established (Bardhan *et al.*, 1978).

It was also shown that glycyrrhizin and glycyrretinic acid inhibited inactivation enzymes of steroidal hormones in the liver and consequently exhibited glucocorticoidlike activity. Thus, inactivation of Δ^4-3-keto-steroidal hormones is initiated by reduction of the C-4—C-5 double bond with Δ^4-5α-reductase and Δ^4-5β-reductase to yield dihydro derivatives that have no more hormonal activity. Glycyrrhetinic acid was found to be a potent inhibitor of Δ^4-5β-reductase *in vitro,* the main counterpart enzyme in human, while the inhibitory activity of glycyrrhizin was found to be much weaker (Tamura *et al.*, 1979). Consequently, it is concluded that glycyrrhizin elicits its activity mainly in the form of its aglycone, glycyrrhetinic acid, through inhibition of the metabolism of Δ^4-3-ketosteroids, leading to potentiation of the activity of the steroids. Such an action results in sodium ion retention and potassium ion excretion and is manifested only by the existence of mineralocorticoids, since variation of the sodium/potassium ratio in the urine occurs in normal animals, but not in adrenalectomized animals. It was suggested that the essential portion of the molecule for enzyme inhibition is the 11-oxo-$\Delta^{12(13)}$-system in the C ring of glycyrrhetinic acid, which seems to be competitive with the 3-oxo-Δ^4-system in the A ring of steroid hormones at an active site of the reducing enzyme. Other mechanisms were also claimed from the findings that (1) glycyrrhetinic acid was shown to bind up to the aldosterone receptor in rat kidney (Ulmann *et al.*, 1975), and (2) in frog skin, a model of the tubuli renales in the kidney, glycyrrhetinic acid had no affection on active transport of sodium but did potentiate the activity of aldosterone (Ishikawa and Sato, 1980).

Glycyrrhizin has been widely used clinically, but the most critical problem is the occurrence of pseudoaldosteronism. On cessation of administration or giving an antialdosterone agent such as spironolactone, this effect can be removed. Efforts to separate the therapeutic actions from the side effects led to the finding of a congener, olean-12-ene-3β,30-diol, which showed lower toxicity than glycyrrhizin, exhibited no inhibition of Δ^4-5α- and Δ^4-5β-reductases, was effective for the suppression of stress ulcer formation, and exerted some antiallergic activity (Takahashi *et al.*, 1980).

Although contact dermatitis due to delayed allergy induced by *p*-phenylenediamine was remarkably suppressed by the i.p. administration of glycyrrhizin, which yielded more of an effect than that caused by prednisolone, glycyrrhizin afforded no reproducible results by means of p.o. administration and, as already mentioned, produced pseudoaldosteronism. Efforts to compensate these shortcomings resulted in the selection of *cis*-1-methyl-4-isohexylcyclohexanecarboxylic acid (IG-10), whose structure comprises most of the E ring and part of the B, C, and D rings of glycyrrhizin. IG-10, by p.o. dosing, inhibited a variety of experimental cutaneous reactions. In contrast to glycyrrhizin, the activity of IG-10 against delayed allergy was found not to be due to inhibition of metabolism of endogenous steroids, but was regarded as being mainly due to the suppression of inflammation induced by lymphokines and partly due to inhibition of the release of lymphokines. Although the inhibitory activity of IG-10 against delayed allergy was inferior to that of steroidal hormones, the activity of the latter was significantly potentiated by concurrent administration of the former. In fact, the dosage of a steroidal hormone could be reduced by IG-10 in clinical steroid therapy (Koda, 1980).

During the course of clinical use since the 1950s, a glycyrrhizin preparation has empirically been found to be effective for chronic hepatitis. Recent evaluation of the therapeutic effects of its i.v. administration on chronic hepatitis afforded significant improvements in the final global improvement rating, liver function improvement rating, and global utility rating (Suzuki *et al.*, 1977). The question then arises as to why a significant improvement in serum transaminase activity is brought about by glycyrrhizin through i.v. injection but not through p.o. administration. The answer to this question may be deduced from the observation that after i.v. injection of glycyrrhizin (80–200 mg) to humans, its concentration in blood rapidly decreased, and glycyrrhetinic acid appeared in the serum at a much later stage and in a very low concentration. On the other hand, p.o. administration of glycyrrhizin (100 mg) to humans gave no glycyrrhizin in the serum, and yielded glycyrrhetinic acid only in a low concentration (several hundred nanograms per milliliter) (Yano and Nakano, 1981).

The mechanism of action of glycyrrhizin in chronic hepatitis has not been completely clarified. It may be that glycyrrhizin elicits effectiveness through its corticosteroidlike activity. However, this possibility seems unlikely because activity is exerted mainly by the aglycone glycyrrhetinic acid, whose blood concentration after administration of glycyrrhizin is very low. Since glycyrrhizin was found to be immunostimulative by the lymphocyte transformation test induced by phytohemagglutinin (PHA) stimulation and dinitrochlorobenzene- (DNCB) induced dermal reaction, a glycyrrhizin preparation was evaluated in patients with chronic liver disease to observe acceleration of lymphotic transformation by PHA stimulation (Yamamoto, 1975). Liver-protective effects, especially the membrane stabilizing effect, of glycyrrhizin (inhibition of histamine release from rat mast cells, of carbon tetrachloride-induced liver lesion, of macrophage-mediated cytotoxicity, and of antibody-dependent cell-mediated cytotoxicity) have also been demonstrated (Mizoguchi *et al.*, 1981). Both glycyrrhizin and glycyrrhetinic acid were also shown to exert antihepatotoxic actions by the *in vitro* assay methods using carbon tetrachloride- and galactosamine-induced cytotoxicity in primary cultured rat hepatocytes (Kiso *et al.*, 1983a,b). In addition to these effects, the stabilizing effect on the lysosome membrane of liver cells (Matsushima, 1978) and the antiviral effect *in vitro* (Pompei *et al.*, 1979, 1980) may also contribute to the therapeutic efficacy of glycyrrhizin.

It was also reported that preparations containing glycyrrhizin induced interferon formation in mice and humans. The induced interferon was from its biophysical properties considered to be immune interferon, and this may play an important role in revealing of a number of therapeutic effects of glycyrrhizin in hepatitis, etc. (Hayashi *et al.*, 1979).

Glycyrrhizin improved hyperlipidemia in high cholesterol diet rats, and increased plasma HDL-cholesterol (Yamamoto *et al.*, 1982). It also increased albumin and the alubumin/globulin ratio significantly, but decreased IgG, IgA, total lipid, phospholipid, cholesterol, the hemolytic activity CH_{50}, C_3 activator, and C_3. This suggests that glycyrrhizin may have favorable effects on the process of arteriosclerosis (Yamauchi and Tsunematsu, 1982).

When *shōsaiko-tō* (containing licorice), one of the most commonly used preparations in Oriental medicine, was dosed p.o. to humans, only glycyrrhetinic acid was detected in blood at very low concentration (~ 10 ng/ml). Therefore, the main role of glycyrrhizin in the *kampō* preparations may be as a corrigent, because the levels of glycyrrhizin in most preparations in Oriental medicine are about the same order of magnitude (Yano and Nakano, 1981).

It was found that a methanol extract was twice as active as a water extract in its antigastric ulcer activity. Separation of the methanol extract led to a fraction (FM100) containing glycyrrhizin (13–19%) and isoflavonoids (liquiritin, isoliquiritin, liquiritigenin, and isoliquilitigenin, total 4–13%). FM100 showed no specific reduction of gastric juice secretion and caused no inhibition of gastric contraction *in situ*. Gastric juice

Liquiritigenin: R = H
Liquiritin: R = D-Glc

Isoliquiritigenin: R = H
Isoliquiritin: R = D-Glc

secretion accelerated by peptone was suppressed by FM100, but that stimulated by gastrin was not, leading to the assumption that the fraction inhibits the secretion of gastrin, as was subsequently proved. FM100 also showed weak but significant inhibitory action on stress ulcer and remarkably accelerated the healing of acetic acid-induced gastric ulcer. However, it was not effective on reserpine-produced ulcers. Further, FM100 given intraduodenally inhibited gastric secretion and prevented aspirin-induced ulcer formation in pylorus-ligated rats. Although glycyrrhizin has no antiulcer activity, it was concluded that the combination of glycyrrhizin and some factors in FM100 shows a synergistic effect on the ulcer (Okabe *et al.*, 1979). On the basis of these results, FM100 was subjected to clinical trials and has now come onto the market based on its gastric ulcer-repairing and antisecretory actions. From its use in Oriental medicine, licorice is expected to exert spasmolytic activity. In fact, however, no significant antispasmodic activity of an aqueous extract was observed in the isolated guinea pig ileum. A weak, noncompetitive anti-cholinesterase action was found in an ethanol extract and in FM100, in which isoliquiritigenin elicited the most intense antispasmodic action. Therefore, it is concluded that licorice does not mediate an appreciable spasmolytic effect clinically.

The antiinflammatory effect of an extract of licorice in the carrageenin-induced rat hind paw edema method was found to be due to liquiritin, whose potency could rationalize half that observed for the extract (Kosuge *et al.*, 1980). As part of a program directed towards the development of antidepressant drugs, fractionation of a methanol extract of licorice was performed by monitoring the inhibitory activity against monoamine oxidase. Liquiritigenin and isoliquiritigenin were identified as the active principles (Tanaka *et al.*, 1980). It is of interest to note that the active principles were glycosides or aglycones depending on physiological actions. A licorice extract suppressed the contractions of guinea pig ileum induced by electrical stimulation and inhibited acetylcholine- and potassium-induced contraction and [^{3}H]quinuclidyl benzoate binding, suggesting that the inhibitory effects of the licorice extract were due to an antimuscular action (Maeda *et al.*, 1982).

Because a licorice extract exhibited immunosuppressive activity on humoral antibody formation in mice, the extract was partially purified to yield a heat-stable fraction (Lx), which was shown to exhibit immunosuppressive activity. It has been revealed that (1) Lx was immunosuppressive in the primary response and elicited an optimal activity when dosed before inoculation of an antigen; (2) Lx had no effect on lymphoid cells already committed to immune response; (3) Lx caused strong suppression of memory cell production; (4) Lx elicited inhibitory effects on the production of both IgG_1- and IgG_2-antibody in guinea pigs and also of IgE-antibody in guinea pigs and rats; (5) Lx completely inhibited the induction of contact sensitivity to DNCB in guinea pigs; (6) Lx partially suppressed the formation of the plaque-forming cells (PFC) against sheep red blood corpuscle (SRBC) in mouse spleen; (7) Lx elicited almost no effect on the response of human peripheral blood lymphocytes to PHA; (8) Lx mediated its immunosuppressive activity at the stage of antigen procession by the macrophages; and (9) Lx suppressed the enhancing effect of human monocytic cells on anti-SRBC PFC response of mouse splenic cells (Yagura *et al.*, 1978). Another polysaccharide fraction (LH-1), containing mainly a glucan, was found to cause immunological enhancement not only in the primary response, but also in the secondary response and in cell-mediated immunity. LH-1 also enhanced memory cell production. However, no effect of LH-1 was found on macrophages or on the response to T-dependent antigen-dinitrophenyl-Ficoll. It was shown that the immunoenhancing actions of LH-1 were exerted by T-cell function (Kumagai and Takata, 1978). Glycyrrhizae Radix (methanol extract, p.o.) has been shown to inhibit significantly 48-hr homologous passive cutaneous anaphylaxis by antidinitrophenylated ascaris–IgE serum in rats (Koda *et al.*, 1982).

In order to perform simple and rapid screening for antiinflammatory drugs, a procedure to determine the release of histamine from rat mast cells was devised. By this screening method, Glycyrrhizae Radix was shown to inhibit histamine release (Hirai *et al.*, 1982). Morphine-induced rigidity of skeletal muscle was inhibited by a licorice extract dosed i.p. or p.o., suggesting it to have centrally acting muscle relaxant activity (Watanabe and Watanabe, 1980).

Further, licorice extracts were shown to be active against bacteria (Saito *et al.*, 1979). During a wide screening on ~1000 crude drugs, an extract of Glycyrrhizae Radix was found to show significant antitumor activity in cultured human cancer cells (JTC-26) *in vitro*. It was noticed, however, that the antitumor potency of this crude drug showed a great variation depending on species, collecting location, storage conditions, processing conditions, and type of tumor cells (Sato *et al.*, 1978).

III. ZINGIBERIS RHIZOMA (GINGER)

Since Zingiberis Rhizoma is consumed by humans as a spice in large quantities, it might be thought that this crude drug has no significant physiological properties. In fact, however, Zingiberis Rhizoma is prescribed in about half of the multiitem prescriptions in Oriental medicine, suggesting that it exerts some therapeutic efficacy. The original plant of Zingiberis Rhizoma is *Zingiber officinale* Roscoe (Zingiberaceae), which gives two different types of preparations: fresh (or air-dried) ginger and processed (steamed and dried) ginger. According to Oriental medicine, fresh ginger is a remedy for vomiting, coughing, abdominal distension, and pyrexia, while processed ginger is used for abdominal pain, lumbago, and diarrhea. As for the pharmacological actions of fresh ginger, central depressant, antiemetic, cardiotonic, antispasmodic, salivary secretory, gastric secretory, antihistaminic, diastase-potentiating, and antibacterial effects have been reported previously. Anticonvulsant activity and hypocholesterolemic activity were announced by Sugaya *et al.*, (1975) and Gujral *et al.* (1978). Ginger was previously shown to be antibacterial, but a ginger extract was found to be practically inactive against bacteria (Saito *et al.*, 1979).

The physiological actions of processed ginger had not been reported until Kasahara and Hikino (1983) showed it to induce analgesia, prolongation of hexobarbital-induced anesthesia, peripheral antivomiting, gastric antisecretion, hypertension followed by hypotension, and heart stimulation. No essential differences in pharmacologic actions were noted from those reported for fresh ginger.

As the main constituents of ginger, cineole, gingiberene, zingiberol (the fragrant principles), gingerone and shogaol (the pungent principles) had been known previously. Detailed examination of the constituents of ginger disclosed the presence of [6]-gingerol, [8]-gingerol, and [10]-gingerol as the pungent principles, dihydrogingerol (later [6]-gingediol) as the bitter principle, and hexahydrocurcumin as the choleretic

[6]-Gingerol: $n = 4$, R^1, $R^2 = O$
[8]-Gingerol: $n = 6$, R^1, $R^2 = O$
[10]-Gingerol: $n = 8$, R^1, $R^2 = O$
[6]-Gingerdiol: $n = 4$, $R^1 = OH$, $R^2 = H$

[6]-Shogaol: $n = 4$

[6]-Dehydrogingerdione: $n = 4$
[10]-Dehydrogingerdione: $n = 8$

[6]-Gingerdione: $n = 4$
[10]-Gingerdione: $n = 8$

principle. Further extentive study of ginger by combined gas chromatography–mass spectrometry revealed the presence of [3]-, [4]-, [5]-, and [12]-gingerol, [4]-, [6]-, [8]-, and [10]-gingediol, [6]-methylgingediol, [4]- and [6]-gingediaceate, and [6]-methylgingediacetate, together with the known gingerone, [6]-, [8]- and [10]-gingerol, [6]-shogaol, and paradol.

The pungent principles, [6]-gingerol and [6]-shogaol, were found to possess central depressant properties (sedative, antipyretic, analgesic, and transient hypotensive actions), and shogaol was more potent than gingerol (Aburada et al., 1982c). When [6]-shogaol was dosed, it elicited both suppression of the gastrointestinal function through the central nervous–adrenal system and the vagolytic mechanism and the direct excitation of the smooth muscles of the digestive tract. On the whole, the former action dominated (Aburada et al., 1981, 1982b). With regard to the effects on humans, powdered ginger rhizome was found to be superior to a reference substance, dimenhydrinate, in preventing the gastrointestinal symptoms of motion sickness (Mowrey and Clayson, 1982).

[6]-Shogaol has both hypotensive activity, due to stimulation of the vagus nerve and cardio-inhibitory action, and hypertensive activity, due to peripheral vasoconstriction and sympathomimetic action (Aburada et al., 1982a). Screening for cardioactive principles in crude drugs reealed activity in zingiberis Rhizoma and Chromatography of an ethyl acetate extract was carried out to yield the pungent principles, [6]-, [8]-, and [10]-gingerol. These substances elicited dose-dependent cardiotonic activity, which was not affected by a specific sodium channel blocker but was inhibited by a calcium antagonist. ATPase in the kidney was also inhibited, which may rationalize the cardioactive action. It is of interest that these constituents are similar in their mechanism of cardiotonic action to the cardioactive steroidal glycosides, although the substances are quite different in structure (Shoji et al., 1981).

In the course of screening crude drugs for inhibitors of prostaglandin (PG) biosynthesis, ginger exhibited inhibitory activity of PG biosynthesis. An extensive analysis for active principles led to the isolation of [6]-gingerol, [6]- and [10]-dehydrogingerdione, and [6]- and [10]-gingerdione, which were shown to be more potent inhibitiors of PG biosynthesis than indomethacin. The presence of a free hydroxyl group is thought to be essential for activity (Kikuchi et al., 1982). Because $PGF_{2\alpha}$ is now known to inhibit absorption of water in the digestive tract, it may be that inhibition of the biosynthesis of $PGF_{2\alpha}$ causes a reduction in bowel activity.

Extracts of Zingiberis Rhizoma displayed significant antitumor activity against cultured human cancer cells (JTC-26) (Sato et al., 1978) and sarcoma 180 ascites in mice (Itokawa *et al.*, 1979). The potency varied depending on the lot (Sato *et al.*, 1978). Juice prepared from ginger strongly inactivated the mutagenicity of tryptophan pyrolysis products (Morita *et al.*, 1978).

IV. HOELEN

Hoelen is the whole fungus, *Poria cocos* Wolf (Polyporaceae), and has been used in Oriental medicine in diuretic, stomachic, and antispasmodic remedies, and in folklore has been used to treat cancer. Although the diuretic activity of Hoelen was observed under certain conditions, it was not found in normal animals. Water and ethanol extracts of Hoelen exerted suppressive effects on guinea pig atria and showed stimulating activity on isolated frog heart; however, no active principles have been reported. It has been reported that Hoelen induced dilation of the isolated rabbit intestine, protected against gastric ulcer formation, and reduced the acidity of the gastric juice in pylorus-ligated rats.

Although Hoelen is an important drug prescribed in one-third of the composite preparations in Oriental medicine, there is no chemical rationale for the uses stated for this crude drug. Reports to date indicate that it contains triterpenoids, eburicoic acid, dehydroeburicoic acid, 3-β-*O*-acetyltumulosic acid, 3-β-*O*-acetyldehydrotumulosic acid, pachymic acid, 7,9(11)-dehydropachymic acid, tumulosic acid, 7,9(11)-dehydrotumulosic acid, and polyporenic C acid, the latter five being cytotoxic *in vitro* against hepatoma (Valisolado *et al.*, 1980).

Another constituent related to physiological activity is one having an antitumor effect. Pachyman, a possible antitumor polysaccharide of Hoelen, was first considered to be a β-(1→3)-linear glucan but was later found to involve a small amount of β-(1→6)-linked branching. Although Hoelen was completely devoid of antitumor activity (Itokawa *et al.*, 1979), severance of the β-(1→6)-linkages in pachyman led to a new polysaccharide possessing a β-(1→3)-linear glucan structure, pachymaran, which exhibited pronounced activity against ascites sarcoma 180 in Swiss albino mice. The active substance is therefore certainly a polysaccharide, and it appears that the structure of polysaccharide is important for the observation of an antitumor effect. In a screening of a number of crude drugs, Yamazaki and Shirota (1981) found Hoelen to have a protective effect on stress-induced ulcer in mice. Koda *et al.* (1982) reported that contact dermatitis produced by picryl chloride in mice was remarkably inhibited by the p.o. administration of Hoelen.

The main reason why the chemical knowledge of active principles of this crude drug is still scanty may be that it has excited no interest for researchers, who considered it to be a block of polysaccharides.

V. PAEONIAE RADIX (PAEONY ROOT)

Paeoniae Radix is prepared from the roots of *Paeonia lactiflora* Pallas (*P. albiflora* Pallas) of the family Paeoniaceae and has been employed as an antispasmodic, an analgesic, a mitigative, and an astringent in Oriental medicine. Concerning the pharmacological actions of the crude drug, the decoction when dosed p.o. induced elevation of blood sugar level in rabbits and showed some antispasmodic effect in the ileum and the uterus of the mouse, rabbit, and guinea pig. Potentiation of respiration, transient hypotension and a transient increase of heart rate in rabbits, and dilatation of the bronchia in guinea pigs were also observed. Paeoniae Radix was also shown recently to exhibit inhibitory effect on adjuvant arthritis in rats (Cho *et al.*, 1982). Antibacterial activity of paeony root extracts is also known (Saito *et al.*, 1979; Namba *et al.*, 1981).

From this crude drug, benzoic acid was isolated in 1907; because benzoic acid possesses some pharmacological actions, it was thought to take part in the therapeutic effectiveness of the crude drug. Since this isolation, no chemical data of interest appeared until paeoniflorin, a monoterpenoid glycoside having a unique cagelike skeleton, was isolated. Further studies led to the isolation of the minor congeners albiflorin, oxypaeoniflorin, and benzoylpaeoniflorin. The contents of these constituents were studied (Yoshizaki *et al.*, 1977; Nishizawa *et al.*, 1979). The toxicity of paeoniflorin was rather low. In animals it showed central effects: loss of righting reflex was produced by i.c. administration in rats, hexobarbital-induced hypnosis was potentiated, acetic acid-induced writhing was inhibited, body temperature was weakly lowered, and pentylenetetrazole-induced convulsions were weakly inhibited in mice. Further, paeoniflorin exhibited a hypotensive effect in guinea pigs, possibly due to its peripheral vasodilation. Vasodilation of coronary vessels and femoral vascular bed by paeoniflorin was found in dogs, and smooth-

Paeoniflorin   Paeoniflorigenone

muscle relaxant activity in both rat stomach and uterus assays were also observed. In isolated organs, it elicited antioxytocic action only in the rat uterus. The relaxing effect on smooth muscle was considered to be mainly musculotropic. Because these effects were not elicited by debenzoyl-paeoniflorin or benzoic acid, they are considered to be specific for paeoniflorin. The content of paeoniflorin is $\sim 1\%$ and the content of the other analogs is less than 0.1%, so that the pharmacological actions of paeoniflorin represent those of the crude drug.

Paeoniae Radix exhibited little effect by p.o. dosing and no remarkable effects even by i.p. injection, indicating that the pharmacological actions of this crude drug are quite moderate. Nevertheless, the results of the above pharmacological studies would support some of the clinical use of Paeoniae Radix.

Paeoniae Radix accelerated phagocytosis of macrophages and nitro-blue tetrazolium reduction, and the latter activity was also shown by paeoniflorin, indicating that this crude drug is participating in the activation of the protective system for infection (Kubo et al., 1979; Nagao et al., 1982). Screening of a crude drug preparation, shōseiryū-tō, against the contact dermatitis produced by picryl chloride in mice revealed it to possess antiallergic activity. Analysis of the component drugs indicated that Paeoniae Radix was one of the active components, and the active principle was again shown to be paeoniflorin. In the homologous passive cutaneous anaphylaxis reaction in guinea pigs and in the adjuvant arthritis assay in rats, paeoniflorin also elicited antiallergic responses. On the other hand, paeoniflorin mediated no apparent activity in nonallergic inflammations. Based on these findings, it was concluded that paeoniflorin contributed to antiallergic potency of the preparation (Fujimura, 1982). Its derivatives, debenzoylpaeoniflorin and paeonone, prepared to find better analogs, were shown to have antiallergic ability about twice that of paeoniflorin in decreasing contact sensitivity to picryl chloride in mice, and in homologous passive cutaneous anaphylaxis reaction in guinea pigs (Fujimura, 1982).

It has been claimed that the previous experiments did not point to an active principle responsible for the muscle-relaxant activity of Paeoniae Radix, which is alleged to be one of its main clinical effects. Consequently, the search for a muscle-relaxant principle was carried out. Monitoring the anticholinesterase activity using the abdominal rectus muscle of frog afforded paeoniflorigenone, showing muscle-relaxant activity (Hayashi et al., 1980; Shimizu et al., 1981). However, this constituent was not regarded as the active principle responsible for the clinical effectiveness of the crude drug, because it is not extractable with hot water. Continuing the survey resulted in the isolation of a second principle exerting anticholinesterase action, and the presence of another principle showing neuromuscular blocking activity was also noted (Hayashi *et al.*, 1981). The active constituent, however, has not yet been characterized.

Paeoniflorigenone, paeoniflorin, oxypaeoniflorin, and benzoylpaeoniflorin inhibited twitch responses of skeletal muscle to direct and indirect stimulation (Kimura and Kimura, 1980; Kimura *et al.*, 1982a,b). Based on the finding that paeoniflorigenone showed the blocking action, which was significantly potentiated by glycyrrhizin at an ineffective concentration, combined action of paeoniflorin and glycyrrhizin in skeletal muscle was examined. As a result, it was found that although paeoniflorin and glycyrrhizin exhibited no neuromuscular blocking effect at lower concentrations, combined administration of both substances at the ineffective concentrations produced significant blocking action *in vitro*, and the combined action largely depends on postsynaptic and musculotropic mechanisms (Kimura *et al.*, 1982c). This combined blocking action of paeoniflorin and glycyrrhizin was later confirmed in mice *in situ* and was found to increase in diabetes mice (Kimura *et al.*, 1983). This observation of the blend effects may accelerate the clarification of the therapeutic efficacy of multiitem preparations in Oriental medicine.

Because Paeoniae Radix gives a positive ferric chloride test and possesses astringent properties, the presence of tannins in this crude drug was suggested. Although no homogeneous gallotannins possessing the polygalloyl side chain had been obtained, six analogs (penta-, hexa-, hepta-, octa-, nona-, and decagalloyl glucose) were isolated homogeneously for the first time from this crude drug by Nishizawa *et al.*, (1980).

Intraperitoneal administration of several crude drugs was found to decrease urea-nitrogen concentration in rat serum (Nagasawa *et al.*, 1978). Examination of one of the active drugs, Paeoniae Radix, revealed 1,2,3,4,6-penta-*O*-galloyl glucose, a typical congener of the gallotannins, as an active principle. Its component parts (gallic acid and glucose) and other constituents [paeoniflorin and (+)-catechin] gave no positive response (Shibutani *et al.*, 1981). The analogs (tetra- to decagalloyl glucoses) isolated from fresh paeony roots afforded similar results (Oura and Nishioka, 1982).

Under the hypothesis that one of the therapeutic effects of Paeoniae Radix is regulation of the blood coagulation process, this crude drug was found to suppress platelet agglutination and to possess inhibitory activity against plasmin. Examination of Paeoniae Radix for antifibrinolytic constituents resulted in the isolation of linoleic acid, palmitic acid, hederagenin, gallic acid, and pentagalloyl glucose (Tani *et al.*, 1982). The crude agglutinins from paeony improved the recovery of rabbits from anemia due to blood loss (Shim *et al.*, 1980).

It was also shown that pentagalloyl glucose from Paeoniae Radix showed antiviral activity against herpes simplex virus. Tannins were considered previously to be responsible for this activity (Takechi and Tanaka, 1982). Although antitumor activity of Paeoniae Radix *in vitro* has been reported, the potency varied markedly with the lot (Sato *et al.*, 1978).

VI. ZIZYPHI FRUCTUS

Zizyphi Fructus is prepared from the fruits of *Zizyphus jujuba* Miller var. *inermis* Rehder of the family Rhamnaceae. Although it has been utilized as an emollient, a sedative, an antitussive, a tonic, a nutrient, and an antianemic, the knowledge that this crude drug contains a large quantity of sugar explains only part of its activity as a nutriment. Ingestion of Zizyphus Fructus is alleged to bring about body wight increase, muscular strength reinforcement, liver protection, and prevention of stress ulcer formation, but no active principles have been deduced.

Together with 11 pentacyclic triterpenoids (Yagi *et al.*, 1978a,b), four triterpenoid glycosides—zizyphus saponins I, II, and III, and jujuboside B—were isolated by Okamura *et al.* (1981). Jujuboside B was shown to exert sedative activity in mice (Wagner *et al.*, 1983). Further, isolation of zizybeoside I and II, zizyvoside I and II, 6,8-di-C-glucosyl-(2S)-naringenin, and 6,8-di-C-glucosyl-(2R)-naringenin, along with vomifoliol and roseoside, from a fraction having hypotensive and sedative activity was reported (Okamura *et al.*, 1981b).

Cyclic adenosine 3′,5′-monophosphate (cyclic AMP) is now well established as an intracellular second messenger in the actions of various hormones and catecholamines. It has also been clarified recently that cyclic AMP in immunological competent cells and target tissues is very concerned with the response and regulation of allergic reactions such as bronchial asthmas for which modern medicine sometimes cannot provide effective therapies. Therefore, the ability of Oriental drugs to increase the intraleukocytic levels of cyclic AMP was examined. This showed that cyclic AMP in the leukocytes was increased by ingestion of certain multiitem preparations clinically employed for asthma and cold in Oriental medicine. Further analysis demonstrated that this increase of cyclic AMP was brought about by the component Zizyphi Fructus (Cyong *et al.*, 1979). The content of cyclic AMP in Zizyphi Fructus was estimated to be 100–500 nmol/g, the value being the highest among those in plant and animal sources ever tested (Cyong and Hanabusa, 1980). In animal tissues, another cyclic nucleotide, cyclic guanosine 3′,5′-monophosphate (cyclic GMP), also plays a part as an intracellular second messenger. The relative concentrations of both cyclic AMP and cyclic GMP have been shown to be important for the regulation of intracellular functions. A high level of cyclic GMP activity was detected in Zizyphi Fructus, and the active principle was determined to be cyclic GMP, whose concentration in Zizyphi Fructus ranged between 30 and 60 nmol/g. Again, this value is the highest ever found either in plant and animal tissues (Cyong and Takahashi, 1982a). Although it is of course difficult to explain all the therapeutic effects of Zizyphus Fructus through the presence of the cyclic nucleotides, the antiallergic action may be thus rationalized. Cyclic AMP in the epithelial cells of the intestinal villi is known to participate in absorption of certain amino acids (Kinzie et al., 1976; Hughes et al., 1978), and it has been suggested that cyclic GMP in the digestive tract takes part in the secretion of digestive fluids. It is therefore probable that the cyclic nucleotides contribute to the therapeutic effectiveness of Zizyphus Fructus on disorders of the digestive tract, since the cyclic GMP concentration in the serum showed no increase when Zizyphus Fructus was administered (Cyong and Takahashi, 1982b). Further survey revealed the presence of a β-adrenergic stimulating substance (Cyong, 1982).

During the course of the examination of Oriental drugs used clinically for antiallergic purposes, an ethanol extract of Zizyphi Fructus was indicated to mediate inhibitory effects on reaginic antibody formation in rats. The ethanol extract was then subjected to bioactivity-directed fractionation to yield ethyl α-D-fructofuranoside, which suppressed reaginic antibody formation but gave no influence on haemagglutinin formation in female rats, and which increased the hemolytic plaque-forming cells in male mice. Although ethyl α-D-fructofuranoside was shown to be an artefact generated during the ethanol extraction of this fructose-rich crude drug, it was suggested to be useful as an antiallergic drug (Yagi et al., 1981). Zizyphus Fructus (aqueous extract, p.o.) was found to inhibit significantly 48-hr homologous passive cutaneous anaphylaxis induced by antidinitrophenylated ascaris–IgE serum in rats (Koda et al., 1982).

A Zizyphus extract was reported to be practically inactive against bacteria (Saito et al., 1979). In vitro antitumor activity of Zizyphus Fructus has been shown (Sato et al., 1978).

VII. CINNAMOMI CORTEX (CASSIA, CINNAMON)

Cinnamomi Cortex has been famous as a drug and a spice for a long time. This crude drug is clinically prescribed for perspiratory, antipyretic, and analgesic purposes in Oriental medicine and is widely used for aromatic stomachic as folkloric remedy throughout the world. Plants for Cinnamomi Cortex, of the family Lauraceae, are distributed widely in the tropical zones of the eastern part of Asia, and important sources are *Cinnamomum cassia* Blume from China, *C. loureiri* Nees and *C. obtusifolia* Nees from Southeast Asia, *C. sieboldi* Meisn from Japan, and *C. zeylanicum* Nees from Ceylon.

Early chemical studies on its constituents were primarily concerned with its essential oil. According to Guenther, the essential oil consists of the major component, *trans*-cinnamaldehyde (80–90%), with cinnamyl acetate and phenylpropyl acetate as minor constituents. Recent gas–liquid chromatographic developments have enabled the detection of over 100 different volatile constituents in *Cinnamomum cassia, C. loureiri,* and *C. zeylanicum,* although the composition varies depending on the lot (Senayake *et al.,* 1978). The most significant feature of the essential oil of *C. cassia* is that it contains no monoterpenoids or sesquiterpenoids, although *C. loureiri* and *C. zeylanicum* do. Pharmacological studies on Cinnamomi Cortex can be divided into two parts, those on the essential oil particularly its main ingredient, cinnamaldehyde, and those on the parts other than the essential oil.

Pharmacological investigations on the essential oil of Cinnamomi Cortex, especially on cinnamaldehyde, have been carried out frequently, particularly during the 1970s (Harada and Yano, 1975; Harada *et al.,* 1976; Harada and Saito, 1978; Harada and Hirayama, 1979; Harada and Yamazaki, 1981). As a result, not only common pharmacological actions, but also a variety of special biological effects have been observed. Because the essential oil contains cinnamaldehyde as the main component and the results of pharmacological testing on the essential oil were found to be similar to those on cinnamaldehyde, it is possible to consider that the pharmacology of cinnamaldehyde represents that of the essential oil. The pharmacological actions of cinnamaldehyde elicited by p.o. administration can be summarized as follows. Although cinnamaldehyde stimulates the central nervous system at low doses and inhibits it at higher doses, its potency is weak. The actions of cinnamaldehyde on the cardiovascular system are complex, since it accelerates release of catecholamines (mainly adrenaline) from the adrenal glands into blood and it also exhibits weak papaverine-like activity. Significant effects of cinnamaldehyde are an increase of peripheral blood flow volume, hypotension, bradycardia, and hyperglycemia. Cinnamaldehyde exerted only weak actions on the digestive system, but it does produce local anesthesia and local stimulation.

Because cinnamaldehyde is subject to facile evaporation and autooxidation, the amount of cinnamaldehyde remaining in a preparation is a balance of loss after storage and processing (decoction, evaporation etc.). In fact, it has been shown that most of cinnamaldehyde in the crude drug is lost during decoction due to evaporation and autooxidation (Takaishi *et al.*, 1979). Therefore, when the weak pharmacological actions are taken into account, the contribution of cinnamaldehyde to the therapeutic effectiveness is doubtful.

Concerning the microbial activity, the essential oil and cinnamaldehyde were reported to show growth-inhibiting and germicidal activity against certain fungi and bacilli. Recently, a new antifungal principle, *o*-methoxycinnamaldehyde, was isolated from *Cinnamomum zeylanicum* and shown to exert stronger activity against fungi rather than bacilli. Thus it inhibited the growth and mycotoxin production of four fungi and re-strained the growth of five dermatophytes (Morozumi, 1978).

The pharmacological actions of an essential oil-free aqueous extract of the crude drug were found to be also rather weak. The only appreciable actions were prolongation of hexobarbital-induced hyponosis and a slight reduction of acetic acid-induced writhing. Administration of an aqueous extract to rats p.o. was reported to have no effect on normal perspiration, but it accelerated pilocarpine-induced perspiration from the sole of rat.

During the course of an investigation for antiinflammatory principles in crude drugs, an aqueous methanol extract of the barks of *Cinnamomum sieboldii* was found to possess intense inhibitory activity on granulation tissue formation in chick embryos. Fractionation for active principles resulted in the isolation of (−)-epicatechin, (+)-catechin, procyanidin B_2, and procyanidin B_4, as well as cinnamonol D_1 and D_2, the activity in-

(−)-Epicatechin: R^1 = OH, R^2 = H
(+)-Catechin: R^1 = H, R^2 = OH

Procyanidin B_2: R^1 = R^2 = OH, R^3 = R^4 = H,
R^5 = α-H
Procyanidin B_4: R^1 = R^4 = OH, R^2 = R^3 = H,
R^5 = β-H

Cinnamonol D_1: R^1 = R^3 = H, R^2 = OH
Cinnamonol D_2: R^1 = R^2 = H, R^3 = OH

creasing in the order of monomers, dimers, and trimers (Otsuka et al., 1982).

In a search for a treatment for nephritis, an aqueous extract of Cinnamomi Cortex was found to inhibit immunological hemolysis conducted by complement in vitro, and to suppress the complement-dependent Forssman shock and Arthus reaction in vivo, a fact that indicated that Cinnamomi Cortex possessed inhibitory activity against complement formation (Nagai et al., 1978). A survey for the active principles responsible for this physiological activity resulted in the isolation of a variety of diterpenoids having complex skeletons, and a series of condensed-type tannin analogs. The diterpenoids can be classified into five groups: the ketal I, II, and III types, the lactone type, and the diketone type. The distribution of these diterpenoids showed variation depending on the lot (origin of plants, collecting location, etc.) (Isogai et al., 1976, 1977; Yagi et al., 1980b; Nohara et al., 1980a,b, 1981; Miyamura et al., 1981; Kashiwada et al., (1981). The physiological actions of these constituents are of quite interest but have not yet been reported. Since Cinnamomi Cortex showed anticomplement activity in vivo, an attempt was carried out to apply it to treat nephritis, which is closely related to complement-dependent immunological reactions (Yagi et al., 1980a). However, anticomplement activity of Cinnamomi Cortex arose through alternative pathways, and consequently it may be possible that Cinnamomi Cortex rather enhances the allergic reaction temporarily (Otsuka, 1982).

Although it is well known that Cinnamomi Cortex contains tannins, the nature of these tannins was not clarified until quite recently. Initiated with the survey for immunologically active components in Cinnamomi

Cortex, a series of polymeric flavanols have been isolated: (−)-epi-catechin as the monomer, procyanidins B-2 and B-5 as the dimers, pro-cyanidin C-1 as the trimer, cinnamtannin I as the tetramer, cinnamtan-nin II as the pentamer, and cinnamtannin III as the hexamer. In the tannin fraction of Cinnamomi Cortex, no hydrolyzable tannin was de-tected (Morimoto et al., 1981, 1982). Tannins are known to elicit some biological actions, pharmacologically, antidiarrhetic, hemostatic, as-tringent, etc., and biochemically, nonspecific inhibition of enzyme ac-tivities have been explained by their ability to combine with protein. The search for a clarification of physiological actions of tannins including those in Cinnamomi Cortex is claimed to be in progress.

Cinnamomi Cortex was found to prevent platelet agglutination and to show antithrombotic activity. These actions together with the antiinflam-matory action of the crude drug were considered to contribute effective-ly to suppression of thrombus formation in certain diseases (Kubo et al., 1981a).

It was also found that a Folin–Ciocalteau reagent positive substance in Cinnamomi Cortex caused a decrease of proteolytic activity of proteases such as pancreatin, providing a practical problem because Cinnamomi Cortex is commonly prescribed with proteases in stomachics and diges-tants (Yamazaki et al., 1981). An extract of Cinnamomi Cortex inhibited adrenaline-induced lipolysis and markedly stimulated lipogenesis from glucose (Ohminami et al., 1982).

Antitumor activity of Cinnamomi Cortex in vivo has also been de-scribed, the intensity varying significantly depending on the lot (Sato et al., 1978).

VIII. ATRACTYLODIS RHIZOMA

The crude drug is prepared from the rhizomes of *Atractylodes* plants of the family Compositae and is empirically classified into two types: Atrac-tylodis Rhizoma from *A. macrocephala* Koidzumi (*A. ovata* De Candolle) and *A. japonica* Koidzumi, and Atractylodis lanceae Rhizoma from *A. lancea* De Candolle and its varieties (*A. lancea* var. *chinensis* Kitamura, etc.). It is used clinically for diuretic and analgesic purposes as well as for stomachic disorders, mainly to improve water metabolism. Interestingly, Atractylodis Rhizoma is said to have antisudorific activity, while Atrac-tylodis lanceae Rhizoma is alleged to possess diaphoretic activity.

When the two types of crude drug are evaluated in light of their sesquiterpenoid constituents, Atractylodis Rhizoma is characterized by an intense reaction in the vanillin–hydrochloric acid test owing to its

Atractylon

Atractylenolide I

Atractylenolide II: R = H
Atractylenolide III: R = OH

β-Endesmol

Hinesol

high content of the furan-containing atractylon and its analogs (Nishikawa *et al.*, 1975, 1976a; Yosioka *et al.*, 1976). From Atractylodis Rhizoma, sesquiterpenoid lactones related to atractylon, atractylenolides I, II, and III, were isolated (Endo *et al.*, 1979; Wang *et al.*, 1980). Atractylodis lanceae Rhizoma, however, is characterized by weak or no reaction in the vanillin–hydrochloric acid test due to its low content or absence of furano analogs. Instead, it contains the alcohols β-eudesmol and hinesol as the main components. The two drug types can also be differentiated from their polyacetylene content: diacetylatractylodiol in the rhizomes of *A. japonica,* and atractylodin and its congeners in the rhizomes of *A. lancea* and its varieties (Nishikawa *et al.*, 1976a,b).

With respect to the diuretic activity of the crude drug, a number of investigations have been performed (Yamahara *et al.*, 1977), however, the results so far obtained lead to no unifying conclusion. Some mechanisms of diuresis were claimed, but further substantiation appears to be required.

Evaluation of the pharmacological actions of the crude drug using the essential oil constituents demonstrated that Atractylodis lanceae Rhizoma had central depressant activity in terms of general behavior, spontaneous movement, antielectroshock convulsion, and prolongation of hexobarbital-induced hypnosis. The active principles were revealed to be primarily β-eudesmol and hinesol. No marked activity was observed with Atractylodis Rhizoma in these actions (Yamahara *et al.*, 1977).

Since prescriptions in Oriental medicine used for antiinflammatory purposes frequently contain Atractylodis Rhizoma as a component, the antiinflammatory activity of *A. japonica* rhizomes was examined. Activity was confirmed, and fractionation for active principles led to the isolation of (+)-eudesma-4(14)-7(11)-dien-8-one and atractylenolide I, which displayed antiinflammatory effects in the increased vascular permeability induced by acetic acid in mice and in granulation tissue formation in chick embryos. The structurally related constituents, atractylenolides II and III, also showed some antiinflammatory activity (Endo *et al.,* 1979). Atractylodis lanceae Rhizoma, however, gave a negative response in the increased vascular permeability (Yamahara *et al.,* 1977; Hikino *et al.,* 1981). Recently, antiinflammatory activity of Atractylodis Rhizoma was observed for adjuvant arthritis in rats (Cho *et al.,* 1982).

Atractylodes lancea rhizomes and *A. lancea* var. *chinensis* rhizomes inhibited formation of aspirin-induced gastric ulcer and pylorus ligation-induced gastric ulcer, indicating that the antiulcerous activity of both *A. lancea* rhizomes is elicited by its gastric antisecretory action. The mechanisms of inhibitory action on gastric juice secretion were reported to involve the antagonistic effect of β-eudesmol against histamine H_1 and H_2 receptors in *A. lancea* rhizomes (Kubo *et al.,* 1981b) and intermediation of the adrenal function (inhibition of release of steroids, which stimulates secretion of gastric juice) by *A. lancea* var. *chinensis* rhizomes (Nogami *et al.,* 1982). *Atractylodes macrocephala* rhizomes and *A. japonica* rhizomes exhibited no marked antiulcerous activity (Kubo *et al.,* 1981b).

Atractylodis Rhizoma is known to reduce blood sugar levels, and this was considered to be due to acceleration of metabolism of glucose *in vivo*. Preventive activity against reduction of glycogen in the liver was also claimed. Atractylodis lanceae Rhizoma was previously reported to elicit hypoglycemic activity in hyperglycemic animals but to exert hyperglycemic activity in normal animals, and the latter was confirmed in mice recently (H. Hikino, unpublished). An extract of *A. japonica* rhizomes, which showed substantial hypoglycemic activity, was fractionated to reveal that the active principle was a peptidoglycan (H. Hikino, unpublished).

The two types of crude drug can therefore be distinguished through their central depressant, antiinflammatory, antiulcerous, and hypoglycemic actions. From these differences, it seems that Atractylodis Rhizoma and Atractylodis lanceae Rhizome are entirely different crude drugs. So far, however, the previously mentioned differences in pharmacological effects are not related to differences in the alleged therapeutic effects.

Atractylodis Rhizoma showed protective activity against carbon tetrachloride-induced liver lesions, with active principle established as atractylon. Atractylodis lanceae Rhizoma, on the other hand, afforded negative responses to this test system (Yamahara *et al.,* 1981). Screening of various preparations of the crude drug from *Atractylodes* rhizomes for antihepatotoxic activity by *in vitro* assay methods using carbon tetrachloride- and galactosamine-induced cycotoxicity in primary cultured rat hepatocytes indicated that some preparations exhibited liver-protective activity but that the potency varied with the lot. Their main sesquiter-penoid components were subjected to screening by these methods to indicate that atractylon, β-eudesmol, and hinesol exerted significant antihepatotoxic effects (Kiso *et al.*, 1983c).

Determination of the inhibitory activity on histamine- and barium chloride-induced contractile responses of the isolated ileum led to recognition of activity in β-eudesmol and hinesol. Further study on the structure–activity relationships demonstrated that the hydroxyl group is important for activity (Itokawa *et al.*, 1980). A number of their derivatives were then subjected to analysis of quantitative structure–activity relationships, demonstrating that the ileum-contracting activity varied parabolically versus the logarithm of the hydrophobic parameter (Morita *et al.*, 1981).

The other significant physiological activity is as an anticoagulant. Thus, repeated ingestion of Atractylodis Rhizoma to rats and humans caused marked prolongation of coagulation time. Nothing is known of the active principles responsible for this physiological activity. Body weight increase and muscle-strength reinforcement have also been reported as pharmacological actions of Atractylodis Rhizoma.

A hot-water extract of Atractylodis lanceae Rhizoma was found to show macrophage spreading ability (Kumazawa *et al.*, 1983). Although Atractylodis Rhizoma and Atractylodis lanceae Rhizoma gave no significant results in antitumor screening with cultured human cancer cells (JTC-26) (Sato *et al.*, 1978) or sarcoma 180 ascites in mice (Itokawa *et al.*, 1979), the essential oil of *A. macrocephala* rhizomes exhibited inhibitory activity against esophageal cancer Ca 19 *in vitro*, the active principles being identified as atractylon and atractylenolide III (Institute of Materia Medica, 1959–1977).

IX. ANGELICAE RADIX

The authentic and original plant of Angelicae Radix is *Angelica sinensis* Diels of the family Umbelliferae, native to China. When imported drugs were scarce in Japan several hundred years ago, *A. acutiloba* Kitagawa, indigenous to Japan, was found to be useful and has been cultivated to date as a substitute for *A. sinensis*. In the early part of this century, a variety of *A. acutiloba* suitable for mass production was selected in Hokkaido, Japan, and is now recognized as *A. acutiloba* var. *sugiyamae* Hikino. As a result, *A. sinensis* of Chinese origin is now excluded from the Japanese crude drug market, while *A. acutiloba* of Japanese origin has no commercial value in China. Angelicae Radix has been employed as a nutrient, an analgesic, a sedative, a menstruation regulator, and an hematopoietic, and as a whole is said to be a homeostatic remedy for women's disorders.

Concerning the analgesic and sedative effects of Angelicae acutilobae Radix, positive and negative results have been reported. On the basis of the observation that a water extract of Angelicae acutilobae Radix showed inhibitory activity in the acetic acid-induced writhing method in mice, fractionation for analgesic principles afforded falcarinol, falcarindiol, falcarinolone, choline, scopoletin, umbelliferone, and vanillic acid. Among the active principles, the three polyeneynes elicited the most intense effects in the writhing test. Falcarindiol and choline also exerted antinociceptive effects stronger than those of aminopyrine in the retrograde injection test of bradykinin into a carotid artery in rats (Tanaka *et al.*, 1977b). Bioassay demonstrated that the analgesic activity was much stronger in *A. acutiloba* rhizomes than in *A. acutiloba* var. *sugiyamae* rhizomes (Tanaka *et al.*, 1977a). This may be the reason why previous evaluations of the analgesic effect of this crude drug varied. Angelicae sinensis Radix has been reported to have weak central depressant activity equivalent to that of *A. acutiloba* in antinociceptive activity assayed by the acetic acid-induced writhing test (Tanaka *et al.*, 1977a).

$$CH_2=CHCH(OH)[C\equiv C]_2\overset{\overset{\displaystyle R}{|}}{C}H—CH=CHCH_2[CH_2]_5CH_3$$

Falcarinol: R = H
Falcarindiol: R = OH

$$CH_2=CHCO[C\equiv C]_2CH(OH)—CH=CHCH_2[CH_2]_5CH_3$$

Falcarinolone

An aqueous extract of Angelicae acutilobae Radix (p.o.) inhibited an acid-induced increase of vascular permeability in mice, and the potency was alleged to be equivalent to that of acetylsalicylic acid. Topical application of an extract of Angelicae Radix suppressed acute inflammation reactions (Hayashi, 1977), supporting the utility of an ointment prepared from it for burns, cuts, and frostbite. However, the antiinflammatory activity of Angelicae acutilobae Radix was not remarkable in other experimental models (Hayashi, 1977; Yamahara *et al.*, 1980). Therefore, Angelicae Radix seems to be effective only on some acute inflammations, since it exhibits no significant actions on chronic inflammation models. However, Angelicae Radix did elicit a suppressive action in adjuvant arthritis in rats (Cho *et al.*, 1982), where the effect was shown to be intense and stable (K. Takahashi, unpublished).

The actions of Angelicae Radix on the cardiovascular system have been subjected to frequent investigations; in summary, the results indicated that it exhibited a negative inotropic effect and a hypotensive effect with a peripheral vasodilation. The active principles for these actions, however, have not yet been determined. An alcoholic extract of Angelicae sinensis Radix (i.v.) showed antiarrhythmic effects in several experimental arrhythmia (Cha *et al.*, 1981).

Effects of Angelicae sinensis Radix on homodynamics and myocardiac oxygen consumption were examined in dogs, and the volatile oil and its ingredient ferulic acid were claimed to be effective on thromboabgiitis obliterans (Chou *et al.*, 1979). The aqueous extract of Angelicae sinensis Radix and its component ferulic acid inhibited blood platelet agrregation and serotonin release by blood platelets of rat, and it was alleged that this action took part in the therapeutic effect of this crude drug on cerebrovascular disorders (Yin *et al.*, 1980).

Based on the clinical use of Angelicae Radix, a hexane extract prepared from *Angelica acutiloba* var. *sugiyamae* was screened for anti-acetylcholine activity using the isolated rat jejunum. Fractionation monitoring this activity yielded a number of phthalides having substituents at C-3, as do other phthalides isolated from umbelliferous plants. Bioassay indicated that the most active congener was ligustilide, followed by *n*-butylidenephthalide. *n*-Butylphthalide was inactive. When a physiologically active phthalide fraction was left standing in air, its activity diminished quite rapidly. This was found to be due to the light-induced transformation: ligustilide→*n*-butylidenephthalide→ phthalic anhydride. However, others view the clinical efficacy of Angelicae Radix on smooth muscles as doubtful, since aqueous extracts of Angelicae Radix were reported to elicit no effect on the isolated ileum.

Ligustilide   3-*n*-Butylidenephthalide

Although an ethanol extract of Angelicae Radix was observed to induce contractions of the uterus, an extract induced no contractions of the isolated guinea pig uterus at a concentration of 50 µg/ml. With respect to Angelicae sinensis Radix, an aqueous extract contracted smooth muscle organs such as the ileum and uterus, while the essential oil inhibited contraction of these organs.

From the use of Angelicae Radix, it may be considered that it partici-

pates in the regulation of sex function. However, long-term administration of an extract of Angelicae Radix inhibited the estrus rate in mice.
Ingestion of Angelicae sinensis Radix for prolonged periods brought
about an increase of uterus weight, a significant increase of nucleic acid
(in particular DNA) and glycogen contents, and an increase in the
glucose consumption in the liver and uterus. The conclusion was made
that Chinese Angelicae Radix may accelerate the growth of the uterus.
An aqueous extract (i.p.) and a light petroleum extract (s.c.) of Angelicae
sinensis Radix showed estrogenic activity in mice.

The immunologic actions of Angelicae Radix have been examined.
Repeated administration of an ethanol extract of Angelicae acutilobae
Radix slightly suppressed the production of reaginic antibody in serum,
and treatment with an aqueous extract also showed activity. Based on
these results, an antiallergic activity of Angelicae Radix was suggested.

From Angelicae acutilobae Radix, a polysaccharide, angelica immunostimulating polysaccharide (AIP), was isolated (Kumazawa et al.,
1981a), which was a polyclonal B-cell activator, induced interferon, had
inhibitory activity against Ehrlich ascites carcinoma, potentiated antibody formation against sheep red blood corpuscle (Kumazawa et al.,
1980, 1981b), and activated helper T cells (Kumazawa et al., 1982a).
Macrophages exuded into the peritoneal cavity by i.p. injection of AIP
remarkably inhibited the growth of EL-4 leukemia cells, activation of
macrophage was not induced by AIP in vitro, and the inhibitory effect of
AIP was shown by coexistence of the macrophage activity factor at a
concentration showing no inhibitory activity. This led to the suggestion
that two signals are necessary for AIP to exhibit inhibitory activity on the
growth of leukemia cells (Kumazawa et al., 1982b). A recent report claimed that AIP-activated macrophages exuded into the peritoneal cavity
inhibited the growth of EL-4 leukemia cells in mice (Kumazawa et al.,
1982a). The properties of AIP have been studied (Matsumoto et al.,
1982; Shukumae et al., 1982). It was reported that mitogenic activity of
AIP was not exerted by its main component, the polygalacturonic acid
unit, but by another glycoprotein moiety (Shukumae et al., 1982), which
was resistant to protease treatment (Ono et al., 1983a). Participation of
carboxyl groups and phosphate groups in the mitogenic activity was also
suggested (Ono et al., 1983b). The water-soluble polysaccharide fraction
of Angelicae acutilobae Radix gave another polysaccharide, an ara-
binogalactan, which showed anticomplement activity mainly through activation of alternative pathway of complement (Yamada et al., 1983a,b).
A hot-water extract of Angelicae acutilobae Radix was shown to exhibit
macrophage spreading ability (Kumazawa et al., 1983). It was also found
that the hot-water extract exerted macrophage-activating activity, as re-

vealed by means of induction of inhibitory action to the growth of tumor cells (Kumazawa *et al.*, 1983).

ACKNOWLEDGMENT

Grateful thanks are due to Professor G. A. Cordell, University of Illinois, for his grammatical correction in preparing this chapter.

REFERENCES

Aburada, M., Ishige, A., Yuasa, K., and Shinbo, M. (1981). *Nippon Yakurigaku Zasshi* **78**, 108P.

Aburada, M., Iizuka, S., Suekawa, M., Ishige, A., and Hosoya, E. (1982a). *Abstr. Pap., 29th Annu. Meet. Jpn. Soc. Pharmacogn.* p. 11.

Aburada, M., Ishige, A., and Yuasa, K. (1982b). *Abstr. Pap., 66th Meet. Kanto Branch Jpn. Pharmacol. Soc.* p. 31.

Aburada, M., Ishige, A., Yuasa, K., Sudo, K., Shinbo, M., and Ikeda, Y. (1982c). *Proc. Symp. Wakan-Yaku* **15**, 162.

Akamatsu, K. (1970). "Sintei Wakan-Yaku (Oriental Medicines, New Ed.)." Ishiyaku Publ. Co., Ltd., Tokyo.

Bardhan, K. D., Cumberland, D. C., Dixon, R. A., and Holdsworth, C. D. (1978). *Gut* **19**, 779.

Bickel, M., and Kauffmann, G. L. (1981). *Gastroenterology* **80**, 770.

But, P. P., Hu, S., and Kong, Y. (1980). *Fitoterapia* **51**, 245.

Cha, L., Chien, C., and Lu, F. (1981). *Yao Hsueh T'ung Pao* **16**, 53.

Cho, S., Takahashi, M., Toita, S., and Cyong, J. (1982). *Shoyakugaku Zasshi* **36**, 78.

Chou, Y., Huang, L., Chen, Y., Fan, L., Chang, L., and Tseng, K. (1979). *Yao Hsueh Hsueh Pao* **14**, 156.

Cyong, J. (1982). *Adv. Pharmacol. Ther. Proc. Int. Congr., 8th, 1981* Vol. 6, p. 251.

Cyong, J., and Hanabusa, K. (1980). *Phytochemistry* **19**, 2747.

Cyong, J., and Takahashi, M. (1982a). *Chem. Pharm. Bull.* **30**, 1081.

Cyong, J., and Takahashi, M. (1982b). *Proc. Symp. Wakan-Yaku* **15**, 150.

Cyong, J., Hanabusa, K., and Otsuka, Y. (1979). *Proc. Symp. Wakan-Yaku* **12**, 1.

Dajani, E. Z., Bianchi, R. G., Casler, J. J., and Weet, J. F. (1979). *Arch. Int. Pharmacodyn. Ther.* **242**, 128.

Derelanko, M. J., and Long, J. F. (1981). *Proc. Soc. Exp. Biol. Med.* **166**, 394.

Endo, K., Taguchi, T., Taguchi, F., Hikino, H., Yamahara, J., and Fujimura, H. (1979). *Chem. Pharm. Bull.* **27**, 2954.

Epstein, M. T., Espiner, E. A., Donald, R. A., and Hughes, H. (1977). *Br. Med. J.* **1**, 488; *Chem. Abstr.* **86**, 166176n (1977).

Fujimura, H. (1982). *Abstr. Pap., Symp. Stud. Dev. Appl. Nat. Occurr. Drug Mater., 4th, 1982* p. 46.

Gujral, S., Bhumra, H., and Swaroop, M. (1978). *Nat. Rep. Int.* **17**, 189.

Harada, M., and Hirayama, Y. (1979). *Abstr. Pap., 60th Meet. Kanto Branch Jpn. Pharmacol. Soc.* p. 34.

Harada, M., and Saito, A. (1978). *J. Pharm. Dyn.* **1**, 89.

Harada, M., and Yamazaki, R. (1981). *Abstr. Pap., 64th Meet. Kanto Branch Jpn. Pharmacol. Soc.* p. 27.

Harada, M., and Yano, S. (1975). *Chem. Pharm. Bull.* **23**, 941.
Harada, M., Fujii, Y., and Kamiya, J. (1976). *Chem. Pharm. Bull.* **24**, 1784.
Hata, T., and Sankawa, U. (1982). *Abstr. Pap., 102nd Annu. Meet. Pharm. Soc. Jpn.* p. 587.
Hayashi, M. (1977). *Nippon Yakurigaku Zasshi* **73**, 177, 193, 205.
Hayashi, T., Kurosawa, S., Shimizu, M., and Morita, N. (1980). *Abstr. Pap., 27th Annu. Meet. Jpn. Soc. Pharmacogn.* p. 41.
Hayashi, T., Shimizu, M., Morita, N., Kimura, I., Kimura, M., and Sankawa, U. (1981). *Abstr. Pap., 101st Annu. Pharm. Soc. Jpn.* p. 484.
Hayashi, Y., Nakata, K., Aso, H., Suzuki, F., and Ishida, N. (1979). *Yakuri to Chiryo* **7**, 3861.
Hikino, H., Taguchi, T., and Endo, K. (1981). *Nippon Toyoigaku Kaishi* **31**, 1.
Hirai, Y., Furubayashi, H., Takase, H., Fujioka, N., Yamasaki, K., and Nakajima, T. (1982). *Abstr. Pap. 102nd Annu. Meet. Pharm. Soc. Jpn.* p. 587.
Hughes, J. M., Murad, F., Chang, B., and Guerrant, R. L. (1978). *Nature (London)* **271**, 755.
Institute of Materia Medica, Chinese Academy of Medical Sciences *et al.* (1959–1977). "Zhong-yao-zhi (Chinese Materia Medica)," Vols. 1–4. People's Medical Publishing House, Beijing.
Ishikawa, S., and Sato, T. (1980). *Endocrinol. Jpn.* **27**, 697.
Isogai, A., Suzuki, A., Tamura, S., Murakoshi, S., Ohashi, Y., and Sasada, Y. (1976). *Agric. Biol. Chem.* **40**, 2305.
Isogai, A., Murakoshi, S., Suzuki, A., and Tamura, S. (1977). *Agric. Biol. Chem.* **41**, 1779.
Itokawa, H., Watanabe, K., and Mihashi, S. (1979). *Shoyakugaku Zasshi* **33**, 95.
Itokawa, H., Watanabe, K., Morita, M., Kitazawa, H., Mihashi, S., and Hamanaka, T. (1980). *Abstr. Pap., 100th Annu. Meet Pharm. Soc. Jpn.* p. 240.
Jiang-su New Medical School (1977). "Zhong-yao Da-ci-dian (Encyclopaedia of Chinese Medicines)." Jiang-su New Medical School, Shanghai.
Kasahara, Y., and Hikino, H. (1983). *Shoyakugaku Zasshi* **37**, 73.
Kashiwada, Y., Nohara, T., Tomimatsu, T., and Nishioka, I. (1981). *Chem. Pharm. Bull.* **29**, 2686.
Kikuchi, F., Shibuya, M., and Sankawa, U. (1982). *Chem. Pharm. Bull.* **30**, 754.
Kimura, I., and Kimura, M. (1980). *Prog. Abstr., World Congr. Chin. Med. Pharm.* p. 2.
Kimura, M., Kimura, I., and Nojima, H. (1982a). *Adv. Pharmacol. Ther. Proc. Int. Congr., 8th, 1981* Vol. 6, p. 245.
Kimura, M., Kimura, I., and Takahashi, K. (1982b). *Planta Med.* **45**, 136.
Kimura, M., Kimura, I., and Nojima, H. (1982c). *Proc. Symp. Wakan-Yaku* **15**, 157.
Kimura, M., Kimura, I., Takahashi, K., and Kanaoka, M. (1983). *Abstr. Pap. 103rd Annu. Meet. Pharm. Soc. Jpn.* p. 330.
Kinzie, J. L., Grimme, N. L., and Alpers, D. H. (1976). *Biochem. Pharmacol.* **25**, 2727.
Kiso, Y., Tohkin, M., and Hikino, H. (1983a). *Planta Med.* **49**, 222.
Kiso, Y., Tohkin, M., and Hikino, H. (1983b). *J. Nat. Prod.* **46**, 841.
Kiso, Y., Tohkin, M., and Hikino, H. (1983c). *J. Nat. Prod.* **46**, 651.
Koda, A. (1980). *Abstr. Pap., Symp. Stud. Dev. Appl. Nat. Occurr. Drug Mater., 3rd, 1980* p. 40.
Koda, A., Nishiyori, T., Nagai, H., Matsuura, N., and Tsuchiya, H. (1982). *Nippon Yakurigaku Zasshi* **80**, 31.
Kosuge, T., Yokota, M., and Kinoshita, T. (1980). *Abstr. Pap., 100th Annu. Meet. Pharm. Soc. Jpn.* p. 181.
Kubo, M., Shin, H., Urata, Y., and Arichi, S. (1979). *Abstr. Pap., 26th Annu. Meet. Jpn. Soc. Pharmacognosy* p. 14.
Kubo, M., Matsuda, H., Nogami, M., Izumi, S., Anase, K., Tani, T., and Arichi, S. (1981a). *Abstr. Pap., 28th Annu. Meet. Jpn. Soc. Pharmacognosy* p. 41.
Kubo, M., Matsuda, H., Nogami, M., Tani, T., Tono, M., and Arichi, S. (1981b). *Abstr. Pap., 28th Annu. Meet. Jpn. Soc. Pharmacognosy* p. 42.

Kumagai, A., and Takata, M. (1978). *Proc. Symp. Wakan-Yaku* **11,** 79.

Kumazawa, Y., Nishimura, C., and Otsuka, Y. (1980). *Abstr. Pap., 100th Annu. Meet. Pharm. Soc. Jpn.* p. 253.

Kumazawa, Y., Watanabe, Y., Nishimura, C., Mizunoe, K., and Otsuka, Y. (1981a). *Proc. Jpn. Sci. Immunol.* **11,** 113.

Kumazawa, Y., Watanabe, Y., Otani, A., Nishimura, C., and Otsuka, Y. (1981b). *Abstr. Pap., 101st Annu. Meet. Pharm. Soc. Jpn.* p. 285.

Kumazawa, Y., Watanabe, Y., Fukumoto, M., Otsuka, Y., Nishimura, C., and Mizunoe, K. (1982a). *Program Abstr., Symp. Wakan-Yaku* **16,** 12.

Kumazawa, Y., Nishimura, C., Fukumoto, M., Watanabe, Y., and Otsuka, Y. (1982b). *Abstr. Pap., 102nd Annu. Meet. Pharm. Soc. Jpn.* p. 231.

Kumazawa, Y., Fukumoto, M., Watanabe, Y., Nishimura, C., and Otsuka, Y. (1983). *Abstr. Pap., 103rd Annu. Meet. Pharm. Soc. Jpn.* p. 397.

Liu, S., ed. (1975). "Zhong-yao Yan-jii Wen-xian Zhai-yao (Summary of Literature on Chinese Medicine Research) 1820–1961." Scientific Publishing House, Beijing.

Liu, S., ed. (1979). "Zhong-yao Yan-jii Wen-xian Zhai-yao (Summary of Literature on Chinese Medicine Research) 1962–1974." Scientific Publishing House, Beijing.

Maeda, T., Shinozuka, K., Okabe, T., Hayashi, M., and Hayashi, E. (1982). *Proc. Symp. Wakan-Yaku* **15,** 83.

Matsumoto, S., Shukumae, T., Suzuki, I., Tanigawa, S., Miyazaki, T., Kumazawa, Y., Watanabe, Y., and Otsuka, Y. (1982). *Abstr. Pap., 102nd Annu. Meet. Pharm. Soc. Jpn.* p. 233.

Matsushima, T. (1978). *Rinsho to Kenkyu* **56,** 3461.

Miyamura, M., Kashiwada, Y., Nohara, T., Tomimatsu, T., and Nishioka, I. (1981). *Abstr. Pap., 28th Annu. Meet. Jpn. Soc. Pharmacognosy* p. 49.

Mizoguchi, Y., Yamamoto, S., and Morisawa, S. (1981). *Minophagen Med. Rev.* **26,** 210.

Morimoto, M., Nonaka, G., and Nishioka, I. (1981). *Abstr. Pap., 28th Annu. Meet. Jpn. Soc. Pharmacognosy* p. 29.

Morimoto, S., Nonaka, G., and Nishioka, I. (1982). *Abstr. Pap., 102nd Annu. Meet. Pharm. Soc. Jpn.* p. 555.

Morita, K., Hara, M., and Kada, T. (1978). *Agric. Biol. Chem.* **42,** 1235.

Morita, M., Nakanishi, H., Mihashi, S., Itokawa, H., and Hamanaka, T. (1981). *Abstr. Pap., 28th Annu. Meet. Jpn. Soc. Pharmacognosy* p. 44.

Morozumi, S. (1978). *Appl. Environ. Microbiol.* **36,** 577.

Mourey, D. B., and Clayson, D. E. (1982). *Lancet* 20, 655.

Nagai, H., Ichikawa, M., Watanabe, S., and Koda, A. (1978). *Proc. Symp. Wakan-Yaku* **11,** 51.

Nagao, K., Mae, H., Nakanishi, J., Kubo, M., Tani, T., Higashino, M., and Arichi, S. (1982). *Abstr. Pap., 29th Annu. Meet. Jpn. Soc. Pharmacognosy* p. 14.

Nagasawa, T., Shibutani, S., and Oura, H. (1978). *Yakugaku Zasshi* **98,** 1642.

Namba, T., Tsunezuka, M., Bae, K., and Hattori, M. (1981). *Shoyakugaku Zasshi* **35,** 295.

Nishikawa, Y., Watanabe, Y., and Seto, T. (1975). *Shoyakugaku Zasshi* **29,** 139.

Nishikawa, Y., Yasuda, I., Watanabe, Y., and Seto, T. (1976a). *Shoyakugaku Zasshi* **30,** 132.

Nishikawa, Y., Yasuda, I., Watanabe, Y., and Seto, T. (1976b). *Yakugaku Zasshi* **96,** 1322.

Nishizawa, M., Yamagishi, T., Horikoshi, T., and Homma, N. (1979). *Shoyakugaku Zasshi* **33,** 65.

Nishizawa, M., Yamagishi, T., Nonaka, G., and Nishioka, I. (1980). *Chem. Pharm. Bull.* **28,** 2850.

Nogami, M., Moriura, S., Nakanishi, J., Kubo, M., Tani, T., and Arichi, S. (1982). *Abstr. Pap., 29th Annu. Meet. Jpn. Soc. Pharmacogn.* p. 14.

Nohara, T., Nishioka, I., Tokubuchi, N., Miyahara, K., and Kawasaki, T. (1980a). *Chem. Pharm. Bull.* **28,** 1969.

Nohara, T., Tokubuchi, T., Kuroiwa, M., and Nishioka, I. (1980b). *Chem. Pharm. Bull.* **28,** 2682.

Nohara, T., Kashiwada, Y., Murakami, K., Tomimatsu, T., Kido, M., Yagi, A., and Nishioka, I. (1981). *Chem. Pharm. Bull.* **29,** 2451.

Ohminami, H., Kimura, Y., Maki, S., Yamanouchi, Y., Doi, R., Okuda, H., and Arichi, S. (1982). *Proc. Symp. Wakan-Yaku* **15,** 9.

Okabe, S., Kunimi, H., Nosaka, A., Ishii, Y., Fujii, Y., and Nakamura, K. (1979). *Oyo Yakuri* **18,** 469.

Okamura, N., Nohara, T., Yagi, A., and Nishioka, I. (1981a). *Chem. Pharm. Bull.* **29,** 676.

Okamura, N., Yagi, A., and Nishioka, I. (1981b). *Chem. Pharm. Bull.* **29,** 3507.

Ono, H., Matsumoto, S., Hashimoto, Y., Suzuki, I., Miyazaki, T., Shukumae, T., Kumazawa, Y., and Otsuka, Y. (1983a). *Abstr. Pap., 103rd Annu. Meet. Pharm. Soc. Jpn.* p. 394.

Ono, H., Hashimoto, Y., Matsumoto, S., Suzuki, I., Miyazaki, T., Shukumae, T., Kumazawa, Y., and Otsuka, Y. (1983b). *Abstr. Pap., 103rd Annu. Meet. Pharm. Soc. Jpn.* p. 395.

Otsuka, H., Fujioka, S., Komiya, T., Mizuta, E., and Takamoto, M. (1982). *Yakugaku Zasshi* **102,** 162.

Otsuka, Y. (1982). *Kampo Igaku* **6,** 12.

Oura, H., and Nishioka, I. (1982). *Abstr. Pap., Symp. Stud. Dev. Appl. Nat. Occurr. Drug Mater., 4th, 1982* p. 61.

Parke, D. V. (1976). *In* "North American Symposium on Carbenoxolone" (I. T. Beck, ed.), p. 5. Excerpta Medica, Amsterdam.

Pompei, R., Flore, O., Marccialis, M. A., Pani, A., and Loddo, B. (1979). *Nature (London)* **281,** 689.

Pompei, R., Pani, A., Flore, O., Marccialis, M. A., and Loddo, B. (1980). *Experientia* **36,** 304.

Rees, W. D. W., Rhodes, J., Wright, J. E., Stamford, I. F., and Bennett, A. (1980). *Scand. J. Gastroenterol.* **14,** 605 (1979); *Chem. Abstr.* **92,** 584t (1980).

Saito, K., Iwasaki, S., Nakajima, Y., Kano, Y., and Konoshima, M. (1979). *Shoyakugaku Zasshi* **33,** 198.

Sato, A., Nakano, Y., and Taguchi, T. (1978). *Abstr. Pap., Symp. Stud. Dev. Appl. Nat. Occurr. Drug Mater., 2nd, 1978* p. 25.

Senanayake, U. M., Lee, T. H., and Wills, R. B. H. (1978). *J. Agric. Food Chem.* **26,** 822.

Shibutani, S., Nagasawa, T., Oura, H., Nonaka, G., and Nishioka, I. (1981). *Chem. Pharm. Bull.* **29,** 874.

Shim, C., Kim, I., and Paik, S. (1980). *Koryo Taehakkyo Uikwa Taehak Chapchi* **17,** 183; *Chem. Abstr.* **93,** 107257m (1980).

Shimizu, M., Hayashi, T., Morita, N., Kimura, I., Kimura, M., Kiuchi, F., Noguchi, H., Iitaka, Y., and Sankawa, U. (1981). *Tetrahedron Lett.* **22,** 3069.

Shoji, N., Iwasa, A., Takemoto, T., Ohizumi, Y., Ishida, Y., Kajiwara, A., and Shibata, S. (1981). *Abstr. Pap., 101st Annu. Meet. Pharm. Soc. Jpn.* p. 505.

Shukumae, T., Matsumoto, S., Suzuki, I., Ono, H., Miyazaki, T., Kumazawa, Y., Watanabe, Y., and Otsuka, Y. (1982). *Program Abstr., Symp. Wakan-Yaku* **16,** 26.

Sugaya, A., Tsuda, T., Takato, M., Takamura, K., and Sugaya, E. (1975). *Shoyakugaku Zasshi* **29,** 160.

Suzuki, H., Ohta, Y., Takino, T., Fujisawa, K., Hirayama, C., Shimizu, N., and Aso, Y. (1977). *Igaku no Ayumi* **102,** 562.

Takagi, K., Kimura, M., Harada, M., and Otsuka, T., eds. (1982). "Wakan Yakubutsugaku (Pharmacology of Medicinal Herbs in East Asia)." Nanzando Co., Ltd., Tokyo.

Takahashi, K., Shibata, S., Yano, S., Harada, M., Saito, H., Tamura, Y., and Kumagai, A. (1980). *Chem. Pharm. Bull.* **28**, 3449.

Takaishi, K., Kuwajima, H., and Kotari, A. (1979). *Abstr. Pap., 99th Annu. Meet. Pharm. Soc. Jpn.* p. 161.

Takechi, M., and Tanaka, Y. (1982). *Abstr. Pap., 102nd Annu. Meet. Pharm. Soc. Jpn.* p. 587.

Tamura, Y., Nishikawa, T., Yamada, K., Yamamoto, M., and Kumagai, A. (1979). *Arzneim.-Forsch.* **29**, 647.

Tanaka, S., Hoshino, C., Ikeshiro, Y., Tabata, M., and Konoshima, M. (1977a). *Yakugaku Zasshi* **97**, 14.

Tanaka, S., Ikeshiro, Y., Tabata, M., and Konoshima, M. (1977b). *Arzneim.-Forsch.* **27**, 2039.

Tanaka, S., Yuwai, Y., and Tabata, M. (1980). *Abstr. Pap., 100th Annu. Meet. Pharm. Soc. Jpn.* p. 239.

Tani, T., Tono, M., Yamada, J., Iwanaga, M., Namba, K., Kubo, M., and Arichi, S. (1982). *Abstr. Pap., 102nd Annu. Meet. Pharm. Soc. Jpn.* p. 581.

Ulmann, A., Menard, J., and Corvol. P. (1975). *Endocrinology* **97**, 46.

Valisolado, J., Bang, L., Beck, J. P., and Ourisson, G. (1980). *Bull. Soc. Chim. Fr.* p. 473.

Van Marle, J., Aarsen, P. N., Lind, A., and Van Weeran-Kramer, J. (1981). *Eur. J. Pharmacol.* **72**, 219.

Wagner, H., Ott, S., Jurcic, K., Morton, J., and Neszmelyi, A. (1983). *Planta Med.* **48**, 136.

Wang, Y., Chang, J., Li, K., Wu, C., Tin, K., and Liu, Y. (1980). *Shan-hsi Hsin I Yao* **9** (4), 47; *Chem. Abstr.* **94**, 20289v (1981).

Watanabe, Y., and Watanabe, K. (1980). *Proc. Symp. Wakan-Yaku* **13**, 16.

Yagi, A., Okamura, N., Haraguchi, Y., Noda, K., and Nishioka, I. (1978a). *Chem. Pharm. Bull.* **26**, 1798.

Yagi, A., Okamura, N., Haraguchi, Y., Noda, K., and Nishioka, I. (1978b). *Chem. Pharm. Bull.* **26**, 3075.

Yagi, A., Nohara, T., Nishioka, I., Koda, A., Nagai, H., Noda, K., and Tokubuchi, N. (1980a). *Proc. Symp. Wakan-Yaku* **13**, 72.

Yagi, A., Tokubuchi, N., Nohara, T., Nonaka, G., Nishioka, I., and Koda, A. (1980b). *Chem. Pharm. Bull.* **28**, 1432.

Yagi, A., Koda, A., Inagaki, N., Haraguchi, Y., Noda, K., Okamura, N., and Nishioka, I. (1981). *Yakugaku Zasshi* **101**, 700.

Yagura, T., Teranishi, T., and Yamamura, Y. (1978). *Proc. Symp. Wakan-Yaku* **11**, 73.

Yamada, Y., Kiyohara, H., Cyong, J., and Shindo, M. (1983a). *Abstr. Pap., 103rd Annu. Meet. Pharm. Soc. Jpn.* p. 207.

Yamada, Y., Kiyohara, H., Cyong, J., Shindo, M., and Otsuka, Y. (1983b). *Abstr. Pap., 103rd Annu. Meet. Pharm. Soc. Jpn.* p. 207.

Yamahara, J., Sawada, T., Tani, T., Nishino, T., Kitagawa, I., and Fujimura, H. (1977). *Yakugaku Zasshi* **97**, 873.

Yamahara, J., Matsuda, H., Watanabe, H., Sawada, T., and Fujimura, H. (1980). *Yakugaku Zasshi* **100**, 713.

Yamahara, J., Matsuda, H., Sawada, T., Kimura, H., and Fujimura, H. (1981). *Abstr. Pap., 101st Annu. Meet. Pharm. Soc. Jpn.* p. 470.

Yamamoto, M. (1975). *Proc. Symp. Wakan-Yaku* **9**, 127.

Yamamoto, M., Uemura, T., Nakama, S., Uemiya, M., Kishida, Y., Kasayama, S., Yamauchi, K., Komuta, K., and Kumagai, A. (1982). *Proc. Symp. Wakan-Yaku* **15**, 29.

Yamauchi, Y., and Tsunematsu, T. (1982). *Proc. Symp. Wakan-Yaku* **15**, 29.

Yamazaki, K., Yokoyama, H., Nunoura, Y., Umezawa, C., and Yoneda, K. (1981). *Abstr. Pap., 28th Annu. Meet. Jpn. Soc. Pharmacognosy* p. 35.

Yamazaki, M., and Shirota, H. (1981). *Shoyakugaku Zasshi* **35**, 96.
Yano, S., and Nakano, N. (1981). *Chiryogaku* **7**, 703.
Yin, Z., Zhang, L., and Xu, L. (1980). *Yao Hsueh Hsueh Pao* **15**, 321.
Yoshizaki, M., Tominori, T., Yoshioka, S., and Namba, T. (1977). *Yakugaku Zasshi* **97**, 916.
Yosioka, I., Nishino, T., Tani, T., and Kitagawa, I. (1976). *Yakugaku Zasshi* **96**, 1229.

3

Gossypol: Pharmacology and Current Status as a Male Contraceptive

DONALD P. WALLER
LOURENS J. D. ZANEVELD
NORMAN R. FARNSWORTH

Program for Collaborative Research in the Pharmaceutical Sciences
College of Pharmacy
Health Sciences Center
University of Illinois at Chicago
Chicago, Illinois, U.S.A.

I. INTRODUCTION

The ability of orally administered gossypol to cause male infertility was brought to the attention of the world in 1978 when the Chinese National Coordinating Group on Male Antifertility Agents published its findings in an English edition of the *Chinese Medical Journal* (Anonymous, 1978). This news was greeted by Western science both with excitement and skepticism. How could a compound that had been extensively investigated in numerous laboratories because of its toxic effects in animals possibly be a safe and effective oral male contraceptive? On the other hand, data presented in the article could not be ignored. More than 4000 men had received gossypol, with claims that this compound was 99.89% effective in reducing sperm counts below 4 million, with a low frequency of

ECONOMIC AND MEDICINAL PLANT RESEARCH
VOLUME 1

side effects. Additionally, withdrawal of gossypol treatment is now known to allow recurrence of sperm output in many men, varying with the length of gossypol treatment.

Numerous animal experiments have been performed since the publication of these data by investigators from several other countries in an attempt to confirm the effects described by the Chinese. Since 1980, conclusive evidence of the ability of gossypol to cause infertility in several species has been produced (e.g., Chang et al., 1980; Hadley et al., 1981; Hahn et al., 1981; Waller et al., 1981a). Additionally, Chinese researchers have extended their investigations into the mechanism of gossypol's action and its effects on men (Xie et al., 1981; Y.-Y. Ye et al., 1981; Lei, 1982; Chen et al., 1982; Fai et al., 1982; Liu et al., 1982; Shao et al., 1982; Shu et al., 1982; Su et al., 1982; Sun et al., 1982; Wang et al., 1982; Yang et al., 1982; S.-J. Ye et al., 1982; W.-S. Ye et al., 1982; Zhou et al., 1982a,b; Zhou, 1982). Gossypol has now been tested in at least 8500 men (Prasad and Diczfalusy, 1981). This large volume of data obtained in the human and in certain animal species identifies gossypol as a potentially safe, effective, reversible, oral male contraceptive.

II. HISTORICAL PERSECTIVE AND TOXICOLOGICAL ASPECTS

Gossypol (1) is a yellow polyphenolic bisesquiterpene found in high concentrations in the pigment glands of the cotton plant (*Gossypium* spe-

CHO OH   HO CHO

HO   OH

HO   OH

1

cies) (Bernardi and Goldblatt, 1969). These pigment glands are present in the leaves, stems, roots, and especially in the seeds of the plant. Cottonseeds from several varieties of *G. hirsutum* are reported to contain between 0.4 and 1.7% gossypol. Gossypol levels vary greatly in different plants depending on the species, variety, and environmental conditions under which the plants are grown. There are some glandless varieties of the cotton plant that contain very small amounts of gossypol.

Cottonseeds are a by-product of cotton production and have been used for medicinal and commercial purposes for centuries. For instance, Alexander of Macedon (356–323 BC) carried cottonseed products during the campaigning of his armies (Altschul *et al.*, 1958). At the turn of the nineteenth century, cottonseed was primarily ground into meal and used as fertilizer. Eventually the abundance of cottonseed meal and its richness in protein led to its use as a feed supplement for livestock. Unfortunately, the animals often became intoxicated and died. At first, the intoxication was thought to be due to a lack of vitamins in the feed (Gallup, 1931). Cows and other ruminants were not affected when adequate vitamin supplements were provided (Reiser and Fu, 1962). However, some species of domestic livestock continued to suffer adverse effects when fed cottonseed meal even with vitamin supplements (Gallup and Reder, 1931–1932). Swine appeared to be the most sensitive to the toxic effects of cottonseed meal. Intoxicated animals had difficulty in breathing and became lethargic due to muscular weakness. The animals exhibited dyspnea with gasping or "thumping" breathing. Frequently a froth, sometimes bloody, was observed in or on the mouth. The cardiovascular system became compromised and edema developed. Generalized convulsions and cyanosis were at times observed just prior to death (Smith, 1957a). On autopsy, generalized edema of the tissues was noted, especially that of the lungs and liver. Additionally, venous congestion, severe myocarditis, necrosis of the liver, nephritis, hemorrhagic areas throughout the intestine with multiple round lesions, and general gastroenteritis were observed (Smith, 1957b).

The toxicity of cottonseed meal was first attributed to gossypol in 1915 (Withers and Carruth, 1915a,b). Subsequently, a large number of publications appeared describing these toxic properties (Bernardi and Goldblatt, 1969). Controversy existed for many years on whether the toxicity of the pigment glands from the cotton plant was entirely due to gossypol. Some investigators reported that the LD_{50} of the glands was greater than that predicted from the measured concentration of gossypol in these glands (El-Nockrashy *et al.*, 1963). For example, glands containing 43% gossypol had an oral LD_{50} of 1.12 g/kg in rats, while gossypol had an LD_{50} of 2.57 g/kg. It is difficult to interpret these earlier results because the "pure" gossypol used as reference may have been contaminated. The methods used to analyze gossypol at that time were not sophisticated enough to identify small amounts of contaminants. Additionally, there are several pigments in cotton seeds with structures similar to gossypol (Bernardi and Goldblatt, 1969) that are also toxic and will add to the toxicity of the pigment glands. The toxicity of some cottonseed pigments has been studied, for instance, gossypurpurin (LD_{50}: 6.68 g/kg) (Eagle *et al.*, 1951) and gossyverdurin (LD_{50}: 0.66 g/kg) (Alsberg and Schwartze, 1919). Some of these pigments, such as gossyverdurin, are much more toxic than gossypol. Although these agents may have contributed to the overall toxicity of the cottonseed products, they cannot alone account for the toxicity caused by the cottonseed extracts.

In contrast, several investigators have demonstrated that a close correlation exists between the toxicity of raw cottonseed and its gossypol content (Withers and Carruth, 1915a,b; Schwartze and Alsberg, 1924a). The toxicity of gossypol was studied extensively during the 1940s and 1950s. Tissue and organ damage following the administration of gossypol was very similar to that observed during cottonseed intoxication. Death was usually due to pulmonary edema or circulatory failure, depending on the duration of dosing and the animal species used. Damage to the kidneys and liver was also noted. In addition, chronic administration of gossypol caused pronounced cachexia and inanition (Alsberg and Schwartze, 1919).

Different species vary greatly in their sensitivity to gossypol. The pig, dog, guinea pig, and rabbit appear most susceptible to gossypol intoxication, while the rat and hamster are more resistant. Daily oral doses of 150 mg per pig (approximately 1.5 mg/kg) cause death after approximately 28 days (Tollet *et al.*, 1957). By contrast, rats and hamsters can be dosed daily with 10 to 15 mg/kg for more than 8 weeks without death. Gossypol is toxic at doses greater than 5 mg/kg when administered to dogs in a variety of dosing regimens, even for short periods of exposure (5–12 consecutive doses; Eagle, 1950). The dog heart is particularly sensitive to gossypol. A striking decrease in body weight occurs in almost all animals intoxicated with gossypol, and some investigators suggested that gossypol could be used to treat obesity. However, Eagle (1950) warned soon thereafter of the cardiac toxicity of gossypol in the dog. It is interesting that the effect of gossypol on the male reproductive tract was only mentioned in two of the more than 100 articles which reported on the toxicity of gossypol. One referred to "breeding failures" (Schwartze and Alsberg, 1924b), while the other mentioned "mild suppression of spermatogenesis" (Ambrose and Robbins, 1951).

In spite of the toxicity of gossypol, the abundance, ready availability, and inexpensive supply of cottonseed meal prompted the development of procedures that removed gossypol during processing (Withers and Carruth, 1918). In some countries, flour is made from cottonseed and is used as an inexpensive flour substitute or extender (Bressani *et al.*, 1965; Srikantia and Sahgal, 1968). No adverse effects from this processed cottonseed-meal flour have been reported (Harper and Smith, 1968). In addition, cottonseed oil is commonly used for cooking and as a salad oil. The toxicity of gossypol did, however,prompt the WHO/FAO/UNICEF Protein Advisory Committee to adopt standards for the maximum allowable amounts of gossypol in food products. This committee established that the level of free gossypol in edible-grade cottonseed flour should not exceed 0.060%. The presence of some gossypol in flour is probably of little concern, since it usually binds to other constituents in the flour during the baking process and is not readily absorbed.

The potential use of gossypol as an oral male contraceptive has raised a question with regard to its toxicity in humans. Little information is at present available on this subject. The most serious problem reported in the Chinese clinical contraceptive studies with humans was a hypokalemia associated with fatigue, lassitude, changes in electroencephalogram (EEG), and, occasionally, with paralysis (Prasad and Diczfalusy, 1981). Hypokalemia occurred in 0.75% of the treated subjects. However, regional differences were noted, and no such problems with hypokalemia were reported in a limited clinical study conducted in Brazil (Coutinho, 1982). This has led to the suggestion that the hypokalemia was due to a dietary deficiency of potassium exacerbated by gossypol rather than to a direct effect of gossypol on body potassium via absorption or excretion.

III. ANTIFERTILITY ACTIVITY OF GOSSYPOL

The events that led to the discovery of the antifertility activity of gossypol in humans are not clear (Farnsworth and Waller, 1982). Several versions appear in the Chinese literature (Liu, 1957; Qian et al., 1981; Zatuchni and Osborne, 1981). Apparently, improperly manufactured cottonseed oil was used for cooking by a small isolated population, resulting in a loss of fertility of the village inhabitants. Studies by Dai et al. (1978) identified gossypol as the most probable agent causing the infertility. Subsequently, a number of studies were performed in humans and animals to investigate the effects of gossypol on male fertility. These efforts were summarized in an English translation of a *Chinese Medical Journal* article describing gossypol as a new antifertility agent (Anonymous, 1978). The investigations that formed the basis for this article have now been published in several Chinese journals. However, since most of these articles are in Chinese, few appear to have been read on a wide scale.

The publication of the 1978 article from China fostered a widespread interest in gossypol as a male antifertility agent. Initial investigations confirmed the ability of gossypol to produce male infertility in several animal models (Zatuchni and Osborne, 1981). The onset of antifertility

TABLE I

SENSITIVITY OF VARIOUS SPECIES TO THE ANTIFERTILITY EFFECTS OF ORALLY ADMINISTERED GOSSYPOL

Species	Daily dose (mg/kg)	Infertility Onset (weeks)	Infertility Return (weeks)	Lethality (number deaths/number treated)	Duration of experiment (weeks)	References
Mouse						
CD-51	20	NO[a]	—	2/10	8	Hahn et al. (1981)
	40	NO	—	7/10	8	Hahn et al. (1981)
Rat						
Sprague–Dawley	10	NO	—	—	14	Hahn et al. (1981)
	20	6	3	—	8	Hahn et al. (1981)
	30	5	4	0/16	5	Hadley et al. (1981)
	40	NO	—	0/8	8	Zatuchni and Osborne (1981)
Wistar	3	NO	—	—	16	Wang and Lei (1979)
	7.5	12	—	—	12	Wang and Lei (1979)
	10	NO	—	—	14	Weinbauer et al. (1982)
	15	NO	—	—	20	Weinbauer et al. (1982)
	15	5	—	—	—	
CD	5	12	—	—	12	Chang et al. (1980)

	10	—	8	—	12	Chang et al. (1980)
	15	—	4	4/6	4	Chang et al. (1980)
Unspecified	15–40	3–5	2–4	—	5	Anonymous (1978)
Hamster						
Syrian	5	—	12	0/8	12	Chang et al. (1980)
	10	—	10	0/6	10	Saksena and Salmonsen (1982)
	10	5	10	0/6	12	Saksena and Salmonsen (1982)
	10	3	6	4/10	8	Hahn et al. (1981)
	15	7	7	0/6	8	Saksena and Salmonsen (1982)
	20	4	4	4/10	8	Hahn et al. (1981)
	40	4	3	2/10	8	Waller et al. (1981a)
Rabbit						
Unspecified	10	—	NO	—	15	Chang et al. (1980)
Dutch belted	1.25	—	NO	—	14	
	10	—	NO	—	5	
	10	—	NO	0/7	10	Saksena et al. (1981)
	20	—	NO	4/6	10	Saksena et al. (1981)
Dog	2–4	—	9	4/6	20	Prasad and Diczfalusy (1981)

aNO, not observed.

93

activity of gossypol is clearly dose-dependent (see Table I). One of the most striking features of gossypol is the large species and, in some cases, strain differences in its antifertility action. The hamster consistently responds to gossypol (Saksena and Salmonsen, 1982). In this species the minimum antifertility dose is about 5 mg/kg administered daily by oral gavage for 6–8 weeks (Chang *et al.*, 1980). Several strains of rats were reported to become infertile when they are orally treated with gossypol, including the Sprague-Dawley, Wistar, and CD strains (Table I), although, in our laboratory, the treatment of Sprague-Dawley rats with 40 mg/kg·day of gossypol for more than 20 days did not affect testicular function. It has become clear that not all strains of rats respond in an identical manner to gossypol (Zatuchni and Osborne, 1981).

Infertility in dogs was noted when gossypol was administered orally at 2–4 mg/kg, but this was accompanied by an increase in the death rate. Gossypol has also been effective in reducing sperm concentrations in male cynomologus monkeys when treated with 5 and 10 mg/kg for 6 months (Shandilya *et al.*, 1982). No fertility studies with primates have been reported so far. The mouse is relatively resistant to the antifertility effects of gossypol (Hahn *et al.*, 1981). Doses as high as 40 mg/kg·day for 8 weeks have been administered with no antifertility effects, although such doses caused a significant mortality rate. The rabbit is very sensitive to the general toxic effects of gossypol and very resistant to its male antifertility effects (Chang *et al.*, 1980; Saksena *et al.*, 1981). As also found in our laboratory, doses of 20–40 mg/kg caused death in rabbits before infertility was induced.

The antifertility activity of gossypol in men has been and is being evaluated through several clinical trials in China. Gossypol acetic acid, gossypol formic acid, and enteric-coated tablets of gossypol formic acid have been tested. The dose commonly used is ~20 mg/day for 60 to 70 days followed by a maintenance dose of about 60 mg/week (Prasad and Diczfalusy, 1981). This dose level (about 0.3 mg/kg·day loading dose and 1.0 mg/kg·week maintenance dose) is much smaller than the doses required for antifertility effects in most animal species, even the most sensitive ones. For instance, the minimal dose in the hamster is about 35 mg/kg·week. It appears that man is either more sensitive to the contraceptive effects of gossypol or has different pharmacokinetic parameters, which lead to an increased response. However, some caution must be observed when comparing clinical trials with animal experiments. The criterion for infertility in the clinical trials is reduction of sperm concentration to less than 4 million/ml, an amount that could still result in pregnancy. No reports on the actual fertility before and after treatment with gossypol in the human have been published. Additionally, although gossypol-treated men were reported to show a clear decrease in sperm count and an increase in abnormal spermatozoa, these data were based on a single ejaculate (Tong *et al.*, 1982). Large variations in semen parameters can occur from ejaculate to ejaculate, and many factors, including stress, can cause the semen parameters to change.

Interestingly, gossypol can cause infertility in the hamster and rat without a large decrease in sperm count (Chang *et al.*, 1980; D. P. Waller, unpublished) (see further). If this is also true in man, lower doses of gossypol than those used during the clinical trials may suffice to cause infertility. Such lower doses would decrease the number of side effects experienced by treated men. Unfortunately, it will be difficult, if not impossible, to evaluate this activity of gossypol without extensive clinical trials that assess the pregnancy rate of the female partners or the capability of the ejaculated spermatozoa to fertilize *in vitro,* not just considering reduced sperm output.

Gossypol obtained from the *Gossypium* species is in the form of optically inactive, racemic (±) material. Throughout this chapter, the term gossypol refers to this optically inactive form, unless otherwise specified. However, it is possible to isolate the (+) optical isomer of gossypol in good yields from the plant *Thespesia populnea* (King and de Silva, 1968). The optical isomerism results from restricted rotation around the C—C bond that connects the two symmetrical naphthalene moieties of gossypol. The (+) isomer does not reduce fertility when orally administered to rats or hamsters (Wang *et al.*, 1979; Waller *et al.*, 1983). For instance, the oral gavage of hamsters with 40 mg/kg·day of (+)-gossypol for 7 weeks caused no change in their fertility (Waller *et al.*, 1983). This dose is more than 8 times the minimal dose required for the antifertility effect of (±)-gossypol in the hamster. Additionally, at 40 mg/kg·day, (±)-gossypol causes infertility within 3 weeks. Although a significant decrease in weight gain occurs when animals are treated with either (+)- or (±)-gossypol, the decrease in weight gain is greater in animals treated with (±)-gossypol. It appears that (+)-gossypol has some of the nonspecific weight gain inhibition effects of (±)-gossypol but has no antifertility action. This apparent stereoselective action may be due to dispositional or pharmacodynamic differences. If (−)-gossypol exists and can be isolated, it will be interesting to determine if it acts in a specific manner on the reproductive tract of the male or if the only difference is in the biotransformation and excretory handling of the various isomers. We hope (−)-gossypol will be more potent as a contraceptive and less toxic than (±)-gossypol, resulting in an improved therapeutic index.

IV. MECHANISMS OF CONTRACEPTIVE ACTION

A. EFFECT ON HORMONES

Gossypol ultimately causes infertility by either preventing the production and/or ejaculation of spermatozoa, or by altering the ability of spermatozoa to be transported through the female genital tract, and/or ability to fertilize the egg. Such effects of gossypol may either be due to a direct effect of gossypol or its metabolites on spermatozoa, sperm maturation in the epididymis, or sperm production in the testis (spermatogenesis), or may be mediated indirectly via its effects on the hormones that regulate reproductive tract function (Fig. 1).

The primary hormone controlling sperm production and function is testosterone, which is produced by the Leydig cells of the testis. In turn, testosterone synthesis is regulated by luteinizing hormone (LH) of the anterior pituitary gland. Follicle-stimulating hormone (FSH) of the ante-

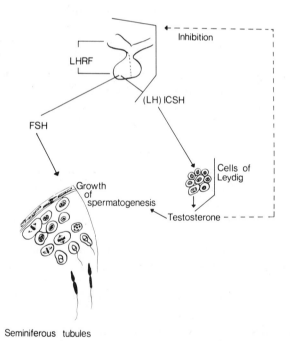

FIG. I Hormonal control of spermatogenesis. Abbreviations: LHRF, LH-releasing hormone; FSH, follicle-stimulating hormone; LH, luteinizing hormone; ICSH, interstitial cell-stimulating hormone.

rior pituitary gland may also influence spermatogenesis via its effect on the Sertoli cell, a cell that envelops all the germ cells (see below). It is known that both FSH and testosterone are required for the initiation of spermatogenesis at puberty, but testosterone stimulation alone appears to suffice for the maintenance of sperm production.

The synthesis and release of the gonadotropins (LH and FSH) from the pituitary gland are regulated by a hypothalamic releasing factor (hormone), by the feedback inhibitory effect of testosterone in the case of LH, and by an as yet poorly identified factor (possibly inhibin) in the case of FSH.

Interference with the hormonal regulation of genital tract function should be reflected by changes in the blood levels of testosterone and possibly also the gonadotrophins LH and FSH. The effect of gossypol on these hormones in animal model systems is somewhat confusing. Hadley et al. (1981) observed a decrease in serum testosterone and LH levels without an accompanying change in serum FSH levels after rats were dosed orally for 5 weeks with gossypol at 30 mg/kg. However, other investigators have noted a decrease in testosterone levels in rats without a change in blood levels of LH (Liang et al., 1981). Thus, one cannot state with certainty that the decreased testosterone levels in blood are due to decreased LH stimulation of the Leydig cells, and it is possible that gossypol affects the Leydig cells directly. The latter possibility is supported by in vitro investigations on isolated Leydig cells. Leydig cells isolated from gossypol-treated rats and stimulated with exogenous LH synthesized testosterone at a decreased rate as compared with LH-stimulated Leydig cells from untreated controls (Lin et al., 1981). Additionally, isolated Leydig cells have a decreased ability to synthesize testosterone in response to LH stimulation when incubated in vitro with gossypol. However, Leydig-cell response may vary between species. Leydig cells from gossypol-treated rabbits secreted normal amounts of testosterone when stimulated with LH (Saksena et al., 1981). It is notable that rabbits are also refractory to the antifertility action of gossypol (as already discussed).

Although these results may lead to speculation that the antifertility action of gossypol parallels its inhibition of Leydig-cell function, gossypol can induce infertility in animal species other than the rat without a concomittant decrease in testosterone levels. Hamsters treated with 40 mg/kg for 8 weeks showed no reduction in testosterone (Waller et al., 1981a). Also, no decrease in testosterone levels was observed in non-human primates after oral dosing for 6 months with gossypol at 10 mg/kg·day (Shandilya et al., 1982). A normal response (release of LH and FSH) to an injection of luteinizing hormone-releasing hormone

(LH-RH) was also observed, indicating that the hypothalamic–pituitary axis was functioning normally. This was in spite of the fact that the ejaculates of the primates possessed poorly motile and morphologically altered spermatozoa. Histologically, no changes occur in the Leydig cells of rabbits and nonhuman primates treated with gossypol (Xue, 1981). Even in the rat, decreases in testosterone may occur only at high dose levels of gossypol, beyond which an antifertility effect is already obtained. Chang *et al.* (1982) observed a decrease in both serum testosterone and LH levels when rats were administered gossypol at 30 mg/kg· day for 6 weeks. There was, however, no significant change in serum testosterone, LH, and FSH levels when animals were dosed at 7.5 or 15 mg/kg·day for 12 weeks, even though the rats were infertile at all three dose levels.

Based on these observations, it appears that the antifertility action of gossypol is not primarily mediated via a direct or indirect effect on testosterone synthesis. Further evidence is that no significant decreases in the weight of the testes and the male accessory organs were observed in any of the nonhuman species after gossypol treatment; these were also not reported for men. Such decreases in the size and weight of the testes are normal sequellae to decreases in testosterone stimulation. Decreased testosterone levels may also cause a loss of libido. However, no significant decrease in libido has occurred in man or the various species of animals studied. In clinical studies, the incidence of loss of libido was about the same as the incidence of increased libido (Anonymous, 1978). Since libido is also psychologically controlled, the few cases of decreased libido may have been caused by psychological factors not directly associated with gossypol.

B. EFFECT ON SPERMATOGENESIS, SPERM MATURATION, AND SPERMATOZOA

After dosing with gossypol, one of the first effects seen in the ejaculate is a loss of sperm motility that is concurrent with or followed by a decrease in sperm numbers. As the number of spermatozoa decrease, the percentage of abnormally shaped sperm and germ (spermatogenic) cells increases (Tong *et al.*, 1982). Even Sertoli cells have been observed in the ejaculate. In later stages of treatment in the human, 90–100% of sperm cells in the ejaculate are germ cells. Numerous spermatids are present with pronounced morphological changes, including multinucleated giant cells with eccentrically placed nuclei. If spermatozoa are present, their structures are generally altered with the spiral mitochondrial sheath being irregularly shaped and frequently showing a loss of cristae.

The acrosomes are not properly formed, and the nuclear chromatin is not condensed. These morphological aberrations of the spermatozoa in the ejaculate, the decrease in sperm numbers, and the presence of spermatogenic cells indicate that spermatogenesis is deranged after prolonged gossypol treatment. Indeed, such altered and/or immotile spermatozoa have also been found in the cauda epididymis and in the testis. Significant ultrastructural changes in epididymal spermatozoa from rats were observed after 3 weeks of treatment with gossypol at 20 and 30 mg/kg (Hoffer, 1982). Flagellar structures of the sperm tail were damaged. On continuous treatment, almost all of the spermatozoa throughout the epididymis showed flagellar damage by the fifth week. Alterations of the sperm head were also observed in the later stages of dosing. However, this could have been the result of death of the spermatozoa and not a direct effect of gossypol.

Spermatogenesis takes place in the seminiferous tubules of the testis. The most basic germ cells, spermatogonia, rest on the basement membrane. These divide mitotically into more advanced spermatogonia and finally into primary spermatocytes. The primary spermatocytes undergo meiosis, first forming secondary spermatocytes and than haploid spermatids. Spermatids do not divide further but undergo biochemical and morphological changes to become spermatozoa. Testicular spermatozoa are immature and generally immotile. They mature, that is, gain the ability to move and fertilize an egg, as they pass through the epididymis.

There is little doubt that gossypol affects the spermatocyte and spermatid stages of spermatogenesis. The earliest onset of infertility in rats after 3 to 4 weeks of gossypol treatment corresponds with the time theoretically required to affect the spermatid stage (Gomes, 1970). Also, ligation experiments indicate that gossypol damage to spermatozoa occurs in the testes (Dai and Dong, 1978). The morphological changes observed in the spermatocytes and spermatids include pyknosis, vacuolation, karyorrhexis, and cytolysis (Xue, 1981). After 4 to 5 weeks of treatment in rats, most of the germinal epithelium is severely altered and the late-stage spermatids and spermatocytes have completely disappeared (Shi *et al.*, 1981). The same pattern of damage appears in the testes of nonhuman primates and in testicular biopsies of men following treatment with gossypol (Xue, 1981).

The organelle that appears to be most sensitive to gossypol treatment is the mitochondrion. Dilated mitochrondria were reported in the midpiece of rat spermatids following gossypol treatment for 4 weeks (Wang and Lei, 1979). Gu *et al.* (1983) examined the testis of rats treated for 8 to 12 weeks with gossypol and found seriously damaged mitochondria in spermatids, including vacuolization, loss of cristae, and complete degeneration of some mitochondria. The mitochondrial sheath was irregular and fragmented, while the other structures, such as the microtubules, coarse fibers, and cell membranes of the spermatids remained basically normal.

Sertoli cells are the nongerm cells of the seminiferous tubules and have a supportive and regulatory function towards the germ cell. Sertoli cells respond to FSH to produce an androgen-binding protein (ABP) that may be required for spermatogenesis. Histologically, only limited damage to Sertoli cells occurs in gossypol-treated animals (Wang *et al.*, 1982), although multiple changes in the mitochondria and lysosomes were evident in rats treated with gossypol at 30 mg/kg for 4 weeks. A decrease in protein synthesis of Sertoli cells treated with gossypol has been reported, suggestive of an effect on Sertoli-cell function (Zhou, 1982). Sertoli cells have been observed in the semen of males treated with gossypol, indicating some disruption of Sertoli-cell attachment in the semeniferous tubules (Tong *et al.*, 1982). Further investigations are necessary before the significance of these observations can be determined and to determine if gossypol's action on spermatogenesis is at least in part mediated via the Sertoli cells.

Loss of motility of spermatozoa in gossypol-treated animals may not only be mediated via an alteration in the spermatogenic process but may also be due to a direct effect of the compound on the epididymis where sperm maturation takes place. Spermatozoa obtained from the epididymis of gossypol-treated rats and hamsters were found to be immotile or poorly motile (Hadley *et al.*, 1981; Chang *et al.*, 1982), and flagellar pathology was present (Hoffer, 1982) (see above). The application of micromolar concentrations of gossypol to the fat pad of the rat epididymis results in a decrease in sperm motility (Hadley and Burgos, 1982). Additionally, mixing gossypol directly with rat, hamster, primate, or human spermatozoa causes immediate sperm death (see next section). Finally, the recovery of sperm motility after withdrawal of gossypol from treated animals is first seen in the caput epididymis, then the corpus, and finally the cauda and the vas deferens (Xue, 1981). Since sperm maturation primarily occurs in the caput, these results further indicate an effect of gossypol on the maturation process.

Although gossypol ultimately causes a spermatogenic defect, a large decrease in the production of spermatozoa may not be necessary for gossypol's antifertility action. Decreases in the numbers of epididymal or vas deferens spermatozoa following the administration of gossypol to either rats or hamsters were not observed by Chang *et al.* (1982). Also, the number of spermatozoa in vaginal smears did not diminish following mating of gossypol-treated male hamsters, although these animals were infertile (unpublished observations by the authors). By contrast, Hadley *et al.* (1981) observed a decrease from 12.2 to 1.6 million spermatozoa/ml collected from vaginal washings within 3 weeks following the initiation of gossypol administration to male rats following mating. A decrease in sperm motility was also noted. After 5 weeks of gossypol administration, the vaginal sperm concentration was further reduced to 0.1 million and the spermatozoa were completely nonmotile. However, these decreases in sperm numbers in the rats occurred much sooner than would have been expected if the effect was only mediated via spermatogenesis (Gomes, 1970). Finally, histological examination of the testis did not reveal any damage to the germ cells of rats following daily treatment with gossypol at 15 mg/kg for up to 12 weeks, even though the animals were infertile (Gu *et al.*, 1983).

These observations, as well as the apparent ability of gossypol to alter sperm maturation in the epididymis, make it possible that the antispermatogenic effects of gossypol are secondary to its antifertility effect. Infertility may occur first by inhibiting sperm motility and/or inactivating certain sperm enzymes necessary for fertilization (see next section) and only later (after long periods of dosing) by altering the spermatogenic process. Severe alterations of spermatogenesis may make recurrence of fertility difficult when gossypol is withdrawn. Indeed, a small but significant number of men who were dosed with gossypol for long periods of time failed to ejaculate spermatozoa after cessation of gossypol treatment (Prasad and Diczfalusy, 1981).

C. EFFECT ON SPERM ENZYMES

Biochemically, gossypol inhibits several sperm enzyme systems. This may, in part, account for its antifertility action. One such enzyme system is involved in energy production by the sperm's midpiece (Myers and Throneberry, 1966). Poso *et al.* (1980) observed an inhibition of glucose degradation to CO_2 by gossypol. Gossypol also has an effect on sperm respiration (Abou-Donia and Dieckert, 1974a; Tso and Lee, 1982c). The sperm-specific enzyme LDH-X is inhibited to a larger degree by gossypol than other LDH isozymes (Maugh, 1981; Giridharan *et al.*, 1982; Tso and Lee, 1982a). Inhibition of these midpiece (energy-producing) enzyme systems, together with alterations seen in the flagellar apparatus (see above), explain the antimotility effect of gossypol when it is added directly to spermatozoa. However, there is some question as to whether the antimotility effect of gossypol *in vivo* is mediated by the same mechanisms. Gossypol chelates with metals or can act as an oxidizing agent.

This could lead to *in vitro* effects that are unrelated to the *in vivo* antifertility activity: e.g., the antimotility effect *in vivo* may be induced via inhibition of sperm maturation rather than by a direct effect on spermatozoa.

Other investigations have been directed to the effect of gossypol on sperm enzymes involved in the fertilization process, such as acrosin. Acrosin is an enzyme located in the acrosome of the sperm head. It performs an important role in the processes leading to sperm penetration of the layers surrounding the egg. Gossypol reversibly inhibits acrosin (Kennedy *et al.*, 1983; Johnson *et al.*, 1982; Tso and Lee, 1982b), but, more importantly, it irreversibly prevents the conversion of the zymogen form of acrosin (proacrosin) to acrosin (Kennedy *et al.*, 1983). This is important because most of the acrosin on the spermatozoon is in the proacrosin form. When gossypol is added to human spermatozoa at dose levels that do not effect sperm motility, *in vitro* fertilization is still prevented. This is at least due in part to the inhibition of proacrosin conversion (Kennedy *et al.*, 1983). Additionally, the vaginal placement of gossypol in amounts that do not cause a decrease in sperm motility still causes a dose-dependent decrease in fertility (Waller *et al.*, 1983). Thus, the antifertility effect of gossypol produced even when spermatozoa are still ejaculated—that is, when no major effects on spermatogenesis have occurred (see previous section)—may not only be due to an effect on sperm motility but also to inhibition of sperm enzymes such as acrosin that are essential for fertilization.

These results provide hope that the dose of gossypol currently being used in humans can be decreased to much lower levels because an absence of spermatozoa or low sperm concentrations may not be essential for contraception. Decreasing the dose of gossypol would result in a significant increase in the therapeutic index. Additionally, gossypol's direct effect on spermatozoa makes the compound potentially useful for vaginal contraceptive purposes, particularly since only low doses need to be used, applied only at intercourse rather than daily. Thus, gossypol should be less toxic as a vaginal contraceptive than as an oral male antifertility agent.

V. DOSING AND PHARMACOKINETICS

Numerous toxicological and antifertility studies have been performed with gossypol, both *in vitro* and *in vivo*. However, little mention is usually made regarding the purity of the gossypol used or the conditions under which it was administered. Gossypol is chemically a highly reactive molecule that is unstable under certain conditions. Early studies on the biological actions of gossypol employed samples of questionable purity. Structurally related pigments normally found in cottonseeds, as well as gossypol degradation products, were probably present in many of the gossypol preparations used in early pharmacological and toxicology studies. Current studies on the antifertility effects of gossypol often utilize samples from a variety of sources, including the U. S. Department of Agriculture, commercial laboratory supply houses, and in-house extracts obtained from cotton seeds. In most cases, limited or no analyses are performed to determine the presence of any contaminants or breakdown products of gossypol in the preparations tested. Additionally, little or no attempt has been made to prevent the breakdown of gossypol in the solutions used during the long dosing periods necessary to induce infertility. In our laboratories, we have demonstrated large variations in the stability of gossypol in organic solvents. It is further known that impurities from degraded gossypol can affect the antifertility properties of gossypol (Waller et al., 1981a). Therefore, it is of utmost importance to determine the purity of gossypol before beginning an experiment and to prepare and store the dosage form under conditions that will assure minimal degradation. Until these conditions have been determined, gossypol should be protected from exposure to light and stored under nitrogen at reduced temperature. In addition, solutions of gossypol should be made up fresh on each day of use.

Gossypol is insoluble in water but will dissolve in alkaline solutions. However, it oxidizes very quickly under alkaline conditions, which may affect its activity (Carruth, 1918). Even so, a number of investigators have used gossypol dissolved in alkaline solution. Various attempts have been made to develop a water-soluble form of gossypol. A water-soluble adduct with glycine was used to study the effects of gossypol on fish (Castillon and Altschul, 1950). Our laboratory has developed a polyvinylpyrrolidone (PVP) coprecipitate that increases the water solubility of gossypol and improves the availability of gossypol to tissues and cells. The PVP-gossypol mixture greatly increases the in vitro action of gossypol on spermatozoa and has also been utilized for vaginal application (Waller et al., 1980, 1981b).

Although gossypol is soluble in most organic solvents, such as ether, benzene, chloroform, and ligroin, these are normally not used as solvents for the administration of gossypol. However, it is possible that a gossypol preparation may contain traces of organic solvents, depending on its method of preparation, which could result in varying degrees of gossypol degradation. Recrystallization of gossypol from various solvents leads to gossypol with different melting points (Campbell et al., 1937), which is in some cases due to polymorphism. However, gossypol also interacts with some solvents to form stable adducts. Gossypol recrystallized from benzene contains 0.5 mol benzene/mol gossypol (Shirley, 1975). Gossypol purified from a mixture of ethanol and 50% acetic acid produces a gossypol–acetic acid adduct (Carruth, 1918). Thus, precautions should be taken to insure that the gossypol being used in biological experiments does not contain unknown organic solvents that could influence its pharmacological effects.

Gossypol is a chemically reactive molecule, mostly due to the presence of six phenolic hydroxyl groups. These hydroxyl groups are strongly acidic and readily react to form esters and ethers. Most importantly, they can react with free amine groups, such as those found in lysine residues of proteins, to form Schiff bases (Lyman *et al.*, 1959). This phenomenon probably leads to inactivation of gossypol in the stomach of ruminants and could be the cause of bioavailability problems in other species when administered orally. Gossypol has been shown to inhibit gastric pepsinogen (Tanksley *et al.*, 1970), probably by binding to the free amino group of the lysine residues (Conkerton and Frampton, 1959). This ability to form Schiff bases may play a role in the mechanism of action of gossypol but may also lead to many nonspecific biological responses.

Gossypol readily chelates with divalent cations, especially iron. This chemical property may be a cause of its toxicity but may also be involved in the mechanism of action of gossypol. Changes in the oxygen-carrying ability of blood from rabbits were observed by Menaul (1923), possibly the result of iron chelation in red blood cells. In addition, gossypol's chelating activity may be the basis for its nonspecific inhibition of biochemical systems that require a divalent cation for activity. Thus, great care must be taken when interpreting *in vitro* results of the activities of gossypol action in extending them to its *in vivo* activity.

The route of gossypol administration should be carefully selected. Leakage around the site of intravenous injection leads to severe edema near the site of injection (Schwartze and Alsberg, 1924b). When administered parenterally, local irritation and edema occur at the site of injection. Intraperitoneal injection of gossypol in peanut oil results in a marked serous exudation in the abdominal cavity, although good absorption of gossypol appears to occur. Ascites and evidence of tissue trauma at or near the injection site take place when gossypol is chronically administered by the intraperitoneal route. This could eventually prevent the absorption of gossypol if repeatedly administered or cause a disturbance to tissues in the area. The reported ability of (+)-gossypol to decrease fertility in female rats (Murthy and Basu, 1981) was probably a result of local tissue trauma and not a specific action of gossypol, because it has been reported that severe trauma alone can decrease fertility. Thus, the chronic use of the parenteral route for the administration of gossypol may result in nonspecific toxic effects.

The oral route of administration is used most frequently in gossypol studies. When administered in this fashion, gossypol may have serious effects on the gastrointestinal (GI) tract. *In vitro,* dilute solutions of gossypol paralyze the muscle of the rabbit intestine and stomach (Schwartze and Alsberg, 1924b). The oral administration of cottonseed meal, cottonseed kernels containing gossypol, or pure gossypol acetate leads to tissue congestion, mucosal sloughing, necrosis of the mucosa, and hemorrhage of the intestinal wall (Gallup and Reder, 1931–1932). These effects are frequently observed when gossypol is administered over long periods of time. The anorexia and weight loss noted in animal studies may reflect these disturbances of the GI tract.

When gossypol is administered orally, the diet of the animals must be carefully controlled. Rats fed a diet containing 500 ppm $FeSO_4$ had greatly reduced tissue levels of gossypol (Abou-Donia et al., 1970). Similarly, swine fed a diet containing 0.1% $FeSO_4$ and gossypol for 15 days had reduced tissue levels of gossypol as compared with swine fed gossypol only (Smith and Clawson, 1965). These effects probably resulted from the chelation of gossypol with iron, forming a complex that was not readily absorbed. Large amounts of iron are found in many sources of water and may dramatically influence the toxic and pharmacological effects of orally administered gossypol.

Protein may also alter the absorption of orally administered gossypol. This is in part due to the Schiff base formation between gossypol and certain proteins (see above). Swine fed rations containing gossypol and 14 or 18% protein for 14 or 28 days had tissue levels of gossypol that were inversely proportional to the levels of protein (Sharma et al., 1966). In most research laboratories, the protein content of animal feed is rigidly controlled. However, in some parts of the world, protein supplements must be administered to ensure healthy animals. The deletion or excessive addition of protein to the diet could significantly alter the effects of gossypol within the body. Effects of iron and protein on gossypol absorption may in part explain the large variations of responses to gossypol that are often reported by different laboratories. These effects may also be a problem when attempting to develop a gossypol formulation that will be consistently absorbed in a large number of individuals, since people will have large variations in diet and living conditions.

The distribution of labeled gossypol administered as a single oral dose has been investigated in rats, swine, chickens, rabbits, dogs, and non-human primates (Anonymous, 1978; Abou-Donia, 1976). High levels of gossypol were found 1 day after dosing in the liver, heart, kidney, spleen, and gastrointestinal tract. Brain levels were very low in all species tested. This is probably due to the blood–brain barrier, which prevents a large polar molecule, such as gossypol, from gaining access to the central nervous system compartment. Large amounts of radioactivity persisted in the gastrointestinal tract and liver for several days. This probably reflects the biliary excretion and possibly enterohepatic cycling of gossypol. High gossypol concentrations in the spleen may be the result of the large amounts of iron available for chelation in that organ.

Interestingly, only very low concentrations of gossypol were found in the testis. As in the brain, the blood–testis barrier probably restricts the movement of gossypol into the testis compartment. It is estimated that less than 0.001% of a single dose ever reaches the testes. The ability of gossypol to selectively disrupt the function of the testes in spite of the small amounts which actually enter the tissue is surprising from this standpoint.

Studies on the distribution of gossypol following chronic administration of cottonseed meal in swine and rats and of pure gossypol in rats have also been performed (Smith and Clawson, 1965; Tone and Jensen, 1975; Sharma *et al.*, 1966; Jensen *et al.*, 1982). The distribution of gossypol was similar to that observed after a single dose. Accumulation of gossypol occurred in all tissues examined for the first 14 days of treatment. The study with swine was terminated after 17 days, while rats were dosed for 5 weeks with cottonseed meal and 8 weeks with pure gossypol. Accumulation of gossypol in the rat continued in most tissues throughout the 5- and 8-week dosing periods. Toward the end of the study with cottonseed meal, gossypol accumulation continued in the liver and spleen but a decrease occurred in lung and kidney. However, in the pure gossypol experiment, gossypol did not accumulate any further in the liver or in the kidney after 6 weeks. This probably indicates that these organs allow the transfer of gossypol in and out of their compartments. Accumulation did continue in the lung, heart, spleen, and testes throughout the 8-week treatment period. Continued accumulation of gossypol in the testicular compartment is probably a result of the blood–testis barrier, which on the one hand inhibits entry but on the other hand prevents the exit of gossypol, leading to an extremely long half-life of gossypol in this particular compartment.

Elimination of gossypol has been studied in several species, including chicks, rats, and swine. In most species, less than 5% of gossypol is excreted in the urine. In all species, the primary route of gossypol elimination is the feces. The molecular weight of gossypol and its acidic hydroxyl groups makes it an ideal molecule for biliary excretion. Indeed, very high levels of gossypol are found in bile (Abou-Donia, 1976; Abou-Donia and Dieckert, 1974b), accounting for the persisting high levels of gossypol in the gastrointestinal tract and feces. Both conjugated and free gossypol have been found in biliary excretions (Abou-Donia and Dieckert, 1974b).

The biotransformation of gossypol has been studied in swine and chicks. Biotransformation products identified in swine include gossypolone, gossypolonic acid, and demethylated gossic acid (Abou-Donia and Dieckert, 1975). Other biotransformation products, such as apogossypol, were proposed but not positively identified. Rats pretreated with phenobarbital had lower blood levels of gossypol than control animals treated with gossypol (Skutches and Smith, 1974). It is not known, however, if phenobarbital causes an increase in the biotransformation of gossypol or an increase in its rate of elimination via the biliary system. Glucuronic acid and sulfate conjugates of gossypol have been identified in bile and feces (Abou-Donia and Dieckert, 1975) and represent a large proportion of the gossypol eliminated via the GI tract. A pathway for gossypol biotransformation has been suggested by Abou-Donia and Dieckert (1975) (see Fig. 2).

From these pharmacokinetic studies, it is easy to understand why several additional weeks are required for the onset of the antifertility effect with low doses of gossypol as compared with high doses. Becuase of gossypol's long half-life in blood (greater than 60 hr), at least a week would be needed for equilibrium to be achieved under multiple dosing conditions. Since gossypol is not distributed rapidly throughout the body and since most of the gossypol remains in the GI tract, the accumulation of gossypol in peripheral tissues such as the testes requires an even greater period of time. For instance, studies by Jensen *et al.* (1982) demonstrated that the testes still accumulate gossypol after 8 weeks. This may be significant to help explain the long onset of infertility at low doses and the possibility of confounding nonspecific effects on the testes due to an excessive buildup of gossypol after long-term dosing. A detailed study of the pharmacokinetics of gossypol with close examination of the parameters in the testicular compartment will help considerably in understanding gossypol's action on the testes and its toxicity.

VI. CONCLUSIONS

The discovery that gossypol has antifertility activity in many species, including humans, when orally administered is extremely exciting. At

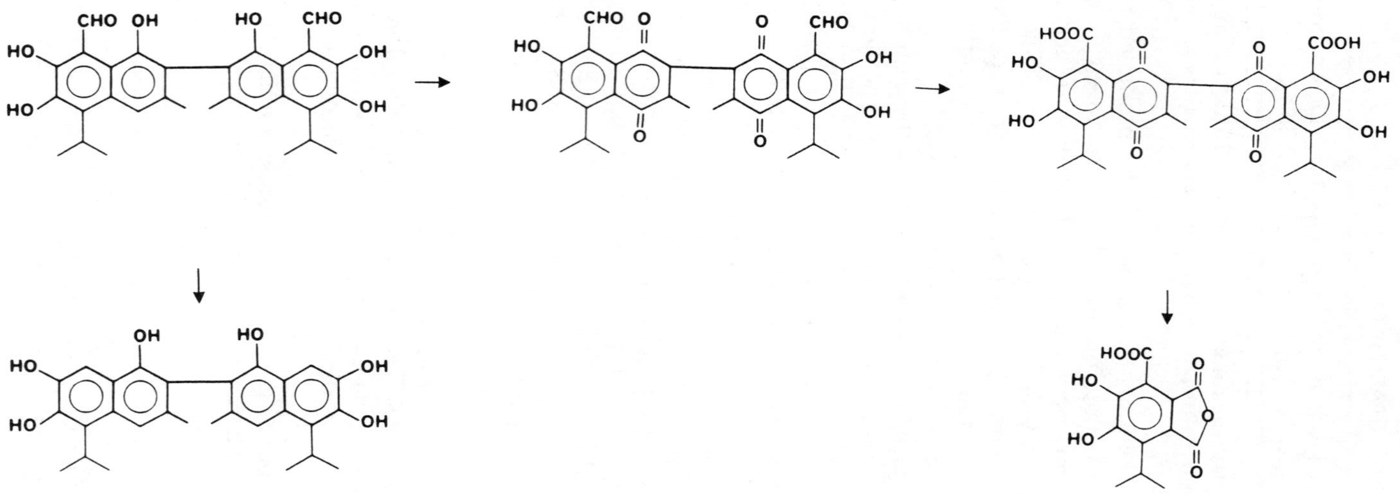

FIG. 2 Suggested gossypol metabolic pathway in the pig liver. From Abou-Donia and Dieckert (1975).

present, some claim that gossypol may well be the only approach to male contraceptive development that has a reasonable chance to reach large-scale clinical testing before the end of the decade (Prasad and Diczfalusy, 1983). Although concern is often expressed regarding the toxicity of gossypol, clinical (human) studies performed so far indicate only a low level of side effects, primarily those associated with hypokalemia. However, this may be associated with regional differences in potassium intake rather than an effect of gossypol. Even so, it is clear that the therapeutic index of gossypol is small, at least for the preparations presently used. It is important that studies with gossypol continue to pinpoint its efficacy and toxicity, because (1) the compound is available in large quantities at a low cost; (2) it is known to be a potent oral male contraceptive; (3) on withdrawal, the antifertility effect is often reversible; and (4) a male contraceptive method is desperately needed. To do so, it is first essential that criteria be defined for gossypol's purity and that only highly purified preparations be used for testing. Second, formulations should be used in which gossypol is stable. Minimally, the purity of gossypol in a stored formulation should be assessed before it is administered. Third, the composition of the formulation should be carefully controlled to minimize gossypol's interaction with any other components. Fourth, when performing animal studies, standards should be set for the diets used, since even protein intake can affect the ultimate toxicity and efficacy of gossypol. Fifth, detailed toxicological studies should be performed, preferably in primates, because the effect of gossypol appears to vary significantly among different species of animals. Sixth, the pharmacokinetics of gossypol should be studied in detail. Seventh, the exact site(s) of gossypol's antifertility action should be determined, keeping in mind its many nonspecific effects when isolated cells or biochemical systems are used for this purpose. When performing contraceptive studies, an absence of spermatozoa in the ejaculate should not be used alone as an endpoint; rather, mating (fertility) studies should be performed, because infertility may well be induced at dose levels of gossypol that do not cause a large decrease in sperm output. Thus, lower dose levels would be required, which would increase the therapeutic index. Finally, derivatives of gossypol should be synthesized, purified, and tested to determine if these cause fewer side effects and have higher antifertility activity than the parent molecule. Since current vaginal contraceptives are not as effective as is desired, detailed studies should also be performed to assess the vaginal contraceptive potency of gossypol. Studies with gossypol may well have a significant impact on overpopulation, probably the largest problem that mankind faces today.

REFERENCES

Abou-Donia, M. B. (1976). *Residue Rev.* **61,** 127.
Abou-Donia, M. B., and Dieckert, J. W. (1974a). *Life Sci.* **14,** 1955.
Abou-Donia, M. B., and Dieckert, J. W. (1974b). *J. Nutr.* **104,** 754.
Abou-Donia, M. B., and Dieckert, J. W. (1975). *Toxicol. Appl. Pharmacol.* **31,** 32.
Abou-Donia, M. B., Lyman, C. M., and Dieckert, J. W. (1970). *Lipids* **5,** 938.
Alsberg, C. L., and Schwartze, E. W. (1919). *J. Pharmacol. Exp. Ther.* **13,** 504.
Altschul, A., Lyman, C. M., and Thurber, F. H. (1958). *In* "Processed Plant Protein Foodstuffs" (A. M. Altschul, ed.), Chapter 17. Academic Press, New York.
Ambrose, A. M., and Robbins, D. J. (1951). *J. Nutr.* **43,** 357.
Anonymous (1978). *Chin. Med. J.* **4,** 417.
Bernardi, L. C., and Goldblatt, L. A. (1969). *In* "Toxic Constituents of Plant Foodstuffs" (I. E. Liener, ed.), pp. 211–267. Academic Press, New York.
Bressani, R., Elias, L. G., and Graham, E. (1965). *Adv. Chem. Ser.* No. 57.
Campbell, K. N., Morris, R. C., and Adams, R. (1937). *J. Am. Chem. Soc.* **59,** 1723.
Carruth, F. E. (1918). *J. Am. Chem. Soc.* **15,** 647.
Castillon, L. E., and Altschul, A. M. (1950). *Proc. Soc. Exp. Biol. Med.* **74,** 623.
Chang, M.-C., Gu, Z.-P., and Saksena, S. K. (1980). *Contraception* **21,** 461.
Chang, M.-C., Gu, Z.-P., and Tsong, Y.-Y. (1982). *Int. J. Fertil.* **27,** 213–218.
Chen, X.-M., Zhou, W.-Y., Ma, X.-X., Feng, J.-B., and Sheih, S.-P. (1982). *Chieh P'ou Hsueh Pao* **13,** 193.
Conkerton, E. J., and Frampton, V. L. (1959). *Arch. Biochem. Biophys.* **81,** 130.
Coutinho, E. M. (1982). *Arch. Androl.* **9,** 37–38.
Dai, R.-X., and Dong, R.-H. (1978). *Shih Yen Sheng Wu Hsueh Pao* **11,** 16.
Dai, R.-X., Pang, S.-N., Lin, X.-K., Ke, Y.-B., Liu, Z.-L., and Dong, R.-H. (1978). *Shih Yen Sheng Wu Hsueh Pao* **11,** 1.
Eagle, E. (1950). *Arch. Biochem.* **26,** 68.
Eagle, E., Hall, C. M., Castillon, L., and Miller, C. B. (1951). *J. Am. Oil Chem. Soc.* **27,** 300.
El-Nockrashy, A. S., Lyman, C. M., and Dollahite, J. W. (1963). *J. Am. Oil Chem. Soc.* **40,** 14.
Fai, Y.-Y., Liang, D.-T., Gou, Y., Liu, Y., Gor, X.-Y., Chow, C.-H., and Xue, S.-P. (1982). *Sheng Chih Yu Pi Yun* **2,** 42.
Farnsworth, N. R., and Waller, D. P. (1982). *Res. Front. Fertil. Regul.* **2,** No. 1.
Gallup, W. D. (1931). *J. Biol. Chem.* **93,** 381.
Gallup, W. D., and Reder, R. (1931–1932). *J. Biol. Chem.* **94,** 221.
Giridharan, N., Bamji, M. S., and Sankaram, A. V. B. (1982). *Contraception* **26,** 607.
Gomes, W. R. (1970). *In* "The Testis" (A. D. Johnson, W. R. Gomes, and N. L. Van Demark, eds.), Vol. 1, pp. 339–432. Academic Press, New York.
Gu, Z.-P., Zong, S.-D., and Chang, C.-C. (1983). *Chung-kuo Yao Li Hsueh Pao* **4,** 40–45.
Hadley, M. A., and Burgos, M. H. (1982). *Ann. N.Y. Acad. Sci.* **383,** 458.
Hadley, M. A., Lin, Y.-C., and Dym, M. (1981). *J. Androl.* **2,** 190.
Hahn, D. W., Rusticus, C., Probst, A., Homm, R., and Johnson, A. N. (1981). *Contraception* **24,** 97.
Harper, G. A., and Smith, K. J. (1968). *Econ. Bot.* **22,** 63.
Hoffer, A. P. (1982). *Arch. Androl.* **8,** 233.
Jensen, D. R., Tone, R. H., Sorensen, R. H., and Bozek, S. A. (1982). *Toxicology* **24,** 65.
Johnson, O., Mas Diaz, J., and Eliasson, R. (1982). *Int. J. Androl.* **5,** 636.
Kennedy, W. P., Van der Ven, H. H., Strauss, W. J., Bhattacharyya, A. K., Waller, D. P., Zaneveld, L. J. D., and Polakoski, K. K. (1983). *Biol. Reprod.* **29,** 999.

King, T. J., and de Silva, L. B. (1968). *Tetrahedron Lett.* **3**, 261.

Lei, H.-P. (1982). *Yao Hsueh Hsueh Pao* **17**, 1.

Liang, S.-X., Pand, S.-N., Dong, R.-H., and Dai, R.-X. (1981). *Sheng Wu Hsueh Pao* **14**, 191.

Lin, T., Morono, E. P., Osterman, J., Nankin, H. R., and Coulson, P. B. (1981). *Fertil. Steril.* **35**, 563.

Liu, B.-S. (1957). *Shang-hai I Hsueh Hsueh Pao* **6**, 43.

Liu, X., Shao, Q., and Liu, H. (1982). *Zhonghua Hayixue Zazhi* **2** (3), 175.

Lyman, C. M., Galiga, B. P., and Slay, M. W.(1959). *Arch. Biochem. Biophys.* **84**, 486.

Maugh, T. H. (1981). *Science* **212**, 314.

Menaul, P. (1923). *J. Agric. Res.* **26**, 233.

Murthy, R. S. R., and Basu, D. K. (1981). *Curr. Sci.* **50**, 64.

Myers, B. D., and Throneberry, G. O. (1966). *Plant Physiol.* **41**, 787.

Poso, H., Wichmann, K., and Luukkainen, T. (1980). *Lancet* p. 885.

Prasad, M. R. N., and Diczfalusy, E. (1983). *In* "Fertility and Sterility" (R. F. Harrison and J. Bonnar, eds.), pp. 255–268. MTP Press, Lancaster, England.

Qian, S.-Z., Hu, Z.-H., Ho, L.-X., Sun, M.-X., Huang, Y.-Z., and Fang, J.-H. (1981). *In* "Clinical Pharmacology and Therapeutics" (P. Turner, ed.), pp. 489–492. Macmillan, New York.

Reiser, R., and Fu, H. C. (1962). *J. Nutr.* **76**, 215.

Saksena, S. K., and Salmonsen, R. A. (1982). *Fertil. Steril.* **37**, 686.

Saksena, S. K., Salmonsen, M. A., Lau, I.-F., and Chang, M. C. (1981). *Contraception* **24**, 203.

Schwartze, E. W., and Alsberg, C. L. (1924a). *J. Agric. Res.* **28**, 173.

Schwartze, E. W., and Alsberg, C. L. (1924b). *J. Agric. Res.* **28**, 191.

Shandilya, L., Clarkson, T. B., Adams, M. R., and Lewis, J. C. (1982). *Biol. Reprod.* **27**, 241.

Shao, T.-S., Zhang, B.-C., Ye, W.-S., and You, M.-M. (1982). *Chieh P'ou Hsueh Pao* **13** (2), 201.

Sharma, M. P., Smith, F. H., and Clawson, A. J. (1966). *J. Nutr.* **88**, 434.

Shi, Q.-X., Zhang, Y.-G., and Yuan, Y.-Y. (1981). *Tung Wu Hsueh Pao* **27** (1), 22.

Shirley, D. A. (1975). *In* "Inactivation of Gossypol with Mineral Salts," pp. 111–146. Nat. Cottonseed Prod. Assoc., Memphis.

Shu, H.-D., Yang, Q.-Z., and Xu, K. (1982). *Chung-kuo Yao Li Hsueh Pao* **3**, 17.

Skutches, C. L., and Smith, F. H. (1974). *J. Nutr.* **104**, 1567.

Smith, F. H., and Clawson, A. J. (1965). *J. Nutr.* **87**, 317.

Smith, H. A. (1957a). *J. Am. Vet. Med. Assoc.* **130**, 300.

Smith, H. A. (1957b). *Am. J. Pathol.* **33**, 353.

Srikantia, S. G., and Sahgal, S. (1968). *Am. J. Clin. Nutr.* **21** (3), 212.

Su, S.-Y., Liu, Y., Zhou, Z.-H., Sheih, S.-P., Zhao, X.-J., Su, M.-Y., and Zhuang, Y.-Z. (1982). *Chieh P'ou Hsueh Pao* **13** (1), 83.

Sun, Y.-B., Chen, Q.-Q., Wang, Y., Su, M., and Lei, H.-P. (1982). *Chung-kuo I Hsueh K'o Hsueh Yuan Hsueh Pao* **4**, 126.

Tanksley, T. D., Neumann, H., Lyman, C. M., Pace, C. N., and Prescott, J. M. (1970). *J. Biol. Chem.* **245**, 6456.

Tollett, J. T., Stepphenson, E. L., and Diggs, B. G. (1957). *J. Anim. Sci.* **16**, 1081.

Tone, J. N., and Jensen, D. R. (1975). *Experientia* **32**, 369.

Tong, S.-M., Zhou, Z.-H., and Zhou, Y.-X. (1982). *Chin. Med. J.* **95**, 355.

Tso, W.-W., and Lee, C.-S. (1982a). *Int. J. Androl.* **5**, 205.

Tso, W.-W., and Lee, C.-S. (1982b). *Arch. Androl.* **8**, 143.

Tso, W.-W., and Lee, C.-S. (1982c). *Contraception* **25**, 649.

Waller, D. P., Zaneveld, L. J. D., and Fong, H. H. S. (1980). *Contraception* **22,** 183.

Waller, D. P., Fong, H. H. S., Cordell, G. A., and Soejarto, D. D. (1981a). *Contraception* **23,** 653.

Waller, D. P., Fong, H. H. S., and Zaneveld, L. J. D. (1981b). U.S. Patent 4,297,431.

Waller, D. P., Martin, A., and Vournazos, C. (1983). *J. Androl.* **4,** 39.

Waller, D. P., Bunapraphatsara, N., Martin, A., Vournazos, C. J., Ahmed, M. M., Soejarto, D. D., Cordell, G. A., and Fong, H. H. S. (1983). *J. Androl.* **4,** 276.

Wang, D.-X., You, M.-M., and Shieh, S.-P. (1982). *Chieh P'ou Hsueh Pao* **13,** 211.

Wang, N.-G., and Lei, H.-P. (1979). *Chung-hua I Hsueh Tsa Chih (Peking)* **59,** 402.

Wang, N.-Y., Luo, V.-D., and Tang, X.-C. (1979). *Yao Hsueh Hsueh Pao* **14,** 662.

Weinbauer, G. F., Rovan, E., and Frick, J. (1982). *Andrologia* **14,** 270.

Withers, W. A., and Carruth, F. E. (1915a). *Science* **41,** 324.

Withers, W. A., and Carruth, F. E. (1915b). *J. Agric. Res.* **12,** 83.

Withers, W. A., and Carruth, F. E. (1918). *J. Agric. Res.* **14,** 425.

Xie, W., Ni, Y., Jiang, Y., Wang, Y., and Wu, Y. (1981). *Sheng Chih Yu Pi Yun* **1** (4), 35.

Xue, S.-P. (1981). *In* "Symposium on Recent Advances in Fertility Regulation" (C.-F., Chang, D. Griffin, and A. Woolman, eds.), pp. 122–146.

Yang, Y., Wu, D., Wang, N., and Chen, X. (1982). *Sheng Chih Yu Pi Yun* **2** (2), 52.

Ye, S.-J., You, M.-M., and Shieh, S.-P. (1982). *Chieh P'ou Hsueh Pao* **13,** 206.

Ye, W.-S., Liu, Y., Shao, T.-S., Liu, Z.-H., You, M.-M., Guo, Y., and Shieh, S.-P. (1982). *Chieh P'ou Hsueh Pao* **13,** 92.

Ye, Y.-Y., Liang, D., Gao, H.-Y., and Ye, G.-Y. (1981). *Yao Hsueh Hsueh Pao* **16,** 390.

Zatuchni, G. I., and Osborne, C. K. (1981). *Res. Front. Fertil. Regul.* **1** (4), 1.

Zhou, L.-F., Lei, H.-P., Gao, Y., Liu, Y., Wang, N.-Y., and Guo, Y. (1982a). *Yao Hsueh Hsueh Pao* **17,** 245.

Zhou, L.-F., Gao, Y., Wang, N.-Y., and Lei, H.-P. (1982b). *Yao Hsueh T'ung Pao* **17** (2), 59.

Zhou, X.-M. (1982). *Sheng Chih Yu Pi Yun* **2,** 39.

4

Immunostimulatory Drugs of Fungi and Higher Plants

H. WAGNER
A. PROKSCH

Institute of Pharmaceutical Biology
University of Munich
Munich, Federal Republic of Germany

I. HISTORICAL ASPECTS

The stimulation of nonspecific defence mechanisms of the human organism as one concept of therapy has a long tradition in medicine. In the German-speaking countries it has been popularized predominantly by physicians such as August Bier, Julius Wagner-Jauregg, and Ferdinand Hoff (Hoff, 1957), and has become known as *Reizkörpertherapie,* "protein shock therapy," *therapeutique de choc,* or *Umstimmungstherapie* (general reorientation or retuning of the organism as a whole). These terms denote the injection of the body's own blood or milk protein (autologous or heterologous proteins), suspensions of inactivated microorganisms such as *Escherichia coli, Plasmodium malariae,* or *Mycobacterium tuberculosis,* the injection of animal organ or plant extracts, as well as the administration of inflammatory agents such as mustard oil or turpentine. These agents were expected to "reorient the organism" and influence favourably chronic inflammatory, allergic, and other diseases considered to be the consequence of impaired host defences or of a vegetative imbalance.

ECONOMIC AND MEDICINAL PLANT RESEARCH
VOLUME 1

Pilot studies to influence positively infectious diseases and malignancies in the same way have also been undertaken.

Subsequently it was discovered that certain minerals (e.g., aluminum hydroxide, magnesium silicate, and beryllium salts) and some plant fatty oils, with or without added Mycobacteria [complete or incomplete Freund's adjuvant (Freund, 1956)], could also be used as adjuvants in immunotherapy, since they increased the production of antibodies triggered off by antigenes.

The object of this chapter is to demonstrate that in the future, apart from microbial preparations (Gram-positive and Gram-negative bacteria), such as BCG vaccines, plant drugs and polyalcohol polymers isolated from fungi may play a role in enhancing immunological host resistance against infections. Although very few of the compounds or preparations described have reached the state of the clinical study, it is suggested that they may be considered as prototypes, the development of which may lead to novel promising compounds.

II. SCOPE AND AIMS OF IMMUNOSTIMULATION

According to Drews (1980), immunostimulation constitutes an attractive alternative to conventional chemotherapy and prophylaxis of infections, especially when the host's defence mechanisms have to be activated under conditions of impaired immune responsiveness. This is of prime importance when infectious diseases, mixed infections, infectious hospitalism, chronic infectious diseases, persistant infections, and their immunopathogenic sequelae, chemotherapy-resistant bacterial and viral infections, have to be treated (Mayr et al., 1979).

A second field of application is the prophylaxis of opportunistic infections of patients at risk. Here, due to a lack of suitable medication, the prophylaxis of viral infections in particular plays a significant role. A further application is the therapy of malignant diseases. It is known, for example, that tumour growth can be inhibited by stimulating specific components of the immune system, such as macrophages or T-killer cells. In addition, the stimulation of the residual mononuclear phagocytotic system remaining intact after treatment with cytostatica may reduce the risk of intercurrent infections.

If chemically defined compounds were available that were capable of specifically stimulating thymus supressor-cell populations, immune stimulation might also be helpful in the therapy of autoimmune diseases such as rheumatoid arthritis, polyarthritis, chronic active hepatitis, myasthenia gravis, multiple sclerosis, psoriasis, and others. So far, the treatment of autoimmune diseases by immunostimulatory drugs lacks a rationale, immunostimulants having been used predominantly on the basis of empirical clinical observations.

III. DEFINITION OF IMMUNOSTIMULANTS

Immunostimulants or immunopotentiators are compounds leading predominantly to a *nonspecific* stimulation of the immunological defence system. They can have either antigenic or nonantigenic properties. The second class of compounds includes the so-called mitogens, which are polycolonal activators of B and/or T lymphocytes (i.e., LPS and lectins). Nonspecific immunostimulants do not affect immunological memory cells, and since their pharmacological efficacy fades comparatively quickly, they have to be administered either in intervals or continuously. Protective immunity due to immunostimulants is brought about quickly and has been termed "paramunity." Compounds capable of inducing paramunity have been termed paramunity inducers (Mayr *et al.*, 1979; Stickl and Mayr, 1979). However, immunostimulants may also stimulate T-suppressor cells and thereby reduce immune resistance. Therefore the terms *immunomodulation* or *immunoregulation,* denoting any effect on or change of immune responsiveness, very often seem to be more appropriate.

Immunoadjuvants are substances that enhance the production of antibodies without acting as antigens themselves. The effects of adjuvants are often thymus-dependent. Their primary target appears to be the macrophage (Waksman, 1980; Allison, 1973). Adjuvants and antigen are administered simultaneously, either subcutaneously or intramuscularly at the same site. Immunostimulants are given either perorally, intravenously, or intraperitoneally. Site and time of administration of immunostimulant and antigen in this case do not coincide.

IV. MECHANISM OF IMMUNOSTIMULATION

Immunological defence is a complicated interplay between nonspecific and specific, cellular and humoral immune responses, stimulation and suppression of immunocompetent cells, and the influence of endocrine and other mechanisms upon the immune system. However, to simplify matters, it may be stated that irrespective of the primary targets of the immunostimulant, be they T or B lymphocytes or the complement system, an increase in phagocytosis by macrophages (macrophagocytosis) and granulocytes (microphagocytosis) plays a central role in immunostimulation (Lohmann-Matthes, 1981).

The activation of macrophages can be brought about by several compounds (stimulants, polyclonal ligands). It is probably important for the stimulating agent to remain in contact with the reactive cell for some time. However, the detailed structural requirements for such a binding are only partially understood. The activated macrophage displays not only increased phagocytosis but is also converted into a secretory cell and acts as a cytotoxic effector cell.

The major role in the amplification of the nonspecific immunological defence is played by lysosomal enzymes secreted by activated macrophages, the components of complement, interferon, lymphokines (Hadden and Stewart, 1981), macrophage migration inhibition factor (MIF), colony-stimulating factor (CSF), prostaglandins, and leukotrienes, to name just the most important factors. The second most important role is the stimulation of T lymphocytes, which can be achieved either directly or indirectly, e.g., via macrophages. The different functions of different T-cell populations have already been mentioned.

An immunostimulation by exogenous mitogens or antigens can also be achieved indirectly by stimulating the organism's own steroid hormones or biogenic amines, such as serotonin, 5-hydroxytryptamine, and histamine. The immunostimulating and suppressive action of estrogens and cortison is well known. Mediator substances secreted by certain immune cells belong to the same type of compounds. Similar mechanisms appear to be responsible for the biological effects of inflammatory substances, either administered or released endogenously.

The details of the molecular biological processes of cell activation are still only known in part. It is assumed that apart from changes in the membrane, surface changes in the cellular metabolism of phospholipids, fatty acids, and prostaglandins, as well as cAMP and cGMP, play an essential role (Hadden, 1980; Ferber and Resch, 1976). There are excellent reviews covering this topic, and the reader is referred to Chédid et al. (1980), Wolstenholme and Knight (1973), Werner and Floc'h (1978), Hadden et al. (1977), Karnovsky and Bolis (1982), and Krakauer and Cathcart (1980).

V. SCREENING METHODS FOR THE DETECTION OF IMMUNOSTIMULATING COMPOUNDS

The most important test systems among the variety of immunological assays available at present are those that allow determination of the functional state and the efficiency of the mononuclear phagocyte system.

Placed second are tests that measure the influence of compounds on T-lymphocyte cell populations:

1. The *carbon clearance test* determines the time-dependent rate of elimination of carbon particles administred intraperitoneally in animal experiments. In humans, soybean emulsions (Lipofundin) are used. The carbon clearance is determined as the ratio of regression coefficients of treated and nontreated groups of animals (see Biozzi *et al.*, 1953; Lemperle and Reichelt, 1973).

2. In another model of phagocytosis, the phagocytotic index is determined *in vitro*, using yeast particles and a granulocyte fraction derived from human serum (see Brandt, 1967). The bioluminescence method developed more recently also uses granulocytes or macrophages. The amounts of O_2 and H_2O_2 liberated during phagocytosis in the presence of zymosan are determined by using this technique (Allen, 1981).

3. The *macrophage migration inhibition test* measures the migration of macrophages and its inhibition in the presence of an antigen (e.g., tuberculin). The migration inhibition factor released by activated lymphocytes attracts macrophages and keeps them at the site of reaction (see Weir, 1978).

4. The so-called *T-lymphocyte transformation test* reflects the degree of blastogenesis of T lymphocytes after having contacted the mitogen. The stimulation rate can be measured by [³H]thymidine incorporation in T lymphocytes, using phytohemagglutinin as control mitogen. Using monoclonal antibodies and fluorescence activated cell sorting, it is even possible to monitor T-lymphocyte subpopulations (subsets) [Ly 1^+, 2^- (helper cells), and Ly 1^-, 2^+, 3^+ (killer/suppressor cells)] (Olsson and Bicker, 1982). Only these measurements allow prediction of whether a substance displays predominantly either stimulating or suppressing activity, and how such a substance might be applied therapeutically.

VI. CLASSIFICATION AND CHARACTERIZATION OF IMMUNOSTIMULATING COMPOUNDS ISOLATED FROM FUNGI AND HIGHER PLANTS

In contrast to typical antigens derived from bacteria and viruses, nonbacterial and nonsynthetic natural compounds with potential immunostimulating activity described up to now can be classified as high and low molecular weight compounds. These compounds may or may not contain nitrogen and belong to different classes of substances. Terpenoids, phenolic compounds, and alkaloids dominate among low molecular weight compounds, and polysaccharides dominate among the high molecular weight compounds. A classification according to their mechanism of action is feasible only in a few cases, because detailed immunological studies are lacking in most cases. If such classifications are used, they are mainly based on medical or clinical empirical knowledge. The pharmacological efficacy for human use cannot be predicted for sure in cases where these compounds have only been tested *in vitro*.

A. LOW MOLECULAR WEIGHT COMPOUNDS

1. ALKALOIDS, NITROGEN-CONTAINING COMPOUNDS

The best studied substance in this group is aristolochic acid, a 3,4-methylendioxy-8-methoxy-10-nitrophenanthrencarbonic acid (AS) (1)

Aristolochic acid
AAI : R = OCH₃
AAII: R = H

isolated from *Aristolochia clematitis*. Extracts of this plant were used in antiquity to treat snake bites and wound infections. Mezger (1966) may have been first in recognizing that the efficacy of the plant extract in treating wounds was not due to a direct antimicrobial activity but directed rather toward the "natural curative power of the organism." Animal experiments carried out by Möse and Lukas (1961), Möse (1963, 1966), Stehr and Wahle (1968), and Lemperle (1972) revealed a pronounced enhancement of phagocytosis of leucocytes and peritoneal macrophages. Tympner (1981) and Henrickson (1970) described an enhancement of granulocytic phagocytosis.

Homeopathic dilutions of D_3 (1 : 10^3) showed maximum activity (Fanselow, 1982). Bartfeld (1977) noted that aristolochic acid increased the production of lymphokines *in vivo*. The rate of recurrent herpes labialis infections could be reduced significantly (Giss, 1980). Exploiting the immunocytoadherence phenomenon, Siering and Müller (1981) demonstrated in rats that aristolochic acid was capable of preventing a reduction of rosette numbers induced by prednisolone. Prednisolone at 5 mg/kg appeared to be equivalent to aristolochic acid at 0.005 mg/kg body weight. The biological activity was interpreted to be the result of a competitive inhibition of lymphocytic surface receptors.

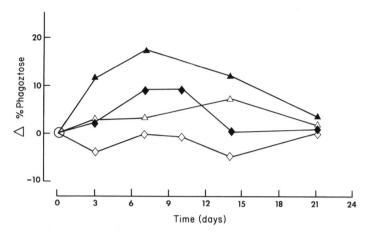

FIG. I Influence of aristolochic acids on phagocytotic activity.

Lemperle (1972) demonstrated that the reduction of phagocytosis induced by chloramphenicol and tetracycline could be compensated by simultaneous administration of aristolochic acid. Kluthe *et al.* (1980) studied the influence of therapeutical doses of aristolochic acid (3 × 0.3 mg/day for 10 days) on phagocytotic activity in two independent doubleblind studies. The highest increase in phagocytotic activity (15–10%) was found between days 7 and 10 (Fig. 1). When treatment was discontinued, phagocytotic activity normalized within a week. Using methylcholanthrene-induced transplantable tumours, Möse (1975) described a reduction in positive "tumour-takes" under treatment with aristolochic acid. Strauch and Hiller (1974) reported activity of aristolochic acid against adenocarcinoma and HeLa cells in tissue culture.

Aristolochic acid has also been suspected of being carcinogenic. This suspicion is based on experiments with mice treated with 0.1–10 mg/kg of aristolochic acid per day for 3 months, after which the formation of gastric tumours was observed (Mengs *et al.*, 1982; Mengs, 1983). Therefore, drugs containing aristolochic acid have been withdrawn from the market following orders of the Bundesgesundheitsamt in Berlin (Federal Board of Public Health).

Among other substances containing nitrogen, the biscoclaurin alkaloid cepharanthine (**2**) isolated from *Stephania cepharantha* and *S. susakii* is reported to stimulate the production of antibodies in animal experiments and to inhibit secretion of biogenic amines histamine and serotonin from mast cells (Sugiyoshi, 1976; Kasajima, 1974). This compound exerts a protective effect on the suppression of hematopoesis induced by cytostatic agents (Mori *et al.*, 1979). This drug and also the pure alkaloid have been used effectively to treat alopecia, pertussis, and

2

Cepharanthine

tuberculosis. The pharmacological efficacy may be the result of immunostimulation (Takahashi, 1956). However, the antiallergic efficacy of
this alkaloid appears to be more pronounced.

The alkaloid tylophorine (**3**), isolated from *Tylophora indica* nec *asth-matica*, is chemically related to aristolochic acid and has also been reported to exert some effects on immunopathological disorders (Gopalak-
rishnan *et al.*, 1978, 1979). Leaf extracts of this plant are used in folk
medicine against bronchial asthma (Nadkarni, 1954) and rheumatism
(Chopra *et al.*, 1958). This initiated a number of experimental studies
that revealed an antianaphylactic effect, the induction of leukopenia,
and inhibition of the Schultz–Dale reaction (Haranath and Shyamala,
1975). As reported by Gopalakrishnan *et al.* (1978, 1979), animal experiments (adjuvant arthritis and antiphlogistic models) revealed a significant suppression of acute, subacute, and chronic inflammations.
Gopalakrishnan *et al.* concluded that this was due to an inhibition of
histamin-release from mast cells. All investigations carried out so far
suggest an immunosuppressive rather than an immunostimulating
process.

3   4

Vidal (1952), Hanisch *et al.* (1966), Del Puerto *et al.* (1968), and
Grollman (1968) reported that emetin (**4**), hitherto known for its activity
as an amoebicide and expectorant only, also possesses antiviral activity.
These observations induced Bauer (1979) to carry out experiments with
embryonic mouse fibroblasts infected with mouse cytomegalovirus. He

demonstrated that emetin in concentrations above 10^{-8} µg/ml applied for 4 days prevented a cytopathic effect. Wagner *et al.* (1985b) observed that emetin increases microphagocytosis significantly at concentrations as low as 0.0001%, a finding that suggests an immunostimulatory mechanism of action. A similar effect was observed by the same author when studying pure oxindole alkaloids isolated from *Uncaria tomentosa* (e.g., isopteropodin) at concentrations of 0.001–0.0001% (Wagner *et al.*, 1985a).

According to Schwarz *et al.* (1974), vincristine, a cytostatic agent utilized for the treatment of leukemias, induces a higher level of antibody production in cells stimulated by antigens prior to the administration of the drug than in control animals. This effect is caused by a partial synchronization of cells. The doses applied were 5, 10, 20, and 30 µg/100 g body weight. At higher doses, the cytostatic activity of vincristine resulted in a reduction of antibody formation.

The well-studied antitumour agent camptothecin has been reported to induce interferon at concentrations of 100 µg (Atherton and Burke, 1978).

2. TERPENOIDS

Similar to tylophorine, several terpenoids possess antiarthritic or antiphlogistic activity. This is especially true for sesquiterpenlactones with a α-methylene-γ-lactone structure in cis position to the cycloheptane ring. Their biological activities appear to be mediated by immunological processes (Hall *et al.*, 1979). The best antiarthritic activity (50–70% inhibition of oedema), among 20 sesquiterpenlactones studied in animal experiments, was observed with helenalin (**5**), tenulin (**6**), and eupahyssopin (**7**) concentrations of 2.5 mg/kg only (indomethacin, 20

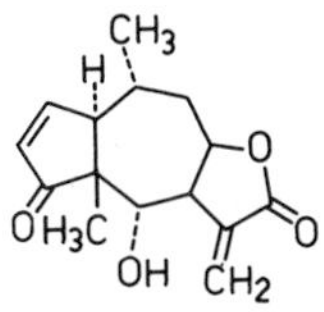

5 Helenalin

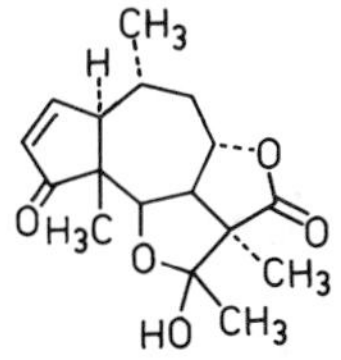

6 Tenulin

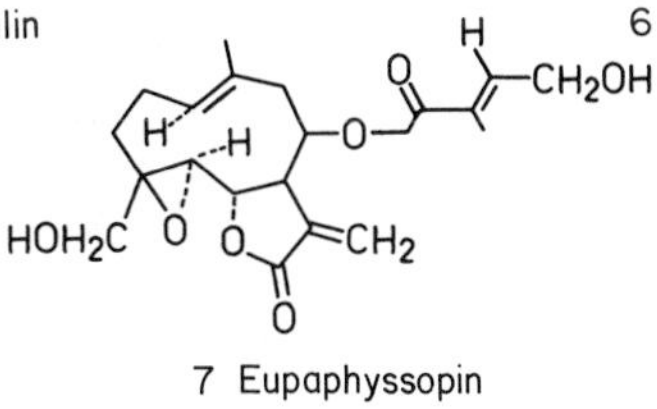

7 Eupaphyssopin

mg/kg). The effects of these compounds on the immune system appear to be twofold: first, the production of antibodies is increased, and second, the production of T lymphocytes is decreased. The mitogenic effect of helenalin would explain the immunostimulating effect observed with B cells.

The results of these studies suggest that the known antiphlogistic activity of *Arnica* drugs is totally or partially due to the presence of the sesquiterpenelactones helenalin, dihydrohelenalin, and its esters, which can be isolated from this plant (Willuhn, 1981).

In contrast, the antitumour activity of several sesquiterpenelactones seems to be the result of a direct alkylation of cellular enzymes, as demonstrated by Wadell *et al.* (1979) for a variety of sesquiterpenelactones of the tenulin series. There are no indications that these terpene structures activate natural killer T cells.

Two germacranolides isolated from *Zexmenia brevifolia*, zexbrevins A and B (**8** and **9**), display a pronounced activity as immunoadjuvants

8   9

(Romo de Vivar *et al.*, 1970; Valdes and Cordoba, 1975). Mice treated with zexbrevin (500 µg per animal) 3–11 days prior to the administration of antigen (sheep erythrocytes, bovine serum albumin, or mouse albumin) revealed a pronounced increase in the number of specific rosettelike cells in the spleen. The same effect was observed when zexbrevin and antigen were administered simultaneously. A mitogenic activity of this compound could be ruled out.

The biological activity of diketocoriolin B (**10**) appears to be different from that of zexbrevin. Diketocoriolin B, an oxidative derivative of (±)-

Coriolin B $\xrightarrow{\text{CrO}_3}$

10

Diketocoriolin B

coriolin, isolated by Takeuchi *et al.* (1969) from the basidiomycete *Coriolus consors*, increases the primary immune response in mice at concentrations of 30–0.01 mg per mouse. At concentrations of 10 μg per mouse, the secondary immune response is increased when the substance is administered intraperitoneally prior to the antigen stimulus (Ishizuka *et al.*, 1972). However, higher doses (125 μg) inhibited the primary immune response. At concentrations of 0.75 μg/ml, Yoshida sarcoma cells were inhibited *in vitro*. When diketocoriolin B was administered at concentrations between 12.5 and 100 μg/day per mouse for 10 days, the survival time of mice carrying implanted L-1210 carcinoma cells was increased.

It is of interest in this context that the well-known tumour promoter 12-*O*-tetradecanoylphorbol 13-acetate (TPA) (**11**) can convert peritoneal

11

macrophages of mice into cytotoxic effector cells at concentrations of 10^{-8}–10^{-7} M. It appears that this effect is not dependent on the presence of lymphocytes (Grimm *et al.*, 1980). The activation was accompanied by a release of plasminogen activator, hydrogen peroxide, and prostaglandin. On the other hand, Abb *et al.* (1979) were able to demonstrate that the same ester (TPA) acts as a mitogen and stimulates the DNA synthesis in primate lymphocytes. It is possible that especially suppressor cells are stimulated, because Goldfarb and Herberman (1981) observed that phorbolesters inhibit the activity of mouse and human natural killer cells.

A steroid glycoside (cynanchoside) has been isolated from the roots of *Cynanchum caudatum*. It increases macrophagocytosis and cellular immunity in mice (Zenyaku Kogyo Co., Ltd., 1980).

In general, saponines are described as immune adjuvants. However, Lutomski *et al.* (1981) demonstrated that a chemically not well characterized mixture of oleanolic acid saponins from *Aralia mandshurica* increased phagocytosis, inhibited the development of certain cancers, and protected the mononuclear system in experimental animals treated with cyclophosphamide.

3. PHENOLS, QUINONES, LIPIDS

Among the aromatic or phenolic compounds from plants, the lignan compound cleistanthin (12) derived from *Cleistanthus collinus* occupies a special position, because the structurally related podophyllotoxin is known to possess cytotoxic activity. In rats, cats, and monkeys, cleistanthin increases the number of granulocytes at a dose of 0.75 mg/kg (Rao and Nair, 1970). This effect has also been observed after oral administration. Studies to reveal the exact mechanism of action, however, are still lacking.

| 12 | 13 |
| Cleistanthin | Dihydroxybenzoic acid |

2,3-Dihydroxybenzoic acid (13) has been reported to stimulate the phagocytotic activity of polymorphonuclear granulocytes and directed cell movement. It has been suggested that the mechanism of action is similar to that of vitamin E and is due to a protection of the leucocyte membrane against autooxidative destruction by H_2O_2. Benzoic acid itself did not display any activity in this test (Boxer *et al.*, 1978). Wacker and Eilmes (1978) reported the antiviral activity of chlorogenic acid, which was suggested to be the result of interferon induction.

The ubiquitous ferulic acid is said to increase phagocytosis in mice, according to studies by Xu *et al.* (1981). Another phenolic compound, anethol, which is present in aniseed oil, increases the number of leucocytes, as has been reported by Ke *et al.* (1980). A stimulation of granulocoytes and an enhancement of the carbon clearance may be measured also with catechins and protocatechuic acid (Wagner and Kreutzkamp, 1985).

Kubo *et al.* (1983) reported that 5-hydroxy-2-*O*-β-D-glucopyranosyl-benzyl-2-*O*-β-D-glucopyranosylbenzyl-2,6-dimethoxybenzoate (curculigoside), isolated from *Curculigo orchioides*, increases phagocytosis. It is not known, however, that this plant is used in folk medicine to enhance resistance nonspecifically.

Preparations of *Rhus* (*Toxicodendron*), which contain urushiolphenols in high dilutions (D_4–D_{12}), are used for the homeopathic treatment of dermatological diseases. Urushiol is a mixture of 3-n-alkyl(ene)catecholes and is the active principle of poison ivy, poison oak, and poison sumac (different *Toxicodendron* species). Less than 2 μg still induces contact allergies. *In vitro*, urushiol induces blastogenesis in peritoneal blood lymphocytes (Byers *et al.*, 1979). The pharmacological effect of such drugs may therefore be due to a nonspecific stimulation of the immune system.

Labzo *et al.* (1980) and Sakykov *et al.* (1980) reported the interferon-inducing activity of gossypol, a 1,1',6,6',7,7'-hexahydroxy-3,3'-dimethyl-5,5'-bisisopropyl-(2,2'-bi-naphthalene)-8,8'-dicarboxyaldehyde **(14)** isolated from cottonseed oil and mainly known for its antispermogenic activity.

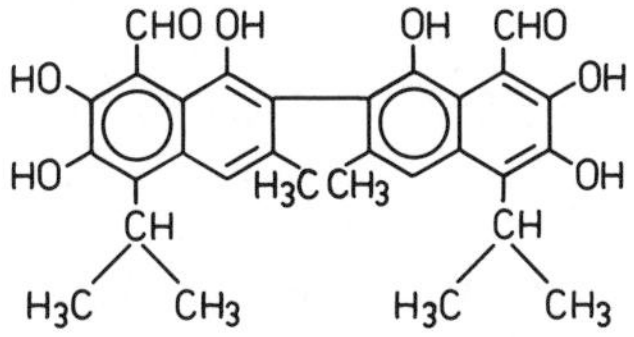

CHO OH   OH CHO
HO   OH
HO   H₃C CH₃   OH
    CH   CH
H₃C   CH₃   H₃C   CH₃

14

Another group of compounds that stimulate the immune system is constituted by the ubiquinones, which are structurally related to urushiol. They are ubiquitous in animals and plants.

According to Block *et al.* (1978) and Mayer *et al.* (1980), ubiquinones Q_7 and Q_8 **(15)** are especially capable of significantly increasing the phagocytosis of mouse macrophages at concentrations of 100 μg/ml.

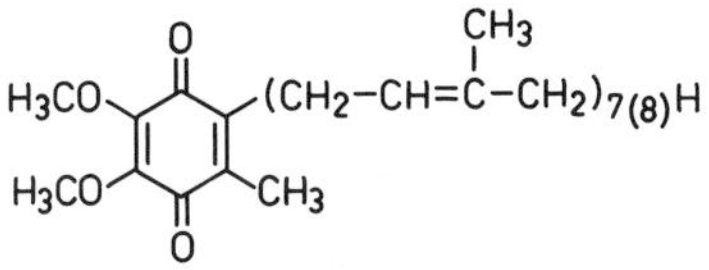

$$H_3CO \quad CH_3$$
$$H_3CO \quad (CH_2-CH=C-CH_2)_{7(8)}H$$
$$CH_3$$

15
Ubiquinone $Q_7 (Q_8)$

The carbon clearance test revealed that ubiquinone Q_7 at 30 mg/kg increased the clearance rate by a factor of 2. In animals pretreated with cyclophosphamide, ubiquinone Q_7 induced the secretion of colony-stimulating factor (CSF) and increased the number of peripheral granulocytes.

The survival rate of granulocytopenic mice with experimentally in-duced bacterial infections (*Klebsiella pneumoniae* and *Staphylococcus aureus*) was significantly increased. In contrast to this, the naphthoquinone plumbagin, which is known for its cytotoxicity, exerts immunstimulating activity when applied in very low concentration (10^{-3}–10^{-5} mg/ml) (Wagner *et al.*, 1985b). Vitamin E, which is structurally related to ubiqui-nones, can also act as an immunopotentiator at low doses (Yasunaga *et al.*, 1982).

Among the lipids that stimulate or modulate the immune response, lysolecithin (2-lysophosphatidylcholine, LPC) (16) and especially its syn-thetic analogues, the alkyllysophospholipids (ALP) (17), deserve special mention. As reviewed by Munder *et al.* (1980), the cytotoxic activity of

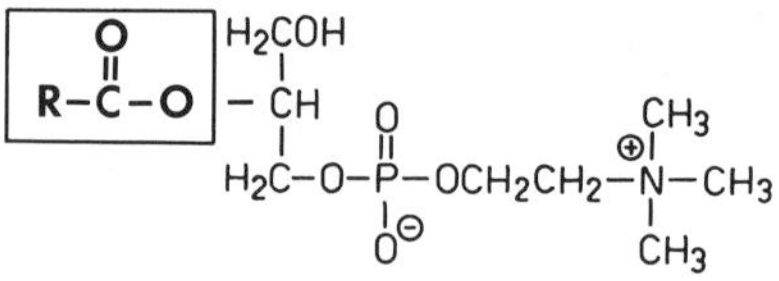

$$16$$
$$R = -(CH_2)_n-CH_3 \quad n = 14\text{-}18$$

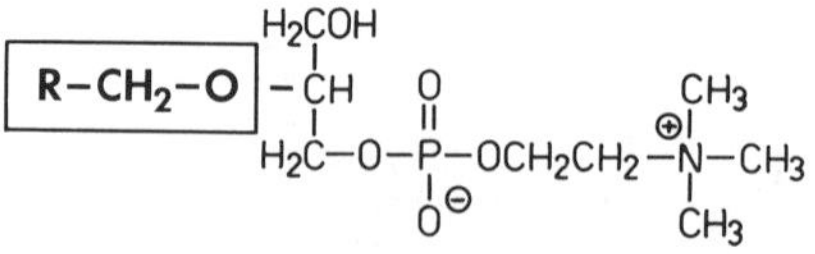

$$R = -(CH_2)_n-CH_3$$
$$n = 14\text{-}18$$
$$17$$

Lysoleithin alkylether

normal peritoneal macrophages and the tumoricidal activity of bone-marrow macrophages are significantly increased by LPC. In addition, ALP appears to possess a direct chemotherapeutic effect selective for tumour cells. When ALP was administered at concentrations of 250–1000 μg per mouse 4–30 days prior to the implantation of Ehrlich ascites cells, tumour growth could be abolished almost completely. When ALP was administered 1–9 days after the tumour had been implanted and was given at concentrations of 10–100 μg/day for 20 days, it still exerted a positive pharmacological effect on tumour growth. This effect was not observable when LPC was used. Lysophosphatidylcholine ana-logues with an ester bond in position S_N1 and no substitution in position S_N2 did not have any pharmacological effect *in vitro*, when the prolifera-tion of human neoplastic cells was studied.

It has been suggested that the cytotoxic activity may be the result of interference in the tumour cell due to the inability of the tumours cells to metabolize ALP (Modolell *et al.*, 1979).

B. HIGH MOLECULAR WEIGHT COMPOUNDS

In this class of compounds, the number of possible polyclonal ligands that may react with surface structures of immunocompetent cells appears to be unlimited at first glance. However, the relatively small number of compounds that can be applied therapeutically demonstrates that the ability of a compound to stimulate the immune system is dependent on several conditions, such as molecular weight, solubility, and structure type.

Polysaccharides, nucleoproteins, and proteins are among the more privileged compounds of biogenic origin. The most important immunostimulants isolated from plants are glycoproteins (lectins) and polysaccharides.

1. LECTINS

Lectins are sugar-binding, carbohydrate-specific proteins or glycoproteins, which agglutinate cells or precipitate glycoconjugates. They are not derived from cells of the immune system. Lectins were first discovered in plants and were therefore called phytohemagglutinins. Later, they were also detected in fungi, bacteria, invertebrates, lower vertebrates, and at last they were also found in higher animals and human organs.

The molecular weights of lectins range between 40,000 and 125,000. They contain one or several polypeptide chains (A and B chains), consisting of several hundred amino acids. These amino acid chains carry carbohydrate ligands as side chains, consisting of approximately several dozens of sugar residues. The binding sites are mainly aspartate and aspartic acid. The predominant sugars are mannose and acetylated glucosamin. "Classical" ricin, like other toxic lectins, is a mixture of a toxic component (ricin) and a nontoxic but agglutinating component (*Ricinus* agglutinin). For the chemical structures of the various known lectins, the reader is referred to the review articles of Lis and Sharon (1973), Rüdiger (1978), and Uhlenbruck (1981).

Apart from their hemagglutinating properties, lectins are of interest because some lectins bind predominantly to lymphocytes, where they induce mitosis. Other lectins inhibit protein synthesis in eucaryotic cells, and some lectins agglutinate malignant cells better than normal cells (Lis and Sharon, 1973). According to Rüdiger (1982), even fragments of lectins may be able to stimulate lymphocytes.

Lymphocytagglutination has been observed with phytohemagglutinin (PHA) of *Phaseolus vulgaris* and pokeweed mitogen, isolated from the root of *Phytolacca americana* (Waxdal, 1974). Concanavalin A, isolated from *Canavalia ensiformis*, and the phytohemagglutinin and lectin from *Lens culinaris* (lentils = Lch) have been reported to exert mitogenic effects on lymphocytes (Nicolson, 1974; Uhlenbruck, 1971).

Tumour-specific lectins—i.e., those only agglutinating tumour cells—are concanavalin A (Inbar and Sachs, 1969), the agglutinin of *Ricinus communis* (RCA$_I$) (Nicolson and Blaustein, 1972), soybean agglutinin (SBA) (Sela *et al.*, 1970), and the agglutinin of white haricot beans (WBA) (Sela *et al.*, 1973).

The cytotoxic activity of several lectins, such as abrin (**18**), ricin, and PHA (Uhlenbruck, 1971), either is due to the inhibition of intracellular

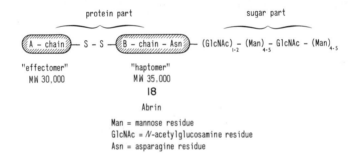

protein synthesis (Sandvig *et al.*, 1976), or is the result of T-killer cell activation [e.g., PHA (Nicolson, 1974)].

Very often, the mitogenic properties of total plant extracts are generally unpredictable, because these extracts contain mixtures of lectins that differ in their mitogenic activity and their specificity (T and B lymphocytes?), as has been observed, for example, for *Phytolacca*. In addition, it is known that lectins, specific for fucose, can saturate the fucose receptors on macrophages and may therefore inhibit migration inhibition factor (Uhlenbruck, 1981).

Apart from some applications in medical diagnosis, lectins have not been used for immunotherapy. Homeopathic dilutions of *Phytolacca* preparations have been reported to have antiarthritic and cytotoxic activity. It is questionable, however, whether this is due to the presence of lectins in these preparations. The same applies to preparations of mistletoe (*Viscum*) that have been used as cancerostatic agents in the treatment of malignancies. However, three fractions containing lectins have been isolated from extracts of *Viscum* (Franz *et al.*, 1981).

2. POLYSACCHARIDES

As compared with bacterial polysaccharides, the number of polysaccharides isolated from fungi and higher plants that have been studied chemically and pharmacologically in detail is relatively small. Most of the latter polysaccharides have been isolated from fungi and lichens (Nakahara et al., 1967; Whistler et al., 1976; Chihara, 1977; Hadden et al., 1977) (Table I). Lentinan, a yeast polysaccharide, isolated from Lentinus edodes, is one of the best studied polysaccharides (Chihara et al., 1969, 1970a; Haba et al., 1976; S. Abe et al., 1982; Moriyama, 1982; Togami et al., 1982). According to Whistler et al. (1976) and Haba et al. (1976), the antitumour activity of this compound is mainly the result of T-lymphocyte activation. It has been suggested that transmitters such as serotonin or 5-hydroxytryptophan are involved in the activation process. An increase in several components of serumproteins, possibly ceruloplasmin and complement 3, has also been observed (Okihara and Maeda, 1982).

When lentinan is given in conjunction with mitomycin and 5-fluorouracil during tumour therapy, immune response and survival time increase as compared to treatment with mitomycin alone or mitomycin and 5-fluorouracil (Taguchi, 1982).

The protein-bound polysaccharide PSK (**19**), isolated from Coriolus

TABLE I

IMMUNOSTIMULATING POLYSACCHARIDES FROM FUNGI, LICHENS, AND ALGAE

Organisms	Polysaccharides	Type	Linkage	M.W.
Fungi	Zymosan	Glucan	$\beta 1 \to 3$	50,000–120,000
	Mannozyme	Mannan	$\beta 1 \to 3$	5,000–20,000
			$\begin{cases} \beta 1 \to 3 \\ \beta 1 \to 4 \end{cases}$	50,000–65,000
	Lentinan	Glucans	$\beta 1 \to 3$	~500,000–1,000,000
	Pachyman		$\begin{cases} \beta 1 \to 2 \\ \beta 1 \to 6 \end{cases}$	
	Pachymaran			
	Schizophyllan	Glucan	$\begin{vmatrix} \beta 1 \to 3 \\ \beta 1 \to 6 \end{vmatrix}$	~400,000
	Krestin (PSK)	Glucan	$\begin{vmatrix} \beta 1 \to 4 \\ \beta 1 \to 6 \end{vmatrix}$	50,000–100,000
			+ covalent proteins	
Lichens and algae	Pustulan	Glucans	$\beta 1 \to 6$	
	Lichenan		$\beta 1 \to 3$	
			$\beta 1 \to 4$	
	Isolichenan		$\alpha 1 \to 3$	100,000–500,000
			$\alpha 1 \to 4$	
	Laminaran		$\beta 1 \to 3$	
			$\beta 1 \to 6$	

19

PSK

versicolor (Ohno *et al.*, 1976), is active against transplanted tumours (Ikekawa *et al.*, 1968; Tsukagoshi, 1975; Mayer and Drews, 1980; Hotta *et al.*, 1982). Likewise, schizophyllan (**20**) from *Schizophyllum commune* (Tabata *et al.*, 1981; Lee *et al.*, 1981; Komatsu *et al.*, 1969; Komatsu, 1976), and pachymaran from *Poria cocos* (see also Chapter 2) display good activity. The latter compound, however, became highly active only after chemical modification (Narni *et al.*, 1980; Chihara *et al.*, 1970b; Hamuro *et al.*, 1971).

20

Schizophyllan

Schizophyllan activates the complement system nonspecifically (Mitani *et al.*, 1981a). It increases not only the concentration of several serum proteins (Mitani *et al.*, 1981b) but also interferon production induced by endotoxin and *Corynebacterium parvum* (Numasaki *et al.*, 1981).

Several models have revealed the activity of schizophyllan against lung carcinoma cells and the development of lung micrometastases (Yamamoto *et al.*, 1981). The yeast polysaccharide zymosan (Bradner *et al.*, 1958; Bradner and Clark, 1959; Diller *et al.*, 1964), and also , other glucans and mannans (Borecký et al., 1967; S. Suzuki et al., 1969; M. Suzuki et al., 1971; Matsumoto et al., 1980) possess stimulating and antitumour activity.

A D-mannan and D-glucan of Saccharomyces cerevisiae inhibits the growth of mouse tumours transplanted subcutaneously into allogenic mice (S. Suzuki et al., 1969; Matsumoto et al., 1980). When given prophylactically, the polysaccharides increase survival rates of mice significantly (90%) when the animals were infected with Staphylococcus aureus. Simultaneously, the activity of the three lysosomal enzymes of peritoneal macrophages was greatly increased. This finding suggests that protection against infection is solely due to increased phagocytosis (Okawa et al., 1982).

Vesiculogen, a polysaccharide from Pezisa vesiculosa, was demonstrated to be an adjuvant and mitogen, mainly activating B cells. Studies of phagocytosis activation revealed that its mechanism of action was mediated by the reticuloendothelial system (I. Suzuki et al., 1982).

Many of the polysaccharides isolated from fungi exhibit a common structural basis element, $(1\rightarrow3)$-β-D-glucan (Whistler et al., 1976; Table I). Several other active polysaccharide fractions from fungi have been described by Kohlmünzer et al. (1980), Miyazaki et al. (1978, 1979), Misaki et al. (1981), Miyazaki and Nishijima (1981), I. Suzuki et al. (1982), Ueno et al. (1982), Mizuno et al. (1981a,b), and Usui et al. (1981). Polysaccharides of algae have rarely been shown to stimulate the immune system or to possess antitumour activity (Mizuno et al., 1980). It is not known whether the antiviral activity of polysaccharides isolated from the Dumontiaceae is the result of interferon induction or another mechanism of action (Ehresmann et al., 1977, 1979). A very good activity has been observed against herpes virus infections (Hatch et al., 1979).

The basic structure of polysaccharides from algae is mainly composed of galactose or mannose, joined together via the C-4 OH and C-3 OH, respectively, to form glycosidic bonds. Frequently, galactose carries sulfate substituents in the C-4 OH position.

At a concentration of 2.5mg/mouse, the polysaccharide traganth has been shown to be active against a variety of tumours (Whistler et al., 1976). Pretreatment of mice with tragacanth also reduced the number of positive "takes" of implanted Ehrlich ascites cells (Osswald, 1968). Since the efficacy of tragacanth depended on time of administration and dosis, it may be assumed that this compound also acts by influencing the immune system. Of all polysaccharides isolated from lichens, lichenan has been particularly well studied because of its pharmaceutical efficacy (Shibata et al., 1968; Nishikawa et al., 1979, 1981). Lichenan was highly active against sarcoma 180 tumours. However, while pustulan and related polysaccharides stimulated the reticuloendothelial system, polysaccharides of the lichenan series appeared to lack this activity (Nishikawa *et al.*, 1981). The lichen polysaccharide isolichenan is a (1→3)-(1→4)-α-D-glucan.

Antitumour activity of polysaccharides of higher plants has also been reported (Tokuzen and Nakahara, 1971; Whistler *et al.*, 1976). The best-studied polysaccharides in pharmacological terms are those derived from rice and wheat straw (Nakahara *et al.*, 1967; Soma *et al.*, 1981), bamboo (Nakahara *et al.*, 1964; Sakai *et al.*, 1963, 1964; Kuboyama *et al.*, 1981), sugar cane bagasse (Sakai *et al.*, 1964), and *Astragalus gummifera* (Roe, 1954; Kojima, 1980; Osswald, 1968; Chen *et al.*, 1981; Wang *et al.*, 1980). Similar activities have been described for polysaccharides isolated especially from *Calendula officinalis* (Manolov *et al.*, 1964), *Solidago* sp., *Trifolium pratense, Arctium lappa, Aucanacua carmizulis, Yucca schidigera, Rumex acetosella,* and *Bryonia alba* et *dioica* (Belkin *et al.*, 1959).

Müller (1962) isolated an acidic polysaccharide (viscic acid) from the berries of *Viscum album* that inhibited the growth of several transplanted tumours. Mathé *et al.* (1963) demonstrated that this compound stimulated the neutrophilic leucocytosis in neutropenic patients. Probably analyzing the same polysaccharide fraction, Bloksma *et al.* (1982) found that this preparation acted as adjuvant on humoral immune response, slightly reduced the number of IgM-producing cells, and markedly increased the number of IgG-producing cells. In addition, an elevated clearance was observed in mice 48 hr after administration of 120 mg/kg of this polysaccharide fraction, while inhibition of clearance was observed after 24 hr. Our own studies (Jordan, 1985) have demonstrated that a lectin-free polysaccharide fraction obtained from *Viscum album* stems strongly increased clearance rate 24 hr as well as 48 hr after administration of 10 mg/kg. However, the sugar composition of this lectin-free polysaccharide differed from that described by Müller, although both preparations contained uronic acid. Exact structural analyses have not been reported for any of these compounds. In most cases, only the sugar composition has been reported. Belkin *et al.* (1959) studied only crude polysaccharide fractions that were obtained by simple solvent precipitations (trichloroacetic acid, phenol, alcohol). However, Wagner and Proksch (1981; Proksch, 1982; Wagner *et al.*, 1984) succeeded in assigning activity to two chemically defined compounds in the case of highly active polysaccharides isolated from *Echinacea purpurea* plant. [Figure 2 shows proposed structures for *Echinacea* polysaccharides I and II (**21,22**).]

One of the polysaccharides isolated from *Echinacea purpurea* could be shown to be a heteroxylan with a mean molecular weight of 35,000, while the other polysaccharide was an arabinorhamnogalactan with a

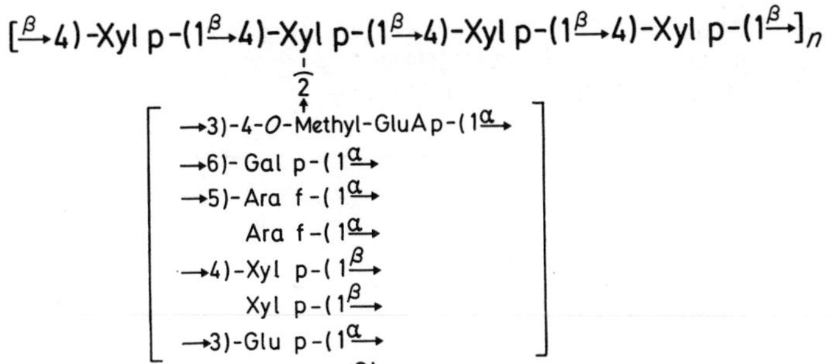

$$[\overset{\beta}{\rightarrow}4)\text{-Xyl p-}(1\overset{\beta}{\rightarrow}4)\text{-Xyl p-}(1\overset{\beta}{\rightarrow}4)\text{-Xyl p-}(1\overset{\beta}{\rightarrow}4)\text{-Xyl p-}(1\overset{\beta}{\rightarrow}]_n$$

$$\uparrow$$
$$2$$

$$\begin{bmatrix} \rightarrow 3)\text{-}4\text{-}O\text{-Methyl-GluA p-}(1\overset{\alpha}{\rightarrow} \\ \rightarrow 6)\text{-Gal p-}(1\overset{\alpha}{\rightarrow} \\ \rightarrow 5)\text{-Ara f-}(1\overset{\alpha}{\rightarrow} \\ \text{Ara f-}(1\overset{\alpha}{\rightarrow} \\ \rightarrow 4)\text{-Xyl p-}(1\overset{\beta}{\rightarrow} \\ \text{Xyl p-}(1\overset{\beta}{\rightarrow} \\ \rightarrow 3)\text{-Glu p-}(1\overset{\alpha}{\rightarrow} \end{bmatrix}$$

$$21$$

$$[\overset{\alpha}{\rightarrow}4)\text{-Gal p-}(1\overset{\alpha}{\rightarrow}2)\text{-Rha p-}(1\overset{\alpha}{\rightarrow}2)\text{-Rha p-}(1\overset{\alpha}{\rightarrow}4)\text{-Gal p-}(1\overset{\alpha}{\rightarrow}]_n$$

$$\uparrow$$
$$4$$

$$\begin{bmatrix} \rightarrow 3)\text{-Glu A p-}(1\overset{\beta}{\rightarrow} \\ \rightarrow 5)\text{-Ara f - }(1\overset{\alpha}{\rightarrow} \\ \rightarrow 4)\text{-Gal p - }(1\overset{\alpha}{\rightarrow} \\ \rightarrow 3)\text{-Glu p - }(1\overset{\beta}{\rightarrow} \\ \rightarrow 4)\text{-Xyl f - }(1\overset{\beta}{\rightarrow} \\ \text{Gal p - }(1\overset{\beta}{\rightarrow} \\ \rightarrow 2)\text{-Rha p - }(1\overset{\alpha}{\rightarrow} \end{bmatrix}$$

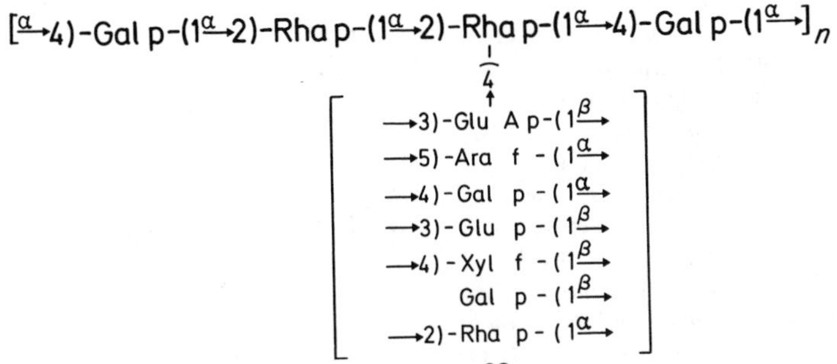

$$22$$

FIG. 2 Proposed structures for *Echinacea* polysaccharides.

mean molecular weight of 450,000. The heteroxylan (**21**) contained arabinose, xylose, and 4-O-methylglucuronic acid as sugar components in a molar ratio of 1 : 4.9 : :0.9. The basic structure resembles that of other xylans found in plants. It consists of $(1\rightarrow4)$-β-bonded xylopyranoses. Branching occurs at about each fifth sugar unit. Side chains are connected to the main chain at the C-2 OH and C-3 OH of the xylose units. Glucuronic acid units appear to be situated directly in the vicinity of the main chain.

The second highly active polysaccharide is composed of rhamnose, arabinose, galactose, glucuronic acid, and galacturonic acid in a molar ratio of 0.8 : 0.6 : 1 : 0.6. The polysaccharide is highly branched and contains chains consisting of rhamnose and galactose in a ratio of 1 : 1. Other chains are connected to these chains via the C-4 OH of about each second rhamnose unit. Linkage of rhamnose units occurs via hydroxyl groups of C-1 and C-2 atoms, while galactose units are joined via C-1 and C-4 atoms.

As determined by the granulocyte and carbon clearance tests, both compounds stimulate phagocytosis to a high degree (Wagner *et al.*, 1984)

TABLE II
INFLUENCE OF POLYSACCHARIDE
FRACTIONS[a] FROM PLANTS ON THE
MICROPHAGOCYTOSIS

Plant/extraction procedure	Mean enhancement of phaocytosis (%)
Echinacea purpurea (1 : 1) (0.5 N NaOH/H$_2$O extract)	45[b]
Eupatorium cannabinum (1 : 2) (0.5 N NaOH/H$_2$O extract)	22[c]
Eupatorium perfoliatum (1 : 0.5) (0.5 N NaOH/H$_2$O extract)	37[b]
Matricaria recutita (1 : 4) (H$_2$O extract)	31[d]
Arnica montana (1 : 4) (H$_2$O extract)	44[b]
Achyrocline saturoides (1 : 4) (H$_2$O extract)	33[b]
Sabal serrulata (1 : 4) (H$_2$O extract)	36[b]
Eleutherococcus senticosus (0.5 N NaOH/H$_2$O extract)	52[d]
Krestin (for comparison)	25[e]

[a]The polysaccharide fractions were prepared from H$_2$O extract or 0.5 N NaOH/H$_2$O extract in a modified procedure according to Caldes et al. (1981). The precepitations were performed with different H$_2$O–alcohol proportions (1 : 0.5, 1 : 1, 1 : 2, 1 : 4). From Wagner et al. (1984). Granulocyte test system according to Brandt (1967).
[b]Test dilution 0.001%
[c]Test dilution 0.05%
[d]Test dilution 0.01%
[e]Test dilution 0.005%

(Tables II and III). For T lymphocytes, the *Echinacea* polysaccharides displayed the same activity as concanavalin A in [³H]thymidine incorporation tests. The B lymphocytes were also stimulated, albeit to a lesser degree. A purified polysaccharide fraction activated macrophages to a pronounced extracellular cytotoxicity against tumor targets. This effect was independent of any cooperative effect with lymphocytes (Stimpl *et al.*, 1984). In infection strain tests, the pharmacological activity of both compounds depended on the time of administration. The same or similar activities were also determined for xyloglucurans from *Eupatorium perfoliatum, E. cannabinum* (Vollmar *et al.*, 1985), and for *Chamomilla re-*

TABLE III
CARBON-CLEARANCE VALUES OF ISOLATED
POLYSACCHARIDE FRACTION[a]

Plant	Regression coefficient		RC_{tr}/RC_c[b]
	Control	Treated	
Echinacea purpurea 1 : 4	−0.0647	−0.1397	2.1681
Echinacea purpurea 1 : 1	−0.0612	−0.1323	2.2152
Eupatorium cannabinum	−0.0597	−0.1087	1.8193
Matricaria recutita	−0.0731	−0.1848	2.5268
Arnica montana	−0.0851	−0.2027	2.3841
Achyrocline saturoides	−0.0583	−0.1350	2.3136
Calendula officinalis	−0.0615	−0.0792	1.2874
Sabal serrulata	−0.0731	−0.2791	3.8163
Althaea officinalis	−0.0597	−0.1306	2.1870
Baptisia tinctoria	−0.0603	−0.0791	1.3105
Eleutherococcus senticosus	−0.0386	−0.1122	2.9040
Krestin[c]	−0.0467	−0.0831	1.400

[a] Tested on mice at 10 mg/kg on one day. From Wagner *et al.* (1985a).
[b] RC_{tr}/RC_c > 1.0 stimulation; < 1.0 suppression
[c] Dose 60 mg/kg

cutita (Odenthal, 1984). They increased phagocytosis by a factor of 1–
2.5, as determined by granulocyte and carbon clearance test (see Tables
II and III). Several of these compounds stimulated T lymphocytes quite
well.

A water-soluble polysaccharide has been isolated from blossoms of
Carthamus tinctorius (safflor) by Caldes *et al..* (1981). It consists mainly of
xylose, fructose, galactose, glucose, arabinose, rhamnose, and uronic
acid, and induces antibodies in mice, following intraperitoneal injection.
These antibodies are similar to those observed after immunization with
type III pneumococcus polysaccharide. *Carthamus* polysaccharide cross-
reacts with antisera specific for *Streptococcus pneumoniae* type III and type
VIII.

Apart from other effects, stimulation of phagocytosis and T lymphocytes and an increase in plasma cell counts has been reported for a
glucoarabinan with an approximate molecular weight of 36,000, isolated
from *Astragalus mongholicus* Bunde (Fang *et al.,* 1982).

Xu *et al.* (1980) isolated two glucose-, galactose-, and arabinose-con-
taining polysaccharides from *Acanthopanax senticosus* Harms, which act as
immunopotentiators (phagocytosis), and also display immunoadjuvant
activity (B lymphocytes). Simultaneous intraperitoneal administration of
125 mg/kg and 0.2 mg bovine serum albumin significantly increased the

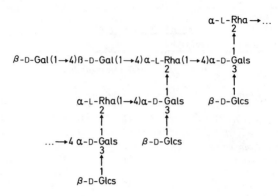

FIG. 3   Undecasaccharide subunit of the *Althaea* mucilage polysaccharide.

serum levels of anti-BSA-IgG and total BSA antibodies in mice. The polysaccharides isolated by us from the same plant differed in molecular weight and sugar composition from those isolated by the Chinese group. However, our isolates were also highly active in the carbon clearance and granulocyte tests (Fang *et al.* 1985). Like most of the other polysaccharides isolated from higher plants, the phagocytosis-enhancing mucous polysaccharide from *Althaea officinalis* (Fig. 3) is a very complex compound. Its structure has been elucidated by Tomoda *et al.* (1980).

Relationships between structure and pharmacological activity of polysaccharides have been discussed several times (Whistler *et al.*, 1976; Miyazaki *et al.*, 1979; Misaki *et al.*, 1981; Gorecka-Tisera *et al.*, 1981). Structural prerequisites could be the nature of bonding, molecular weights, and special molecular conformations.

The highest activity against sarcoma 180 has been demonstrated for (1→3)-β-D-glucans and (1→3)-β-D-glucans possessing β-D-glucopyranosyl groups (1→6)-linked to every third or fourth residue of the main chain. β-D-Glucans containing mainly (1→6) linkages have less activity. In addition, there is evidence that the β configuration of the D-glucosyl residues, which gives rise to a helical chain conformation of the D-glucosyl residues, is essential. X-Ray analysis and viscosity studies, e.g., on lentinan and schizophyllan, indicated that they have triple-strand helices stabilized by hydrogen bonding between (1→3)-linked chains of D-glucosyl residues.

The chemical modification of polysaccharides affects their activity in a variable manner. When O-carboxymethylated, glucans of the lichenan type from *Gyrophora esculenta* yielded products with greatly reduced anti-tumour activity (Nishikawa and Ohno, 1981). However, when glucans from *Auricularia auricula-judae* were subjected to a controlled periodate oxidation, these modified glucan fractions exhibited antitumour effects significantly higher than those of the original glucans (Misaki *et al.*, 1981). On the other hand, when native schizophyllan, which has not yet been put into clinical use as a chemotherapeutic agent because of its high viscosity, was subjected to controlled ultrasonic depolymerization, its antitumour activity was neither reduced nor increased (Tabata *et al.*, 1981). Chihara *et al.* (1970a) have claimed that a structure of higher order or a micelle structure of a polysaccharide, rather than the primary structure, plays an important role in the antitumour activity. Dextranes with a mean molecular weight of 90,000 were highly active, while those with mean molecular weights of less than 50,000 were rather inefficient immunogens. This behaviour has also been observed when polysaccharides of *Echinacea* and *Eupatorium* were fractionated. Fractions with very low molecular weights were more or less inactive (Wagner *et al.*, 1984).

Immunogenic activity appears to be highly influenced by dose level. In antitumour tests, lentinan was most active at a dose of 1–2 mg/kg, administered for 10 days. A dose of 5 mg/kg for 2 days or single doses of 10 and 20 mg/kg were also active. An increase of the daily doses beyond these values resulted in a markedly decreased pharmacological activity. Similar observations have been made for *Echinacea* polysaccharides. The doses of individual polysaccharides tested are given in Table IV.

A new aspect of structure–activity relationship was revealed by the discovery of some polysaccharides of nonmicrobial origin, such as agar, carrageenans, plant gums, and polysaccharides from higher plants. These compounds with proven immunostimulating activity were capable of cross-reacting with antipneumococcal sera and other antisera (Heidelberger, 1960, 1982; Rao, 1975; Caldes *et al.*, 1981). Although the gross chemical composition of these polysaccharides differs from those isolated from microorganisms, their behaviour can be explained if one assumes that these polysaccharides contain chemically identical or similar immunodeterminant regions. It is conceivable, therefore, that these polysaccharides might also become valuable sources for the preparation of future vaccines against some infections.

3. PEPTIDES/PROTEINS

There is only a limited number of defined, immunostimulating proteins that are not sugar-specific. Kojima *et al.* (1980a,b) have isolated a protein from leaves of *Artemisia princeps* by ultrafiltration and Sephadex chromatography. This protein has a molecular weight of 500,000–

TABLE IV

DOSAGES TESTED OF VARIOUS POLYSACCHARIDES

Polysaccharides	Dosage	References
Antitumour test		
Lentinan	1–2 mg/kg × 10 days	
CM-Pachymaran	25 mg/kg × 10 days	Chihara (1977)
Zymosan	5 mg/kg × 10 days	
Schizophyllan	1 mg/kg × 10 days	
Auricularia auricula-judae	8–10 mg/kg × 10 days	Misaki et al. (1981)
Grifora umbellata	1–5 mg/kg × 10 days	Miyazaki et al. (1979)
Gummi arabicum	400 mg/kg × 1 day	
Tragacanth	100 mg/kg × 1 day	Osswald (1968)
Partly 0-acetylated pustulan	150 mg/kg × 10 days	
Glucan (lichenan type)	150 mg/kg × 10 days	Nishikawa et al. (1979)
PSK (krestin)	10 mg/kg × 10 days	Miyazaki and Nishijima (1981)
Ganoderma lucidum	20 mg/kg × 10 days	
Carbon clearance test:		
Glucan (lichnan type)	100 mg/kg × 10 days	Nishikawa and Ohno (1981)
Yeast polysaccharide (glucan, mannan)	150 mg/kg × 10 days	M. Suzuki et al. (1971)
PSK (krestin)	120 mg/kg × 1 day	Mayer and Drews (1980)
Polysaccharides of Echinacea purpurea, Eupatorium canabinum, Calendula officinalis, Achyrocline saturoides, Baptisia tintoria, Eleutherococcus senticosus	10 mg/kg × 1 day	Wagner et al. (1984)

1,000,000 and induces interferon when administered orally and parenterally. As has been reported by Atherton and Burke (1978), α-amanitin, the liver-toxic cyclooctapeptide isolated from *Amanita muscaria* induces interferon.

4. CHEMICALLY UNDEFINED OR STANDARDIZED PLANT PREPARATIONS

Many studies have demonstrated that various extracts or fractions of higher plants possess immunostimulatory activity (Tables V–VII). However, it is difficult to compare the results since different test assays have been used for *in vitro* and *in vivo* studies. The carbon clearance test has been applied frequently. In several cases, the phagocytotic index of leu-

TABLE V
PLANT DRUGS WITH STIMULANT
ACTIVITY ON THE RES[a,b]

Plant	Plant portion
Anthurium wagnerianum	Root
Caulophyllum thalictroides	Rhizome
Centella asiatica	Root
Combretum micranthum	Leaves
Euphorbia hirta	Upper part
Gelsemium sempervirens	Rhizome, leaves
Guatteria spruceana	Bark, leaves
Gymnema sylvestre	Leaves
Herpestis monniera	Upper part
Houstonia purpurea	Whole plant
Jacaranda rhombifolia	Bark, leaves
Leucothoë cutesbaei	Leaves
Manilkara surinamensis	Bark, leaves
Maytenus ficiformis	Bark, leaves
Maytenus laevis	Bark, leaves
Miconia stephananthera	Leaves
Moringa oleifera	Root
Nectandra globosa/truxillensis	Bark
Piper marginatum	Fruits
Rhynchosia phaseoloides	Stems
Tabebuia barbata	Bark

[a] From Di Carlo *et al.* (1964).
[b] Intravenous injection of 15–100 mg MeOH extracts/kg to mice and measurement of the carbon cleareance (carbon clearance index).

cocytes in the presence of *Salmonella typhosa* has been determined. In other cases the influence on interferon release has been measured. Descriptions of pharmacological activities cover a broad range between "nonspecific resistance stimulation," antigenic activity, phagocytosis-stimulant activity, and "interferon induction" activity. Extracts have been administered intraperitoneally, intravenously, subcutaneously, but also perorally. The doses p.a. cover a range up to 100 mg, while up to 1.5 g/kg have been administered p.o. In most of the studies, exact dose–response data are lacking.

Table VIII lists all those drugs used in the German-speaking countries as drugs for nonspecific stimulating therapy for "retuning the organism." *Aristolochia* and *Echinacea* drugs play a prominent role. Aristolochic acid has to be regarded as the active principle in the *Aristolochia* drugs (see Section IV,A,1). Recently, Wagner and Proksch (1981) isolated two poly-

TABLE VI

PLANT DRUGS WITH NONSPECIFIC RESISTANCE STIMULANT ACTIVITY

Plant	Part	Literature
Perilla frutescens var. *crispa*[a]	Leaves	Kojima *et al.* (1980a)
Artemisia princips[a] and other *Artemisia* sp.	Leaves	Kojima et al. (1980b)
Angelica sinensis[b]	—	Yin *et al.* (1980)
Fagopyrum cymosum	Upper part	Liu *et al.* (1981)
Periplaneta australiensis	—	Cheng *et al.* (1981)
Aralia manshurica	Root bark	Halasa *et al.* (1978)
Astragalus membranaceus	Root	
Asarum europaeum	Root	Möse and Lukas (1961)
Withania somnifera	Root	Gupta *et al.* (1977)
Schizandra chinensis	Fruits, leaves	Shipochliev and Ilieva (1967)

[a] Interferon induction.
[b] Phagozytose stimulation.

TABLE VII

PLANT EXTRACTS AND
FRACTIONS WITH PHAGOZYTOSE
STIMULANT ACTIVITY[a,b]

Plant		
Aucoumea klaineana	—	U
Boswellia carterii	G	U
Byrsonima verbascifolia	G	U
Glycine max	—	U
Manilkara achras	—	U
Myroxylum toluiferum	—	U
Persea gratissima	—	U
Pistacia lentiscus	G	—
Tussilago farfara	—	U
Vinca minor	—	U
Canarium luzonicum	G	U
Commiphora abyssinica	G	U
Dyera costulata	G	—
Petiveria alliacea	—	U
Styrax tonkinensis/benzoin	G	U

[a] According to Delaveau *et al.* (1980).
[b] Degree of stimulation of phagocytosis as determined by the survival rate of mice infected with *Escherichia coli* 111B$_4$ (Institut Pasteur, Paris). Out of 50 extracts of angiosperm species, a survival rate of 80% was observed for the 16 species listed in this table. G = total extract; U = nonsaponifiable contents.

TABLE VIII
MEDICINAL PLANTS MOST
FREQUENTLY USED FOR
REIZKÖRPERTHERAPIE AND
UMSTIMMUNGSTHERAPIE (GENERAL
REORIENTATION OR RETUNING OF THE
ORGANISM)[a]

Chamomilla recutita[b]
Achyrocline saturoides[b]
Aconitum napellus
Althaea officinalis
Aristolochia clematitis
Arnica montana[b]
Baptisia tinctoria
Bryonia dioica
Calendula officinalis[b]
Carex flavus/elongens/vesicans
Centaurium minus[b]
Cynanchum vincetoxicum (Vincetoxicum)
Echinacea purpurea, angustifolia[b]
Eupatorium cannabinum, perfoliatum[b]
Euphrasia officinalis
Gelsemium sempervirens
Marrubium vulgare
Peumus boldus
Phytolacca americana
Scrophularia nodosa
Thuja occidentalis

---

[a] According to the Federal Association of the Pharmaceutical Industry (1982).
[b] Family Compositae

saccharides with immunostimulating activity from *Echinacea purpurea* (see Section IV,B,2). In addition, *Echinacea* extract appears to be directly virostatic (Wacker and Hilbig, 1978). Previous animal and clinical studies have shown that *Echinacea* extracts increase predominantly phagocytosis (Choné, 1965; Kuhn, 1953; Mund-Hoyen, 1979; Sickel, 1971), possess antiinflammatory activity, and accelerate wound healing. In addition, there are numerous studies describing the therapeutic use of *Echinacea* preparations in urology, gynecology, internal medicine, and dermatology (for review, see Harnischfeger and Stolze, 1980).

Kumazawa *et al.* (1982) isolated a peptide-bound water-soluble heteroglycan, rich in uronic acid, from the roots of *Angelica acutiloba* Kitagawa (Yamato Tohki). This compound possessed antitumour activity, activated polyclonal B lymphocytes, and enhanced antibody response against sheep erythrocytes. In contrast, the aqueous root extract of *Angelica polymorpha* appears to contain an immunosuppressive substance (inhibition of IgE antibody production) (Sung *et al.*, 1982).

It must be emphasized at this point that it is not known whether the antiviral or antitumour activity of plant extracts described in most studies is caused by a direct action of the drug on viruses and tumours, respectively, or whether the pharmacological effects are totally or partially due to induction of the immune system's effector cells (May and Willuhn, 1978; May, 1981; Barbar *et al.*, 1979).

The phytodrugs customary in trade are mainly mixed preparations. The extracts or squeezed juices are partly administered in homeopathic dilutions (D_1–D_6 = 10^{-1}–10^{-6}), either alone or in conjunction with animal venoms (formic acid, snake venom). These preparations have not been standardized chemically or biologically, in any case. Medical indications for the use of these drugs comprise a general enhancement of resistance, prophylaxis of flu, febrile trivial infections, support of antibiotic, cytostatic, and X-ray treatment, therapy of light or chronic throat or bladder infections, or support of tumour resistance. Beuscher *et al.*. (1977, 1978; Beuscher, 1982) studied a mixed preparation, containing highly diluted extracts of *Echinacea angustifolia* and *E. purpurea, Baptisia tinctoria*, and *Thuja officinalis*, snake and bee venom (Esberitox), which is administered perorally and intramuscularly. In several experiments, this drug was tested for its action on the mononuclear phagocytotic system. All tests revealed a significant increase in phagocytosis and raised lysosomal hydrolase activity in the peritoneal macrophages utilized in the test systems.

Additional information about the influence of these preparations on parameters of the immune system stems from Tympner (1981), who compared the activity of several preparations in granulocyte tests with those of immunoglobulin preparations. The phytodrugs increased the phagocytosis index to the same degree as the immunoglobulins. Similar observations were made by Wagner *et al.* (1985a) in extended studies with the same and other preparations, using the same test systems. These authors even found pharmacological activities when higher dilutions (0.1–0.001%) were tested.

A *Eupatorium perfoliatum* extract in dilution D_2 was subjected to a controlled clinical trial with 53 flu patients. Subjective and clinical symptoms were compared with those occuring after treatment with acetylsalicylic acid. Both drugs were found to be equally active (Gassinger *et al.*, 1981).

With the exception of *Echinacea, Eupatorium, Eleutherococcus, Calendula, Arnica, Sabal,* and *Achyrocline,* the active principles of all other drugs are unknown.

4. ADAPTOGENOUS DRUGS

The term "adaptogenous drug" was coined by Brekhman (1980). It comprises drugs leading to an increase of unspecific resistance against stresses other than those provoked by bacterial and viral infections. These stresses comprise those induced, for example, by physical, chemical, and biological factors. An adaptogen should normalize the organism without directly influencing the pathological processes themselves.

Concentrating mainly on *Eleutherococcus*, Brekhman was able to demonstrate that extracts of this plant and also saponins and lignanglycosides isolated from the extracts increased performance and efficiency in general. The extracts also possessed "antistress" activity (Brekhman, 1980). In contrast, the polysaccharides isolated by a group of Chinese scientists (Xu *et al.*, 1980), and independently by Fang *et al.* (1985) (see Section II) appear to have true immunostimulating properties.

The mechanisms of action of adaptogens are not yet well understood. Studies carried out so far seem to suggest an action primarily on the endocrine system and transmitter substances. Since it is known that minute quantities of cortisone or estrogens can stimulate the immune system, an additional immunostimulating effect of adaptogens might well be the result of indirect action on the endocrine and transmitter system.

VII. CONCLUSION

This survey reveals that there is a multitude of nonmicrobial compounds with potential immunostimulating activity. As far as the class of low molecular weight compounds is concerned, no structure–activity relationships can be recognized at present. However, there appear to be suitable candidates for the development of useful immunostimulatory drugs in the class of high molecular weight compounds, especially polysaccharides of a particular structural type.

However, a critical evaluation of the efficacy of many of the compounds that have been described as immunostimulants is hindered because the results obtained by various *in vitro* and *in vivo* tests frequently employed cannot be compared easily. This is particularly true for investigations carried out with crude plant extracts. Since it is known that a considerable number of T-lymphocyte stimulators also possess immunosuppressive activities, it will be necessary in these cases to study which subpopulations of lymphocytes are influenced in particular and to which degree. A great deal of attention will have to be paid to this particular feature, because immunostimulatory and immunosuppressive activities appear to be strongly dose-related.

In developing useful immunostimulants, a successful compound should meet several criteria that have been laid down by the WHO. A suitable immunostimulatory compound should be (1) chemically defined, (2) easily degradable biologically, (3) non(co)carcinogenic or nonmutagenic, (4) neither too strong nor too weak in its stimulatory efficacy, and (5) nontoxic and without any side or cascade effects.

When developing clinically applicable immunostimulating drugs, it seems to be important to have not only a clear concept about the range of therapeutic indications but also concepts about whether a compound is to be used predominantly as a prophylactic drug or therapeutically or as adjuvant only.

ACKNOWLEDGMENT

The authors are grateful to Horst Ibelgaufts (Institute of Biochemistry, University of Munich) for translating the manuscript.

REFERENCES

Abb, J., Bayliss, G. I., and Deinhardt, F. (1979). *J. Immunol.* **122,** 1639.
Abe, N., Ebina, T., and Ishida, N. (1982). *Mikrobiol. Immunol.* **26,** 539.
Abe, S., Yoshioka, O., Masuko, Y., Tsubouchi, J., Kohno, M., Nakajima, H., Yamazaki, M., and Mizono, D. (1982). *Gann* **73** (1), 91.
Allen, R. C. (1981). *In* "Bioluminescence and Chemiluminescence: Basic Chemistry and Analytical Applications" (M. A. De Luca and W. D. McElroy, eds.), p. 63. Academic Press, New York.
Allison, A. C. (1973). *In* "Immunopotentiation" (G. E. W. Wolstenholme and J. Knight, eds.). Ciba Found. Symp., Excerpta Medica, Amsterdam.
Atherton, K. T., and Burke, D. C. (1978). *J. Gen. Virol.* **41,** 229.
Bähr, V., and Hänsel, R. (1980). *Arch. Pharm. (Weinheim, Ger.)* **313,** 653.
Barbar, O. P., Bajpai, S. K., Chowdhury, B. L., and Khan, S. K. (1979). *Indian J. Exp. Biol.* **17,** 451.
Bartfeld, H. (1977). *Arzneim.-Forsch.* **27,** 2297.
Bauer, J. (1979). Ph.D. Dissertation, Chirurgische Universitätsklinik München, Munich.
Belkin, M., Hardy, W. G., Perrault, A., and Sato, H. (1959). *Cancer Res.* **19,** 1050.
Beuscher, N. (1982). *Arzneim.-Forsch.* **32,** 134.
Beuscher, N., Otto, B., and Schäfer, B. (1977). *Arzneim.-Forsch.* **27,** 1655.
Beuscher, N., Beuscher, H., and Schäfer, B. (1978). *Arzneim.-Forsch.* **28,** 2242.
Biozzi, G., Benacerraf, B., and Halpern,, B. N. (1953). *Br. J. Exp. Pathol.* **34,** 441.
Block, L. H., Georgopoulos, A., Mayer, P., and Drews, J. (1978). *J. Exp. Med.* **148,** 1228.
Bloksma, N., Schmiermann, P., de Reuver, M., van Dijk, H., and Willers, J. (1982). *Planta Med.* **46,** 221.
Borecký, L., Lackovič, V., Blaškavič, D., Masler, L., and Šike, D. (1967). *Acta Virol.* **11,** 264.
Boxer, L. A., Allen, J. M., and Baehner, R. L. (1978). *J. Lab. Clin. Med.* **92,** 730.
Bradner, W. T., and Clark, D. A. (1959). *Cancer Res.* **19,** 673.
Bradner, W. T., Clarke, D. A., and Stock, C. C. (1958). *Cancer Res.* **18,** 347.

ADDENDUM

FURTHER IMMUNOSTIMULATING SUBSTANCES AND EXTRACTS FROM HIGHER PLANTS AND FUNGI

Substance	Plant	Activity	Reference
Maesanin (chinon-derived)	*Maesa lanceolata*	Nonspecific	Kubo et al. (1983)
Glycyrrhicin Glycyrrhicinic acid	*Glycyrrhiza glabra*	Activation of macrophages and of NK cells *in vivo*	N. Abe et al. (1982)
Saponins	*Aralia mandshurica*	Stimulation of the phagocytic system	Lutomski et al. (1981)
Pilocarpine	*Pilocarpus* sp.	Influence on proliferation of T lymphocytes	Bähr and Hänsel (1980)
Colchicin	*Colchicum autumnale*	Increase of antibodies *in vivo*	Shek and Coons (1978)
D-Manno-D-glucan	Extracellular polysaccharide of a cellculture of *Microcellobosporia grisea*	Antitumour (implanted Ehrlich carcinoma and MM46 adenocarcinoma)	Inoue et al. (1983)
Arabinogalactan	*Angelica acutiloba*	Interferon induction, anticomplementary activity, mitogenic activity	Yamada et al. (1984a)
Aqueous extracts	*Bupleurum falcatum*	Influence on T lymphocytes	Nakashima et al. (1978)
Aqueous extract	*Panax ginseng, Astragalus membranaceus*	Influence on lymphocytes	Sun et al. (1983)
Aqueous extract	*Ligustrum lucidum*		
Aqueous extract	*Astragalus seu hedysari*	Interferon induction	Hou et al. (1981)

(continued)

ADDENDUM (*Continued*)

Substance	Plant	Activity	Reference
D-Glucan	*Cordyceps ophioglossoide*		Yamada *et al.* (1984b)
Mannan, Glucan Glucan	*Dictyophora indusiata* *Ganoderma japonicum*	Antitumour (implanted sarcoma 180)	Ukai *et al.* (1983)
Glucuronoxylomannan Polysaccharide–protein fraction	*Auricularia* sp. *Naematoloma fasciculare*		Lee *et al.* (1981)
Polysaccharide (arabinogalactan)	*Epimedium violaceum*	Antiinfectious, immunostimulating activity	Mitsuhashi *et al.* (1982)
Polysaccharide–protein fraction	*Laetiporus sulphureus*	Antitumour (implanted sarcoma 180), stimulation of the humoral immune respons	Chang *et al.* (1982)
Glucuronoxylans	*Eupatorium perfoliatum and cannabinum*	Stimulation of the phagocytic system	Wagner *et al.* (1984)
	Matricaria recutita	Stimulation of the phagocytic system	Wagner *et al.* (1984)
	Eleutherococcus senticosus (Acanthopanax senticosus)	Stimulation of the phagocytic system	Wagner *et al.* (1984)
Arabinoglucuronoxylan, galactan	*Sabal serrulata*	Stimulation of the phagocytic system	Wagner *et al.* (1984)
Polysaccharide	*Polystictus versicolor*	Antitumour (implanted hepatoma)	Wang and Jai (1982)
Polysaccharide I (Glu, Gal, Ara)	*Astragalus mongholicus*	Stimulation of the phagocytosis of peritoneal macrophages	Fang *et al.* (1982)
Protein–polysaccharide fraction (Glu, Gal, Man, Xyl, Fuc)	Cell cultures of *Hydnum repandum*	Neoplasma inhibition, antitumour (implanted sarcoma 180)	Chung *et al.* (1983)

Brandt, L. (1967). *Scand. J. Haematol., Suppl.* **2.**

Brekhman, I. I. (1980). "Man and Biologically Active Substance," p. 58. Pergamon, Oxford.

Byers, V. S., Epstein, W. L., Castagnoli, N., and Baer, H. (1979). *J. Clin. Invest.* **64,** 1437.

Caldes, G., Prescott, B., Thomas, C.-A., and Baker, P. J. (1981). *J. Gen. Appl. Microbiol.* **27,** 157.

Chang, C. K., Lee, C. O., Chung, K. S., Chai, E. C., and Kim, B. K. (1982). *Arch. Pharm. Res.* **5,** 39.

Chédid, L., Miescher, P. A., and Mueller-Eberhard, H. L., eds. (1980). "Immunstimulation." Springer-Verlag, Berlin and New York.

Chen, L. J., Shen, M. L., Wang, M. Y., Zhai, S. K., and Lin, M. Z. (1981). *Chung-kuo Yao Li Hsueh Pao* **2,** (3), 200; *Chem. Abstr.* **95,** 185472.

Chihara, G. (1977). *Gan no Rinsho* **23** (7), 645.

Chihara, G., Maeda, Y. Y., Hamuro, J., Sasaki, T., and Fukuoka, F. (1969). *Nature (London)* **222,** 687.

Chihara, G., Hamuro, J., Maeda, Y. Y., Arai, Y., and Fukuoka, F. (1970a). *Cancer Res.* **30,** 2776.

Chihara, G., Hamuro, J., Maeda, Y. Y., Arai, Y., and Fukuoka, F. (1970b). *Nature (London)* **225,** 934.

Choné, B. (1965). *Arzneim.-Forsch.* **11,** 611.

Chopra, R. N., Chopra, I. C., Handa, K. L., and Kapur, L. D. (1958). "Indigenous Drugs of India," p. 43. Dhur & Sons Ltd., Calcutta.

Chung, K. S., Choi, E. C., Kim, B. K., and Kim, Y. S. (1983). *Hangguk Kyunhakhoechi* **11,** 91; *Chem. Abstr.* **100,** 064589 (1983).

Delaveau, P., Lallouette, P., and Tessier, A. M. (1980). *Planta Med.* **40,** 49.

Del Puerto, B. M., Tato, J. C., Koltan, A., Bures, O. M., De Chieri, P. R., Garcia, A., Escaray, I. I., and Lorenzo, B. (1968). *Prensa Med. Argent.* **55,** 818.

Di Carlo, F. I., Haynes, L. J., Silver, N. J., and Phillips, G. E. (1964). *RES, J. Reticuloendothedial Soc.* **1,** 224.

Diller, I. C., Fisher, M. E., and Gable, D. (1964). *Proc. Soc. Exp. Biol. Med.* **117** (1), 107.

Drews, J. (1980). *Swiss Pharma* **2,** No. 9.

Ehresmann, D. W., Deig, E. F., Hatch, M. T., Di Salvo, L. H., and Vedros, N. A. (1977). *J. Phycol.* **13,** 37.

Ehresmann, D. W., Deig, E. F., and Hatch, M. T. (1979). *In* "Marine Algae in Pharmaceutical Science" (H. A. Hoppe, T. Levring, and Y. Tanaka, eds.), p. 293. de Gruyter, Berlin.

Fang, S., Chen, Y., Xu, X., Ye, C., Zhai, S., and Shen, M. (1982). *You Ji Hua Xue* **2,** 26.

Fang, J., Proksch, A., and Wagner, H. (1985). *Phytochem.* (in press).

Fanselow, G. (1982). Thesis, Universität München, Munich.

Federal Association of the Pharmaceutical Industry (1982). "Rote Liste (List of Registered Drugs)." FAPI, Frankfurt am Main.

Ferber, E., and Resch, K. (1976). *Naturwissenschaften* **63,** 375.

Franz, H., Ziska, P., and Kindt, A. (1981). *Biochem. J.* **195,** 481.

Freund, J. (1956). *Adv. Tuberc. Res.* **7,** 130.

Gassinger, C. A., Wünstel, G., and Netter, P. (1981). *Arzneim.-Forsch.* **31,** 732.

Giss, G. (1980). *In* "Der Chronische Infekt" (W. Rothenberg and H. Sieck, eds.), p. 59. Madaus, Vienna.

Goldfarb, R. H., and Herberman, R. B. (1981). *J. Immunol.* **126,** 2129.

Gopalakrishnan, C., Shankaranavayanan, D., Nuzimudeen, S. K., and Kameswaran, L. (1978). *Indian J. Med. Res.* **71,** 940.

148

H. WAGNER AND A. PROKSCH

Gopalakrishnan, C., Shankaranavayanan, D., Kameswaran, L., and Naturajan, S. (1979). *Indian J. Med. Res.* **69**, 513.
Gorecka-Tisera, A., Proctor, J. W., Yamamura, Y., Harnaha, J., and Meinert, K. (1981). *JNCI, J. Natl. Cancer Inst.* **67** (4), 911.
Grimm, W., Bärlin, E., Leser, H. G., Kramer, W., and Gemsa, D. (1980). *Clin. Immunol. Immunopathol.* **17**, 617.
Grollman, A. P. (1968). *J. Biol. Chem.* **243**, 4089.
Gupta, O. P., Singh, B., and Atal, C. K. (1977). *Indian J. Pharm.* **39**, 163A.
Haba, S., Hamaoka, T., Takatsu, K., and Kitagawa, M. (1976). *Int. J. Cancer* **18** (1), 93.
Hadden, J. W. (1980). *In* "Immunstimulation" (L. Chédid, P. A. Miescher, and H. L. Mueller-Eberhard, eds.), p. 35. Springer-Verlag, Berlin and New York.
Hadden, J. W., and Stewart, W. E., II. (1981). "Contemporary Immunology—The Lymphokines—Biochemistry and Biological Activity." Humana Press, Clifton, New Jersey.
Hadden, J. W., Delmonte, L., and Oettgen, H. F. (1977). *In* "Comprehensive Immunology" (R. A. Good and S. B. Day, eds.), Vol. 3, p. 273. Plenum, New York.
Halasa, J., Pietrzak-Nowacka, M., Giedrys-Galant, S., and Lutomski, J. (1978). *Herba Pol.* **24**, 233.
Hall, I. H., Lee, K. H., Starness, C. O., Sumida, Y., Wu, R. Y., Wadell, T. G., Cochran, J. W., and Gerhart, K. G. (1979). *J. Pharm. Sci.* **68**, 537.
Hamuro, J., Maeda, Y. Y., Arai, Y., Fukuoka, F., and Chihara, G. (1971). *Chem.-Biol. Interact.* **3**, 69.
Hanisch, J., Jarfas, K., and Orban, T. (1966). *Klin. Oczna* **36**, 565.
Haranath, P. S. R. K., and Shyamala, K. (1975). *Indian J. Med. Res.* **63**, 661.
Harnischfeger, G., and Stolze, H. (1980). *In* "Notabene Medici," Vol. 10, p. 484. Verlag Pharmedolingua, Melsungen.
Hatch, M. T., Ehresmann, D. W., and Deig, E. F. (1979). *In* "Marine Algae in Pharmaceutical Science" (H. A. Hoppe, T. Levring, and Y. Tanaka, eds.), p. 343. de Gruyter, Berlin.
Heidelberger, M. (1960). *Fortschr. Chem. Org. Naturst.* **18**, 503.
Heidelberger, M. (1982). *Fortschr. Chem. Org. Naturst.* **42**, 288.
Henrickson, C. U. (1970). *Z. Immunitaetsforsch.* **5**, 425.
Hoff, F. (1957). "Fieber, unspezifische Abwehrvorgänge. Unspezifische Therapie." Thieme, Stuttgart.
Hotta, T., Enomoto, S., Yoshikumi, C., Ohara, M., and Ueno, S. (1982). U. S. Patent 4,289, 688; *Chem. Abstr.* **95**, P185569.
Hou, Y., Ma, G., Wu, S., Li, Y., and Li, H. (1981). *Chin. Med. J.* **94**, 35.
Ikekawa, T., Nakanishi, M., Uehara, N., Chihara, G., and Fukuoka, F. (1968). *Gann* **59**, 155.
Inbar, M., and Sachs, L. (1969). *Proc. Natl. Acad. Sci. U.S.A.* **63**, 1418.
Inoue, K., Nakajima, H., Kohno, M., Ohshima, M., Kadoya, S., Takahashi, K., and Abe, S. (1983). *Carbohydr. Res.* **114**, 164.
Ishizuka, M., Iinuma, H., Takeuchi, T., and Umezawa, H. (1972). *J. Antibiot.* **25**, 320.
Jordan, E. (1985). Thesis, Universität München, Munich.
Karnovsky, M. L., and Bolis, L. (1982). "Phagocytosis—Past and Future." Academic Press, New York.
Kasajima, T. (1974). *Jpn. J. Reticuloendothelial Soc.* **14**, 1.
Ke, M. C., Wang, S. N., Yao, Y. C., and Liu, C. F. (1980). *Hu-Nan I Hsueh Yuan Hsueh Pao* **5**, 277; *Chem. Abstr.* **94**, 109240 (1981).
Kluthe, R., Vogt, A., and Batsford, S. (1982). *Arzneim.-Forsch.* **32**, 443.
Kohlmünzer, S., Grzybek, J., and Tanaka, M. (1980). *Planta Med.* **39** (3), 231.

Kojima, Y. (1980). *Nippon Monaikei Gakkai Kaishi* **19** (6), 261.

Kojima, Y., Konno, S., Tamamura, S., Sano, Y., and Hashimoto, T. (1980a). Ger. Offen. DE 2,947,646; *Chem. Abstr.* **93**, 146618z (1980).

Kojima, Y., Konno, S., Tamamura, S., and Hashimoto, T. (1980b). Ger. Offen. DE 3,000,521; *Chem. Abstr.* **94**, 52919s (1981).

Komatsu, N. (1976). *Mushroom Sci.* **9** (1), 867.

Komatsu, N., Okubo, S., Kikumoto, K., Saito, G., and Sakai, S. (1969). *Gann* **60** (2), 137.

Krakauer, R. S., and Cathcart, M. K. (1980). "Immunregulation and Autoimmunity," Elsevier/North-Holland, New York.

Kubo, I., Kamikawa, T., and Miura, J. (1983). *Tetrahedron Lett.* **24**, 3825.

Kubo, M., Numba, K., Nagamoto, N., Nagao, T., Nakanishi, J., Uno, H., and Nishimura, H. (1983). *Planta Med.* **47**, 52.

Kuboyama, N., Fuji, A., and Tamura, T. (1981). *Nippon Yakurigaku Zasshi* **77** (6), 579.

Kuhn, O. (1953). *Arzneim.-Forsch.* **3**, 194.

Kumazawa, Y., Mizunoe, K., and Otsuka, Y. (1982). *Immunology* **47**, 75.

Labzo, S. S., Novokhatskii, A. S., Kabirov, S. K., and Knyazeva, V. F. (1980). *Vopr. Virusol.* No. 1, p. 81.

Lee, C. O., Choi, E. C., and Kim, B. K. (1981). *Arch. Pharm. Res.* **4**, 117.

Lee, S. A., Chung, K. S., Shim, M. J., Choi, E. C., and Kim, B. K. (1981). *Hanguk Kyunhakhoe Chi* **9** (1), 25; *Chem. Abstr.* **95**, 54979.

Lemperle, G. (1972). "Funktion des retikuloendothelialen Systems bei chirurgischen Erkrankungen." Habil-Schrift der Med. Fakultät der Universität Freiburg, Freiburg.

Lemperle, G., and Reichelt, M. (1973). *Med. Klin. (Munich)* **68**, 48.

Lis, H., and Sharon, N. (1973). *Annu. Rev. Biochem.* **42**, 541.

Liu, W. F., Song, Y. M., Wang, L. Z., Yang, J. L., and Yin, D. X. (1981). *Yao Hsueh Hsueh Pao* **16** (4), 247; *Chem. Abstr.* **95**, 126068.

Lohman-Matthes, M. L. (1981). *Biol. Unserer Zeit* **11**, 135.

Lutomski, I., Garecki, P., and Halasa, J. (1981). *Planta Med.* **42**, 116.

Manolov, T., Boyadzhiev, T., and Nikolov, P. (1964). *Eksp. Med. Morfol.* **3**, 41.

Mathé, G., Schneider, M., Amiel, J.-L., Cattan, A. Schwarzenberg, L., and Berno, M. (1963). *Rev. Fr. Etud. Clin. Biol.* **8**, 1017.

Matsumoto, T., Takanohashi, M., Okubo, Y., Suzuki, M., and Suzuki, S. (1980). *Carbohydr. Res.* **83**, 363.

May, G. (1981). *Z. Angew. Phytother.* **5**, 187.

May, G., and Willuhn, G. (1978). *Arzneim.-Forsch.* **28** (1), 1.

Mayer, P., and Drews, J. (1980). *Eur. Z. Klin. Ther. Infekt.* **8** (1), 13.

Mayer, P., Hamberger, H., and Drews, J. (1980). *Infection (Munich)* **8**, 256.

Mayr, A., Raettig, H., Stickl, H., and Alexander, M. (1979). *Fortschr. Med.* **97**, 1205.

Mengs, U. (1983). *Arch. Toxicol.* **52**, 209.

Mengs, U., Lang, W., and Poch, J.-A. (1982). *Arch. Toxicol.* **51**, 107.

Mezger, J. (1966). "Gesichtete Homöopathische Arzneimittellehre," Vols. I and II. Haug-Verlag, Ulm.

Misaki, A., Kakuta, M., Sasaki, T., Tanaka, M., and Miyaji, H. (1981). *Carbohydr. Res.* **92**, 115.

Mitani, M., Matsuo, T., Arika, T., Nagumo, N., Suzuki, H., Komatsu, Y., and Kamatsu, N. (1981a). *Yakuri to Chiryo* **9** (6), 2265.

Mitani, M., Matsuo, T., and Arika, T. (1981b). *Yakuri to Chiryo* **9** (6), 2273.

Mitsuhashi, S., Takase, M., Yasui, S. Washizawa, I., and Yoshioka, K. (1982). Patent: PCT International; WO 8203771 A 1; *Chem. Abstr.* **98**, 149583 (1982).

Miyazaki, T., and Nishijima, M. (1981). *Chem. Pharm. Bull.* **29** (12), 3611.

Miyazaki, T., Oikawa, N., Yamada, H., and Yadomae, T. (1978). *Carbohydr. Res.* **81**, 235.
Miyazaki, T., Oikawa, N., Yadomae, T., Yamada, H., Yamada, Y., Hsu, H.-Y., and Ito, H. (1979). *Carbohydr. Res.* **69**, 165.
Mizuno, T., Usui, T., Tomoda, M., Shinkai, K., Shimizu, M., Arakawa, M., and Tanaka, M. (1980). *Shizuoka Daigaku Nagakubu Kenkyu Hokoku* No. 30, p. 41.
Mizuno, T., Hayashi, K., Arakawa, M., Shinkai, K., Shimizu, M., and Tanaka, M. (1981a). *Shizuoka Daigaku Nagakubu Kenkyu Hokoku* No. 31, p. 49.
Mizuno, T., Hayashi, K., Iwasaki, Y., Shitano, A., Arakawa, M., Shinkai, K., Shimizu, M., and Tanaka, M. (1981b). *Shizuoka Daigaku Nagakubu Kenkyu Hokoku* No. 31, p. 65.
Modolell, M., Andreesen, R., Pahlke, W., Brugger, U., and Munder, P. G. (1979). *Cancer Res.* **39** (11), 4681.
Mori, M., Nakamoto, S., Arashima, Y., and Seno, S. (1979). *Gan to Kagaku Ryoho* **6** (1), 175.
Moriyama, M. (1982). *Acta Med. Okayama* **36** (1), 49.
Möse, J. R. (1963). *Planta Med.* **11**, 72.
Möse, J. R. (1966). *Arzneim.-Forsch.* **16**, 118.
Möse, J. R. (1975). *Oesterr. Z. Onkol.* **2**, 151.
Möse, J. R., and Lukas, G. (1961). *Arzneim.-Forsch.* **11**, 33.
Müller, J. (1962). Ger. Offen. DE 1,130,112.
Munder, R. G., Modolell, M., Andreesen, R., Weltzien, H. U., and Westphal, O. (1980). *In* "Immunstimulation" (L. Chédid, P. A. Miescher, and H. L. Mueller-Eberhard, eds.), p. 177. Springer-Verlag, Berlin and New York.
Mund-Hoyen, W. D. (1979). *Aerztl. Praxis* **31**, 14.
Nadkarni, A. K. (1954). *In* "Indian Materia Medica." p. 1252. Popular Book Depot, Bombay.
Nakahara, W., Fukuoka, F., Maeda, Y. Y., and Aoki, K. (1964). *Gann* **55** (4), 285.
Nakahara, W., Tokuzen, R., Fukuoka, F., and Whistler, R. L. (1967). *Nature (London)* **216**, 374.
Nakashima, S., Umeda, Y., and Sakai, Y. (1978). *Proc. Symp. Wakan Yaku* **11**, 56.
Narni, T., Takahashi, K., Kobayashi, M., and Shibata, S. (1980). *Carbohydr. Res.* **87**, 161.
Nicolson, G. L. (1974). *Int. Rev. Cytol.* **39**, 91.
Nicolson, G. L., and Blaustein, J. (1972). *Biochim. Biophys. Acta* **266**, 543.
Nishikawa, Y., Yoshimoto, K., Horiuchi, R. (née Murakami), Michishita, K., Okabe, M., and Fukuoka, F. (1979). *Chem. Pharm. Bull.* **27** (9), 2065.
Nishikawa, Y., and Ohno, H. (1981). *Chem. Pharm. Bull.* **29** (11), 3407.
Numasaki, Y., Iwano, K., Yamamoto, M., and Kikuchi, M. (1981). *Kinki Daigaku Igaku Zasshi* **6** (3), 383.
Ohno, R., Yokomaku, S., Wakayama, K. Sugiura, S., Imai, K., and Yamada, K. (1976). *Gann* **67**, 97.
Okawa, Y., Okura, Y., Hashimoto, K., Matsumoto, T., Suzuki, S., and Suzuki, M. (1982). *Carbohydr. Res.* **108**, 328.
Okihara, G., and Maeda, Y. Y. (1982). *Curr. Chemother. Immunother., Proc. Int. Congr. Chemother., 12th, 1981* Abstract 31.
Olsson, L., and Bicker, U. (1982). *J. Immunepharmacol.* **3** (3/4), 277.
Osswald, H. (1968). *Arzneim.-Forsch.* **18**, 1495.
Proksch, A. (1982). Dissertation, Universität München, Munich.
Rao, C. V. N. (1975). *Pure Appl. Chem.* **42** (3), 479.
Rao, R. R., and Nair, I. B. (1970). *Pharmacology* **4**, 347.
Roe, E. M. F. (1954). *Nature (London)* **184**, 1891.
Romo de Vivar, A., Guerrero, C., Diaz, E., and Ortega, A. (1970). *Tetrahedron* **26**, 1657.
Rüdiger, H. (1978). *Nuturwissenschaften* **65**, 239.

Rüdiger, H. (1982). *Planta Med.* **45,** 3.

Sakai, S., Saito, S., Sugayama, J., Kamasuka, T., Takeda, S., and Takano, T. (1963). *J. Antibiot., Ser. B* **16** (6), 387.

Sakai, S., Saito, G., Sugayama, J., Kamasuka, T., Takano, T., and Takeda, S. (1964). *Gann* **55** (3), 197.

Sakykov, A. S., Ershov, F. J., Aslanov, K. H., Novokhatskii, A. S., Ismailov, A. J., Auelbekov, S. A., Bikitirov, L., and Baram, N. J. (1980). USSR Patent 721,103.

Sandvig, K., Olsnes, S., and Pihl, A. (1976). *J. Biol. Chem.* **251,** 3977.

Schwarz, J. A., König, P., and Scheurlen, P. G. (1974). *Verh. Dtsch.-Ges. Inn. Med.* **80,** 1597.

Sela, B., Lis, H., Sharon, N., and Sachs, L. (1970). *J. Membr. Biol.* **3,** 267.

Sela, B., Lis, H., Sharon, N., and Sachs, L. (1973). *Biochim. Biophys. Acta* **310,** 273.

Shek, P. N., and Coons, A. H. (1978). *J. Exp. Med.* **147,** 1213.

Shibata, S., Nishikawa, Y., Tanaka, M., Fukuoka, F., and Nakanishi, M. (1968). *Z. Krebsforsch.* **71,** 102.

Shipochliev, T., and Ilieva, S. (1967). *Farmatsiya (Sofia)* **17** (3), 56; *Chem. Abstr.* **68,** 1870z (1968).

Sickel, K. (1971). *Aerztl. Praxis* **23,** 201.

Siering, H., and Müller, H. J. (1981). *Arzneim.-Forsch.* **31,** 1260.

Soma, E., Kobayashi, K., Karakawa, T., Kato, S., and Uchida, K. (1981). European Patent Appl. EP 25,123.

Stehr, K., and Wahle, H. (1968). *Arch. Kinderheilkd.* **119,** 183.

Stickl, H., and Mayr, A. (1979). *Fortschr. Med.* **97,** 1781.

Stimpl, M., Proksch, A., Wagner, H., and Lohmann-Matthes, M.-L. (1984). *Infect. Immun.* **46,** 845.

Strauch, R., and Hiller, K. (1974). *Pharmazie* **10/11,** 656.

Sugiyoshi, K. (1976). *Allergy* **25,** 685.

Sun, Y., Hersh, E. M., Talapaz, M., Lee, S.-L., Wong, W., Loo, T.-L., and Mavligit, G. M. (1983). *Cancer* **52,** 70.

Sung, C.-P., Baker, A. P., Holden, D. A., Smith, W. J., and Chakrin, L. W. (1982). *J. Nat. Prod.* **45,** 398.

Suzuki, I., Yadomae, T., Yonekubo, H., Nishijima, M., and Miyazaki, T. (1982). *Chem. Pharm. Bull.* **30** (3), 1066.

Suzuki, M., Chaki, F., and Suzuki, S. (1971). *Gann* **62,** 553.

Suzuki, S., Hatsukaiwa, H., Sunayama, H., Suzuki, T., Uchiyama, M., Fukuoka, F., Nakanishi, M., and Akiya, S. (1969). *Gann* **60** (3), 273.

Tabata, K., Ito, W., Kojima, T., Kawabata, S., and Misaki, A. (1981). *Carbohydr. Res.* **89,** 121.

Taguchi, T. (1982). *Curr. Chemother. Immunother., Proc. Int. Congr. Chemother., 12th, 1981* Abstract 974.

Takahashi, R. (1956). *Kagaku Ryoho Kenkyusho Hokoku* **10,** 55.

Takeuchi, T., Iinuma, H., Iwanaga, J., Takahashi, S., Takita, T., and Umezawa, H. (1969). *J. Antibiot.* **22,** 215.

Togami, M., Takeuchi, I., Imaizumi, F., and Kawakami, M. (1982). *Chem. Pharm. Bull.* **30** (4), 1134.

Tokuzen, R., and Nakahara, W. (1971). *Arzneim.-Forsch.* **21,** 269.

Tomoda, M., Satch, N., and Shimada, K. (1980). *Chem. Pharm. Bull.* **28** (3), 824.

Tsukagoshi, S. (1975). "Host Defense against Cancer and its Potentiation" (D. Mizuno *et al.*, eds.), p. 365. Univ. of Tokyo Press, Tokyo.

Tympner, K. D. (1981). *Z. Angew. Phytother.* **5,** 181.

Ueno, Y., Okamoto, Y., Yamauchi, R., and Kato, K. (1982). *Carbohydr. Res.* **101** (1), 160.
Uhlenbruck, G. (1971). "Immunologie, eine Einführung." Goldmann Verlag, München.
Uhlenbruck, G. (1981). *Naturwissenschaften* **68**, 606.
Ukai, S., Kiho, T., Hara, C., Morita, M., Goto, A., Imaizumi, N., and Hasegawa, Y. (1983). *Chem. Parm. Bull.***31**, 741.
Usui, T., Iwasaki, Y., Hayashi, K., Mizuno, T., Tanaka, M., Shinkai, K., and Arakawa, M. (1981). *Agric. Biol. Chem.* **45** (1), 323.
Valdes, R., and Cordoba, F. (1975). *Agents Actions* **5**, 64.
Vidal, J. (1952). *Hospital (Rio de Janeiro)* **40**, 305.
Vollmar, A., Schäfer, W., and Wagner, H. (1985). *Phytochem.* (in press).
Wacker, A., and Eilmes, H.-G. (1978). *Erfahrungsheilkunde* **27** (6), 346.
Wacker, A., and Hilbig, W. (1978). *Planta Med.* **33**, 2.
Wadell, T. G., Austin, A. M., Cochran, J. W., Gebhart, K. G., Hall, I. N., and Lee, K. H. (1979). *J. Pharm. Sci.* **68**, 715.
Wagner, H., and Fang, J. N. (1984). Lecture on the 2. Symp. on *Eleutherococcus* 18–19. April, Moscow.
Wagner, H., and Kreutzkamp, B. (1985). *Planta Med.* (in press).
Wagner, H. and Proksch, A. (1981). *Z. Angew. Phytother.* **2**, 166.
Wagner, H., Proksch, A., Riess-Maurer, I., Vollmar, A., Odenthal, S., Stuppner, H., Jurcic, K., Le Turdu, M., and Fang, J. N. (1984). *Arzneim.-Forsch.* **34**, 659.
Wagner, H., Kreutzkamp, B., and Jurcic, K. (1985a). *Planta Med.* (in press).
Wagner, H., Proksch, A., Vollmar, A., Kreutzkamp, B., and Bauer, R. (1985b). *Planta Med.* (in press).
Waksman, B. H. (1980). *In* "Immunstimulation" (L. Chédid, P. A. Miescher, and H. J. Mueller-Eberhard, eds.), p. 5. Springer-Verlag, Berlin.
Wang, D.-Y., Yang, W.-Y., Zhai, S.-K., and Shen, M.-L. (1980). *Sheng Wu Hua Hsueh Yu Sheng Wu Wu Li Hsueh Pao* **12** (4), 343; *Chem. Abstr.* **94**, 196201.
Wang, S., and Jai, R. (1982). *Baichiuen Yike Daxue Xuebao* **8**, 23; *Chem. Abstr.* **100**, 061451 (1983).
Waxdal, M. J. (1974). *Biochemistry* **13**, 3671.
Weir, D. M. (1978). "Handbook of Experimental Immunology," 3rd ed., Vol. 3, Chapter 27. Blackwell, Oxford.
Werner, G. H., and Floc'h, F. (1978). "The Pharmacology of Immunoregulation." Academic Press, New York.
Whistler, R. L., Bushway, A. A., Singh, P. P., Nakahara, W., and Tokuzen, R. (1976). *Adv. Carbohydr. Chem. Biochem.* **32**, 235.
Willuhn, G. (1981). *Pharm. Unserer Zeit* **10**, 1.
Wolstenholme, G. E. W., and Knight, J., eds. (1973). "Immunopotentiation," Ciba Found. Symp. Excerpta Medica, Amsterdam.
Xu, R. S., Feng, S. C., and Fan, Z. Y. (1980). *Planta Med.* **39**, 278.
Xu, L. N., Ouyang, R., Yin, Z. Z., Zhang, L. Y., and Ji, L. X. (1981). *Yao Hsueh Hsueh Pao* **16** (6), 411; *Chem. Abstr.* **97**, 16815.
Yamada, H., Kiyohara, H., Cyang, J.-C., Kojima, Y., Kumezawa, Y. (1984a). *Planta Med.* **45**, 164.
Yamada, H., Kawaguchi, N., Ohmori, T., Takeshita, Y., Taneya, S.-L., and Miyazaki, T. (1984b). *Carbohydr. Res.* **125**, 107.
Yamamoto, T., Yamashita, T., and Tsubura, E. (1981). *Invasion Metastasis* **1** (1), 71.
Yasunaga, T., Kato, H., Ohgaki, K., Inamoto, K., and Hikasa, Y. (1982). *J. Nutr.* **112** (6), 1075.

Yin, Z. Z., Zhang, L. Y., and Xu, L. N. (1980). *Yao Hsueh Hseuh Pao* No. 6, p. 321; *Chem. Abstr.* **94**, 266 (1981).

Zenyaku Kogyo Co., Ltd. (1980). *Jpn. Kokai Tokkyo Koho* **80 167,300;** *Chem. Abstr.* **94,** 214604 (1981).

5

Siberian Ginseng (*Eleutherococcus senticosus*): Current Status as an Adaptogen

NORMAN R. FARNSWORTH
A. DOUGLAS KINGHORN
DJAJA D. SOEJARTO
DONALD P. WALLER

Program for Collaborative Research in the Pharmaceutical Sciences
College of Pharmacy
Health Sciences Center
University of Illinois at Chicago
Chicago, Illinois, USA

ECONOMIC AND MEDICINAL PLANT RESEARCH
VOLUME 1

I. INTRODUCTION AND NOMENCLATURE

The Far Eastern plant *Eleutherococcus senticosus* (Rupr. and Maxim.) Maxim. (Araliaceae, ginseng family), formerly termed *Hedera senticosa* and *Acanthopanax senticosus*, is known commonly as "Siberian ginseng," "touch-me-not," "devil's shrub," "eleutherococc," "spiny eleutherococc," "wild pepper," "eleuthero," "eleuthero ginseng," and "devil's bush" (Baranov, 1982; Brekhman, 1970). Soejarto and Farnsworth (1978) clarified the taxonomic position and scientific name of this plant as *Eleutherococcus senticosus*, although most Chinese reports on this plant utilize *Acanthopanax senticosus* as the preferred binomial.

The plant is most abundant in the Khabarovsk and Primorsk Districts of the Soviet Union, found in forests covering some 10 million hectares, but its distribution extends to the Middle Amur region in the north, Sakhalin Island and Japan in the east, and South Korea and the Chinese Provinces of Shansi and Hopei in the south (Soejarto and Farnsworth, 1978).

Eleutherococcus senticosus first drew the attention of Soviet pharmacologists in a systematic study to identify a suitable, abundant, and inexpensive substitute for *Panax ginseng* C. A. Meyer (Araliaceae), a plant widely used in the Soviet Union as a "tonic" and/or "adaptogen."

Experimental studies on extracts of *Eleutherococcus senticosus* have been carried out almost exclusively in the Soviet Union, starting in the late 1950s, reaching their peak in the late 1960s and early 1970s, and have continued since that time. *In vitro, in vivo,* and human studies utilizing a 33% ethanolic extract of *E. senticosus* roots have been primarily reported by two groups in the Soviet Union, headed by I. I. Brekhman and I. V. Dardymov.

A 33% ethanolic extract of *Eleutherococcus senticosus* roots was approved by the Pharmacological Committee of the U.S.S.R. Ministry of Health for human use in 1962 (Brekhman, 1970). In 1976 it was estimated that as many as 3 million people were using this extract regularly in the Soviet Union (I. I. Brekhman, personal communication, 1976).

Because the "adaptogen" is uncommon to Western medicine or science, a few words should be said concerning this term. "Adaptogen" was first introduced by N. V. Lazarev in 1947 to describe the unique action of dibazole (2-benzylbenzimidazole), which was claimed to increase "nonspecific" resistance of an organism to adverse influences (Brekhman and Dardymov, 1969b). In 1958, Brekhman defined "adaptogen" as a substance that (1) must be innocuous and cause minimal disorders in the physiological functions of an organism, (2) must have a nonspecific action (i.e., it should increase resistance to adverse influences by a wide range of physical, chemical, and biochemical factors), and (3), usually has a normalizing action irrespective of the direction of the pathologic state.

A number of reviews have been published on the subject of plant adaptogens; those involving discussions on *Eleutherococcus senticosus* have been authored or coauthored by Amirov and Abdulova (1966), Baranov (1982), Brekhman (1960, 1963b, 1964, 1966, 1968, 1977), Brekhman and Dardymov (1969b), Curtze (1975), Dardymov (1976a,b), Koramova (1967), Lazarev (1965a), Rozin (1966), Sandberg (1973), Saratikov (1966), and Suprunov and Samojlov (1967).

In 1971, Imedex International, Santa Monica, California, began to import and market the U.S.S.R. pharmacopoeial liquid extract of *Eleutherococcus senticosus* root as "Siberian Ginseng Liquid Extract" in the United States. It is now widely promoted in health food stores in the United States and in Europe, not always as the pharmacopoeial product. Unfortunately, it has been formulated in a number of products that have no scientific rationale, and in recent years it seems to be losing its popularity, at least in the United States. This does not detract, however, from its potential value, if properly used, based on available scientific studies of *E. senticosus*.

Almost all scientific information on the biological evaluation of extracts of *Eleutherococcus senticosus* has been published in the Russian language, and a great deal of difficulty is encountered in obtaining copies of the reports.

II. CONSTITUENTS

The initial phytochemical report on *Eleutherococcus senticosus* was published in 1965 by members of the Institute of Biologically Active Substances, in Vladivostok, U.S.S.R. (Ovodov *et al.*, 1965b). Seven compounds, termed eleutherosides A–G, were isolated from a physiologically active fraction obtained from a methanol extract of *E. senticosus* roots (Ovodov *et al.*, 1965b). The occurrence of eleutherosides A–G, where known, as well as other identified constituents of several plant parts of this taxon, are shown in Table I. The *E. senticosus* constituents discovered to date are shown as structures **1–25**.

Eleutherosides A–G are obtained from *Eleutherosoccus senticosus* roots in a ratio of approximately 8 : 30 : 10 : 12 : 4 : 2 : 1, respectively (Ovodov *et al.*, 1965b). Total eleutheroside yields in the roots of this plant have been determined as being in the range 0.6–0.9% (w/w), and in the stems in the range 0.6–1.5% (w/w) (Lapchik *et al.*, 1969; Lapchik and Ovodov, 1969). Eleutheroside A proved to be the ubiquitous phytosterol, dauco-

TABLE I

COMPOUNDS PRESENT IN *ELEUTHEROCOCCUS SENTICOSUS*

Compound name	Chemical type	% Yield (w/w) in dried plant part				Reference
		Roots	Stems	Leaves	Fruits	
Eleutheroside A (daucosterol) (1)	Sterol[a]	~0.1	+			Lapchik and Ovodov (1970); Ovodov et al. (1965b, 1967)
Eleutheroside B (syringin) (2)	Phenylpropanoid[a]	~0.5	+			Lapchik and Ovodov (1970); Ovodov et al. (1965b, 1969)
Eleutheroside B₁ (isofraxidin-7-O-α-L-glucoside) (3)	Coumarin[a]	Traces				Ovodov et al. (1967)
Eleutheroside B₂	Unknown (chromone)	Traces				Suprunov and Dzizenko (1971)
Eleutheroside B₃	Unknown (coumarin)	Traces				Suprunov and Dzizenko (1971)
Eleutheroside B₄ [(−)-sesamin] (4)	Lignan	0.023				Suprunov and Dzizenko (1971)
Eleutheroside C (methyl-α-D-galactoside) (5)	Sugar	~0.40	+			Lapchik and Ovodov (1970); Ovodov et al. (1965b, 1967)
Eleutheroside D [(−)-Syringaresinol di-O-β-D-glucoside] (6)	Lignan[a]	~0.08	+			Lapchik and Ovodov (1970); Ovodov et al. (1965b, 1967)
Eleutheroside E[c] (different crystal form of eleutheroside D) (6)	Lignan[a]	~0.10	+			Lapchik and Ovodov (1970); Ovodov et al. (1965a,b, 1967)
Eleutheroside F	Unknown[a]	~0.05				Ovodov et al. (1965b)
Eleutheroside G	Unknown[b]	~0.025				Ovodov et al. (1965b)
Eleutheroside I (= mussenin B) (7)	Triterpene[a]			+		Frolova and Ovodov (1971); Frolova et al. (1971)
Eleutheroside K (8)	Triterpene[a]			+		Frolova and Ovodov (1971); Frolova et al. (1971)
Eleutheroside L (9)	Triterpene[a]			0.16		Frolova and Ovodov (1971); Frolova et al. (1971)

Compound	Type	Amount (%)	References
Eleutheroside M (= hederasaponin B) (10)	Triterpene[a]	0.26	Frolova and Ovodov (1971); Frolova et al. (1971)
Senticosides A–F[d]	Triterpene[a]	+	Suprunov (1970)
Vitamin E (11)	Benzofuran	<1.0 mg %	Zorikov and Burii (1974)
β-Carotene (12)	Carotenoid	+	Zorikov and Burii (1974)
Isofraxidin (13)	Coumarin	+	Wagner et al. (1982)
Coumarin X (unknown)	Coumarin	+	Wagner et al. (1982)
Complex mixture	Essential oil	0.8	Ovodov et al. (1965b)
Copper	Inorganic	+	Grinkevich et al. (1967); Grinkevich and Gribovskaya (1968)
(–)-Syringaresinol (14)	Lignan	+	Wagner et al. (1982)
Caffeic acid (15)	Phenylpropanoid	+	Wagner et al. (1982); Lee et al. (1978)
Caffeic acid ethyl ester (16)	Phenylpropanoid	+	Wagner et al. (1982)
Coniferyl aldehyde (17)	Phenylpropanoid	+	Wagner et al. (1982)
Sinapyl alcohol (18)	Phenylpropanoid	+	Wagner et al. (1982)
β-Sitosterol (19)	Sterol	+	Wagner et al. (1982)
Polysaccharides	Sugar[a]	2.3–5.7	17–20 Ovodov and Shibaeva (1969); Solov'eva et al. (1968); Xu et al. (1980)
Polysaccharides PES A	Sugar[a]	+	Xu et al. (1982)
Polysaccharides PES B	Sugar[b]	+	Xu et al. (1982)
Galactose (20)	Sugar	+	Lee et al. (1979)
Glucose (α and β forms) (21, 22)	Sugar	+	Lee et al. (1979)
Maltose (α and β forms) (23, 24)	Sugar	+	Lee et al. (1979)
Sucrose (25)	Sugar	+	Lee et al. (1979)
Oleanolic acid (26)	Triterpene	+	Wagner et al. (1982)
	Essential oil	0.8	Ovodov et al. (1965b)

[a] Present in the form of glycosides.

[b] Also known as β-calycanthoside.

[c] Also known as acanthoside D.

[d] It is quite possible the "senticosides," which are incompletely characterized oleanolic acid glycosides, are identical with some of the eleutherosides I, K, L, and/or M. This point has not been made clear in any of the publications by the Soviet workers.

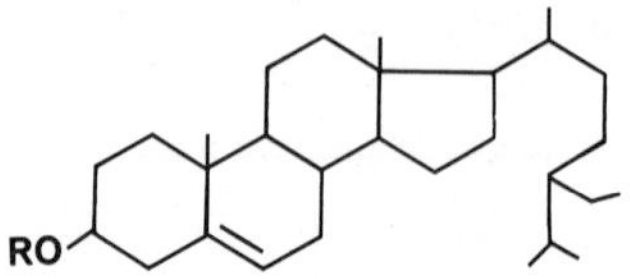

1 R = β-D-Glu
19 R = H

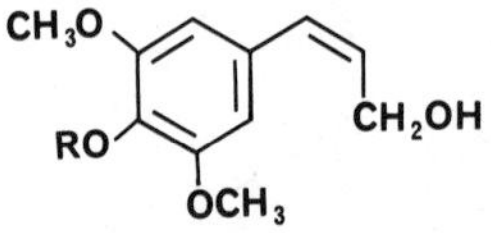

CH₃O
RO — CH₂OH
OCH₃

2 R = β-D-Glu
18 R = H

CH₃O
RO — O — O
OCH₃

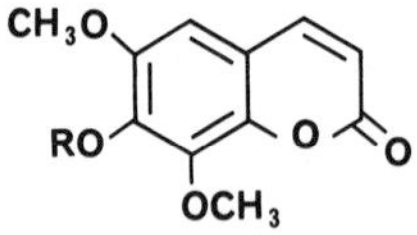

3 R = α-L-Glu
13 R = H

sterol (**1**) (β-sitosterol-3-*O*-β-D-glucoside) (Ovodov *et al.*, 1965b, 1967). Eleutheroside E (**6**) was found to be acanthoside D, a diglucoside of the lignan (−)-syringaresinol (**14**), and eleutheroside D (**6**) is apparently a different crystalline form of eleutheroside E (Ovodov *et al.*, 1967). Early work also permitted the identification of eleutheroside C as methyl-α-D-galactoside (**4**) (Ovodov *et al.*, 1967). "Eleutheroside B" was originally isolated as a complex mixture that was finally resolved and was found to be the known phenylpropanoid derivative eleutheroside B (**2**) (syringin) (Ovodov *et al.*, 1965b, 1967). Trace amounts of eleutheroside B_1 (**3**) were isolated, and this substance was found to be the coumarin glycoside isofraxidin-7-*O*-α-L-glucoside (Ovodov *et al.*, 1967). Eleutherosides B_2 and B_3 were assigned as incompletely identified chromone and coumarin derivatives, respectively (Suprunov and Dzizenko, 1971). Eleutheroside B_4 (**7**) was characterized as the known lignan, (−)-sesamin (Suprunov and Dizenko, 1971). Somewhat inconsistent with the results

	R	R_1	R_2	R_3	R_4	R_5
4	H	O-CH_2-O		H	O-CH_2-O	
6	OCH_3	O-β-D-Glu	OCH_3	OCH_3	O-β-D-Glu	OCH_3
14	OCH_3	OH	OCH_3	OCH_3	OH	OCH_3

5

of previous work (Ovodov *et al.*, 1967), Suprunov and Dzizenko (1971) clearly state that isofraxidin was obtained as an isolate in their studies, and refer to this as "substance B_1." This seems to imply that it is identical with eleutheroside B_1. However, in more recent work, Wagner and co-workers (1982) have established isofraxidin (**13**) as a true constituent of *E. senticosus* roots. Eleutherosides F and G, originally isolated by Ovodov *et al.* (1965b) in their initial study on this species, remain to be characterized.

A series of oleanolic acid glycosides, eleutherosides I,K,L and M (**7–10**) were isolated from the leaves of *Eleutherococcus senticosus*. Eleutheroside H was later found to be a mixture of eleutherosides I and H (Frolova and Ovodov, 1971; Frolova *et al.*, 1971). Eleutheroside I (**7**) was identified as the known compound mussenin B, and eleutheroside M (**10**) is hederasaponin B (Frolova and Ovodov, 1971; Frolova *et al.*, 1971). Senticosides A–F are a series of oleanolic acid derivatives with

	R	R_1
7	α-L-rhamnose-(1→4)-α-L-arabinose-1→	H
8	α-L-rhamnose-(1→2)-α-L-arabinose-1→	H
9	α-L-rhamnose-(1→4)-α-L-arabinose-1→	←1-β-glucose-(6←1)-β-D-glucose-(4←1)-α-L-rhamnose
10	α-L-rhamnose-(1→2)-α-L-arabinose-1→	←1-β-glucose-(6←1)-β-D-glucose-(6←1)-α-L-rhamnose
26	H	H

uncharacterized glycone moieties that have also been isolated from the leaves of *E. senticosus* (Suprunov, 1970).

Miscellaneous minor constituents of *Eleutherococcus senticosus* include vitamin E (**11**) and β-carotene (**12**) (Zorikov and Burii, 1974); the lignan, (−)-syringaresinol (**14**) (Wagner *et al.*, 1982); and the phenylpropanoids caffeic acid (**15**) (Lee *et al.*, 1978; Wagner *et al.*, 1982), caffeic acid ethyl ester (**16**) (Wagner *et al.*, 1982), coniferyl aldehyde (**17**) (Wagner *et al.*, 1982), and sinapyl alcohol (**18**). The last compound is the aglycone of eleutheroside B (Wagner *et al.*, 1982). Additional constituents reported in this plant include undetermined polysaccharides; free sugars (**20–25**); sitosterol (**19**) (the aglycone of eleutheroside A); and oleanolic acid (**26**) (the aglycone of eleutherosides I–M) (Table I). Thin-layer chromatographic analysis has shown that the ginsenosides characteristic of *Panax* species (American, Chinese, Korean, Sanchi ginsengs) are not present in the roots of *E. senticosus* (Wagner and Wurmboeck, 1977; Lui and Staba, 1980).

It may thus be noted that no unusual compounds or compound types,

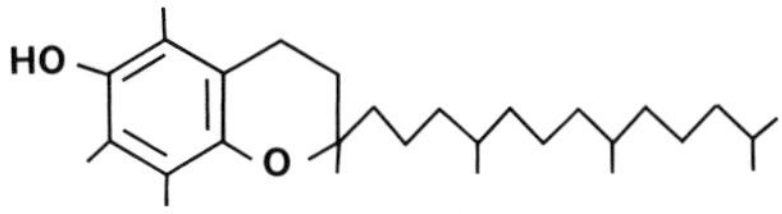

11

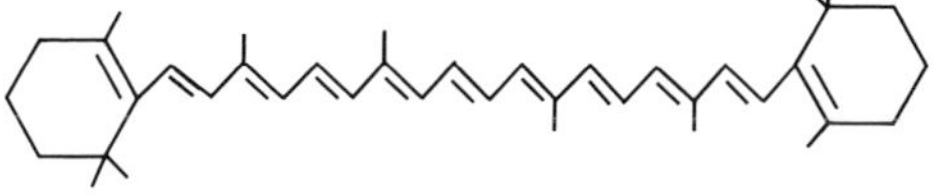

12

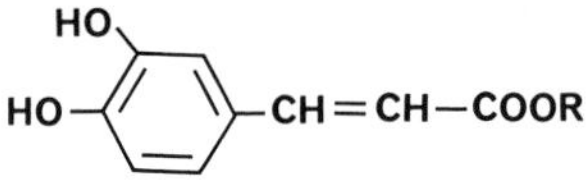

15 R = H
16 R = CH_2CH_3

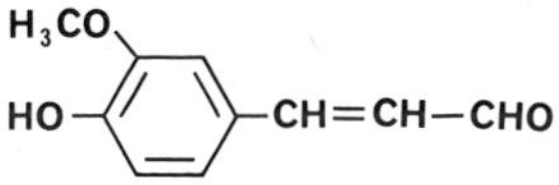

17

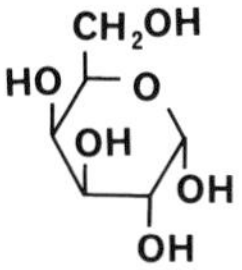

20

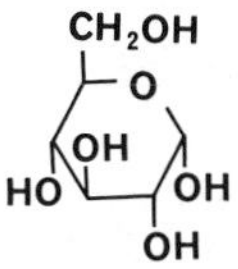

21 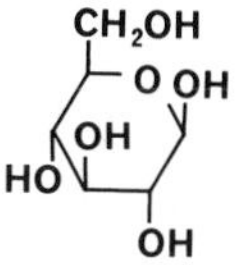   **22**

characteristic only of *Eleutherococcus senticosus*, have been isolated in appreciable amounts from any of the organs of this plant (Elyakov and Ovodov, 1972; Sandberg, 1973).

III. TOXICITY STUDIES IN ANIMALS

A 33% ethanol extract of *Eleutherococcus senticosus* roots has been reported to have an oral LD_{50} (acute) of 14.5 g/kg in mice (Brekhman, 1960; 1963a). The oral acute LD_{50} of powdered *E. senticosus* roots in mice is reported as 31.0 g/kg (Brekhman, 1970). Rats administered a 33% ethanol extract of *E. senticosus* roots daily for 320 days in their drinking water at a dose of 5.0 ml/kg · day showed no toxic manifestations or deaths after 800 days from birth (Golotkin *et al.*, 1972).

Studies involving the administration of a 33% ethanol extract[1] of *Eleutherococcus senticosus* roots by gastric intubation daily for 2 months to groups of male and female rats have been reported. Data relative to animal weight, blood studies [hemoglobin, red blood cells (RBC), white blood cells (WBC), and differential counts], organ weights (adrenals,

---

[1] The type of extract used in these studies was not clearly defined. Doses were stated to be "equivalent to" 10.0 mg/kg of total eleutheroside (0.2 ml/100 g of a 5.0% solution), which was claimed to contain eleutherosides B (syringin), B_1, C, D, and E, derived from *E. senticosus* roots by extraction with ethanol. This dose was stated to be twice that normally employed (Dardymov *et al.*, 1972b).

liver, thymus, spleen, heart, thyroid, gonads, and kidneys), adrenal gland ascorbic acid and cholesterol levels, and liver glycogen and urine analyses in *E. senticosus*-treated rats versus controls were compared (Dardymov *et al.*, 1972b). Results from these studies showed that after 2 months of oral administration of the total eleutherosides, there was no significant effect on urine output or content. Blood studies revealed no change in hemoglobin or RBC count relative to controls. Although there was claimed to be a tendency for low WBC counts in *E. senticosus*-treated rats, these were stated to be statistically insignificant. Similarly, there was no significant change in animal weights in the *E. senticosus*-treated groups, relative to controls. Adrenal ascorbic acid and cholesterol levels were also unaffected by the *E. senticosus* treatment, as was liver glycogen (Dardymov *et al.*, 1972b).

A freeze-dried aqueous extract of *Eleutherococcus senticosus* roots did not produce deaths in mice with single oral doses of 3.0 g/kg (Medon *et al.*, 1981).

Feeding studies with *Eleutherococcus senticosus* root extracts mixed with standard ration to minks, lambs, rabbits, and piglets showed no adverse effects on animal growth (Judin, 1967; Suprunov, 1967; Gorshkov and Antrushin, 1967).

An additional piece of evidence relative to the nontoxicity of *Eleutherococcus senticosus* is a report that spotted deer feed on this plant in the spring (Gorovi, 1963).

An additional study designed to determine the acute LD_{50} of *Eleutherococcus senticosus* was carried out in dogs. The product tested was a commercial extract from the U.S.S.R. Using eight mongrel dogs, four males and four females, weighing from 5 to 12 kg, graded oral doses of the intact extract were administered. At doses exceeding 7.1 ml/kg, the extract produced the following symptoms: sedation, ataxia, loss of righting reflex, hypopnoea, tremors, increased salivation, and vomiting. The symptoms can all be attributed to the 33% ethanol content of the extract. The acute oral LD_{50} was stated to be > 20.0 ml/kg of the liquid extract of *E. senticosus* roots (containing 33% ethanol). Lack of sample precluded determination of the exact LD_{50} (L. M. Strelkoff, personal communication, 1980).

Dardymov and co-workers (1972b) have reported on the absence of teratogenic effects in offspring from male and female Wistar rats given 10.0 mg/kg of total eleutherosides from *Eleutherococcus senticosus* daily for 16 days.

A 10% extract of *Eleutherococcus senticosus* roots (10.0 ml/kg) was fed to pregnant minks and also to the offspring from these animals, from the first day of lactation until the pups were 45 days old (Judin, 1967).

Control groups of animals were also studied. No adverse effects on the pregnant mothers or on the pups were reported (Judin, 1967).

Similar studies, showing a lack of toxicity and/or teratogenicity for *E. senticosus* extract, were carried out with pregnant lambs (Suprunov, 1967).

An unpublished study carried out in West Germany with the *Eleutherococcus senticosus* 33% ethanol extract involved an evaluation for potential teratogenic effects in (1) 48 female New Zealand rabbits (plus 12 controls), in which the extract was administered by gastric intubation daily between the 6th and 18th day of pregnancy at doses of 0.5, 1.5, 4.5, and 13.5 ml/kg and (2) 72 female Sprague-Dawley rats (plus 24 controls) in which the extract was administered by gastric intubation daily between the 6th and 15th day of pregnancy at doses of 1.5, 4.5, and 13.5 ml/kg (L. M. Strelkoff, personal communication, 1980).

In both (1) and (2) above, animals were laparotomized 1 day prior to expected delivery, and the fetuses were removed and examined.

There were no adverse effects on the rabbits administered *Eleutherococcus senticosus* extract at dose levels of 0.5, 1.5, or 4.5 ml/kg, and no teratogenic effects in the offspring. In the mothers receiving 13.5 ml/kg of the extract, all rabbits died between the fourth and tenth day of dosing. Premature death precluded evaluation of the fetuses for teratogenic effects. There was no macroscopic pathology in the parent female rabbits that died in the course of the study.

Since the toxic symptoms described in the rabbits receiving 13.5 mg/kg of the extract are typically of ethanol intoxication, it is most likely that this high dose of ethanol was responsible for the deaths.

There were no teratogenic effects in any of the offspring from the *Eleutherococcus senticosus*-treated pregnant rats at any dose level (1.5, 4.5, or 13.5 ml/kg) (L. M. Strelkoff, personal communication, 1980).

IV. ANTITOXIC EFFECT

It is well known that "citrovorum factor" is used to decrease the toxicity of methotrexate when methotrexate is administered in toxic doses to cancer patients. There is evidence from the literature that *Eleutherococcus senticosus* root extract has a similar detoxifying effect.

For example, *Eleutherococcus senticosus* extract (20 ml/kg, i.p.) administered to guinea pigs treated with the antitumor antibiotic rubromycin C resulted in better tolerance of this toxic antibiotic (Gol'berg *et al.*, 1971). Combined use of Thio-TEPA (Lazarev and Brekhman, 1967; Tsirlina, 1965), Dopan (Lazarev and Brekhman, 1967; Tsirlina, 1965), 6-mercaptopurine (Mironova, 1963), cyclophosphan (Monakhov, 1965, 1967a,b), ethymidine (Monakhov, 1967a,b), benzo-TEPA (Monakhov, 1967a,b), and sarcolysin (Stukov, 1966) with orally administered *E. senticosus* liquid extract decreased the toxicity of these agents and improved their antitumor effect.

The toxic effect of chlorofos (s.c.) was decreased by *Eleutherococcus senticosus* extract (El'kin, 1972b), as was the hemic hypoxia induced in mice by $NaNO_2$ administration or CO_2, or the tissue hypoxia induced by malonic acid (Mikhailova and Fruentov, 1972). In combination with either cytarabine or $N^6(\Delta^2$-isopentenyl)-adenosine, aqueous extracts of *E. senticosus* gave additive antiproliferative effects against L1210 leukemia cells *in vitro* (Hacker and Medon, 1984).

Along similar lines, *Eleutherococcus senticosus* root extract (33% ethanol) administered orally to rodents has been reported to delay tumor take, delay or prevent metastasis, prevent or delay spontaneous mammary tumor or spontaneous leukemia information, and delay the induction of chemically induced or spontaneous tumors in mice or prevent metastasis (Lazarev, 1962, 1963, 1965b; Dzhioev, 1965; Yaremenko and Moskalik, 1967; Karimov, 1969; Leshchev, 1966; Malyugina, 1964, 1966, 1969; Moskalik, 1970a,b, 1973; Ronichevskaya, 1967; Stukov, 1965, 1967; Yaremenko, 1966; Yaremenko and Moskalik, 1971).

V. STUDIES OF SIBERIAN GINSENG IN HUMANS

A. USE IN NORMAL AND STRESSED HUMAN SUBJECTS

Table II presents a summary of studies in which *Eleutherococcus senticosus* root extract (33% ethanol) was administered orally to more than 2100 human subjects having no pathology.

These studies were designed to measure the "adaptogenic" effects of *Eleutherococcus senticosus* and involved administration to determine the ability of humans to withstand adverse conditions (heat, noise, motion, work load increase, exercise, decompression); to improve auditory disturbances; to increase mental alertness and work output; and to improve the quality of work under stress conditions and athletic performance.

Both male and female subjects were involved in the studies, ranging in age from 19 to 72 years. Doses of the 33% ethanol extract of *Eleutherococcus senticosus* roots ranged from 2.0 to 16.0 ml, one to three times a day

TABLE II

SUMMARY OF REPORTS ON ADMINISTRATION OF *ELEUTHEROCOCCUS SENTICOSUS* LIQUID EXTRACT[a] ORALLY TO NORMAL OR STRESSED HUMAN SUBJECTS

Number of subjects	Sex	Age (years)	Dose (ml)	Dosing frequency	Dosing duration	Number of courses[b]	Remarks	References
4	Male	24–28	8.0	Daily	21 days	1	Subjects adapted to high-temperature environment plus 4 hr of light daily work load for 9 days (37.7–38.5°). Extract of *E. senticosus* produced a negative water balance and increased oxygen consumption. No side effects were reported in the subjects receiving the *E. senticosus* extract.	Afanas'ev *et al.* (1973)
5	Male	24–31	4.0 or 8.0	Once	—	1	Subjects were healthy military personnel from mountain rescue units. Oxygen consumption was measured during forced work periods (pedaling a veloergometer).	Brandis and Pilovitskaya (1966a,b)
5	Male	24–31	4.0	Daily	2 days	1	The 8.0 ml dose always caused increase in pulmonary ventilation, discharge of CO_2, secretion of sweat, and minute volume of heart. However, the 4.0 ml dose was judged more favorable. Blood chemistries and several physiological parameters were reported.	Brandis and Pilovitskaya (1966a,b)

N	Sex	Age	Dose	Frequency	Duration		Comments	Reference
20	Male	20–23	2.0	Daily	30 days	1	No side effects in the subjects receiving the *Eleutherococcus* extract were reported.	Baburin (1966a,b)
							Subjects receiving *E. senticosus* extracts were exposed to loud noises in order to determine if hearing capability could be increased. No side effects due to *Eleutherococcus* extract administration were reported.	
18	Male[c]	24–39	4.0	Once	—	1	Subjects receiving the *Eleutherococcus* extract were subjected to a variety of conditions designed to induce motion sickness (vestibular disturbances), with complex set of tests being applied to ascertain positive or negative effects. No side effects due to *E. senticosus* extract administration were reported.	Baburin (1976b)
10	Male[d]	30–45	12.0–16.0	Once	—	1		
46	Male	20–52	2.0	Daily	20–30 days	1	Subjects receiving *Eleutherococcus* extract in this group were seamen with auditory disorders.	Baburin (1976a)
19	Male	20–52	4.0	Daily	4 days	1	This group was made up of telegraph operators with minor hearing losses with rhinolaryngeal catarrh.	Baburin (1976a)

(continued)

169

TABLE II (*Continued*)

Number of subjects	Sex	Age (years)	Dose (ml)	Dosing frequency	Dosing duration	Number of courses[b]	Remarks	References
31	Both	20–52	2.0–3.0	Daily	25–30 days	1	This group was made up of patients having varying degrees of deafness due to a variety of causes.	Baburin (1976a)
8	Both	20–52	2.0–3.0	Daily	25–30		Tests were conducted to measure hearing competency. No side effects due to *Eleutherococcus* extract were reported.	Baburin (1976a)
21	Both	?	1.0	Twice daily	3 weeks	1	Subjects receiving *Eleutherococcus* extract were workers in noisy departments and airplane pilots with hearing losses. Hospitalized patients improved more than outpatients.	Oleinichenko (1966)
							No side effects due to *Eleutherococcus* extract were reported (abstract).	Oleinichenko (1966)
19	Male	21–33	2.0	Daily	30 days	1	Subjects receiving *Eleutherococcus* extract were healthy radiotelegraphers who were subjected to work under controlled high noise conditions. Work speed and number of errors were recorded. The number and speed of radiogram receptions were increased.	Baburin (1966b)

No.	Sex	Age	Dose	Frequency	Duration	No.	Results	Reference
5	Male	22–30	2.0 4.0	1–1.5 hr before bedtime + in a.m.	30 days	1	No side effects due to *Eleutherococcus* extract were reported.	Baburin (1966b)
18	Male	21–33	2.0	Once	—	1	Subjects receiving *Eleutherococcus* extract were healthy radiotelegraphers who were studied relative to work speed and number of errors. No side effects due to *Eleutherococcus* extract were reported.	Medvedev (1966)
?	?	?	8.0	Daily	15 days	1	Subjects receiving *Eleutherococcus* extract were noted to increase physical performance.	Gordeicheva et al. (1975)
76	?	20–25	4.0	Once	—	1	Subjects were skiers with different amounts of training. Single dose of *Eleutherococcus* extract 1–1.5 hr before start of race (20–50 km) increased resistance of body to hypoxemia and afforded good tolerance to heavy physical burdens when skiers had not been appropriately prepared beforehand.	Dalinger (1966b)

(continued)

171

TABLE II (*Continued*)

Number of subjects	Sex	Age (years)	Dose (ml)	Dosing frequency	Dosing duration	Number of courses[b]	Remarks	References
35	Female	53–72	1.5–2.0	Twice daily	12 days	1	Subjects were workers in jobs involving physical labor in publishing house. Improvement noted in activity of cardiovascular system, ability to work, appetite, and general being in those without marked hypertension. Extract was not recommended for persons whose blood pressure is 180/90 mm Hg or higher.	Dalinger (1966a)
20	Male	?	2.0	Once daily	30 days	1	Improvement in quality of proofreading in healthy subjects was noted after administration of *Eleutherococcus* extract.	P. P. Golikov (1963)
98	Male	19–23	20.0[f]	Once	—	1	Stimulatory effect on mental working ability of normal healthy subjects was noted after administration of *Eleutherococcus* extracts. No side effects were reported in the subjects receiving the *Eleutherococcus* extract.	Egorov and Baburin (1966)

5	Male	24–35	2.0	Once daily 49 days	1	Tests conducted on healthy mountain rescuers to measure O_2 consumption and CO_2 expiration under forced work conditions (pedaling) while inhaling gas mixtures enriched with 50 and 96% O_2. A large number of physiologic (blood pressure, heart rate, pulse body temperature) and biochemical (blood sugar, pH, NH_4, N, etc.) parameters, diuresis, etc. were recorded.	Brandis and Pilovitskaya (1966a,b)
5	Male	28–31	4.0	Once weekly	—	No significant adverse effects attributable to the *Eleutherococcus* extract were reported.	Brandis and Pilovitskaya (1966a,b)
32	Male	N.S.[g]	2.0	Once	—	Athletes (pole vaulters, boxers, gymnasts, rowers, etc.) were given *Eleutherococcus* extract and subjected to a variety of work tasks (running, lifting etc.). No side effects attributable to the *Eleutherococcus* extract were reported.	Blokhin (1966a)
9	Male	N.S.	N.S.	Once	—	Weightlifters, wrestlers and gymnasts were subjected to static work taking *Eleutherococcus* extract and quantitative assessments were made.	Blokhin (1966b)

(continued)

TABLE II (Continued)

Number of subjects	Sex	Age (years)	Dose (ml)	Dosing frequency	Dosing duration	Number of courses[b]	Remarks	References
209	Male	N.S.	N.S.	N.S.	N.S.	N.S.	A combination of physical training and *E. senticosus* extract was used to increase work capability and normalization of body functions in sailors subjected to long periods of elevated temperature on long sea voyages in the tropics.	Berdyshev (1970)
16	Male	28–34	4.0	Once	—	1	*Eleutherococcus senticosus* extract was administered to healthy deep sea divers who had not been diving for ~1 year. They were then placed into decompression chambers and stressed. Various types of work were assigned to the subjects and the results were evaluated. No side effects attributed to the *E. senticosus* extract were reported.	Baburin *et al.* (1970)
40	Male	24–38	4.0 ml	Once daily	N.S.	N.S.	*Eleutherococcus senticosus* extract was administered to healthy males who were subjected to conditions designed to induced motion sickness. The data	Baburin *et al.* (1972)

174

						Reference
					suggest that the extract had a beneficial effect. No side effects attributable to the *E. senticosus* extract were reported.	Baburin *et al.* (1972)
117	N.S.	N.S.	1.0 ml	Once	30 days	Dardymov *et al.* (1966)
					1	

Eleutherococcus senticosus extract was administered to healthy subjects and capillary resistance was determined at various periods of time by means of immunological skin tests. Increased capillary resistance was attributed to the *E. senticosus* group and an ascorbic acid-treated group, over controls receiving neither. No side effects attributed to *E. senticosus* were reported. There was no significant change in levels of formed blood elements, with the exception of the appearance of slight increase in neutrophils and a mild eosinophilia in some subjects. There was a decrease in total serum protein, but no significant changes in the albumin fractions or sialic acid content of the serum in subjects receiving the *E. senticosus* extract.

(continued)

175

TABLE II (*Continued*)

Number of subjects	Sex	Age (years)	Dose (ml)	Dosing frequency	Dosing duration	Number of courses[b]	Remarks	References
N.S.	Male	N.S.	N.S.	N.S.	N.S.	N.S.	Blood donors were given *E. senticosus* extract after alternate bleedings. Effects on several hematological parameters were briefly presented, which were claimed to be favorable. No side effects attributed to the *E. senticosus* extract were reported.	Belonosov *et al.* (1965)
9	Male	19–22	2.0	Once	N.S.	1	In subjects given the *Eleutherococcus senticosus* extract, basal metabolism was reported to decrease by 5%. No side effects attributed to the *E. senticosus* extracts were reported.	Dardymov (1963)
19	Both	20–23	2.0	Once daily	30 days	1	The article suggest that *E. senticosus* extract administered to normal subjects enhanced excretion of vitamins B₁, B₂ and C, but only when added vitamins were given at the same time (relative to controls). Given alone, *E. senticosus* did not appear to affect the excretion of	Padkin and Baburin (1966)

176

N.S.	N.S.	18–25	N.S.	N.S.	N.S.	1	water-soluble vitamins. No side effects were reported in the subjects taking the *E. senticosus* extract. Enhanced amount of mental work, as well as increased quality of work output (proofreading) was measured after administration of *Eleutherococcus senticosus* extract.	Kuntsman (1967)
29	Male	N.S.	2.0	Daily	30 days	1	Healthy sailors on a sea voyage in the tropics were given *Eleutherococcus* extract in order to determine whether or not their performance was improved. A number of physiological functions were monitored (see references for details). No adverse effects were reported.	Berdyshev (1977)
48	Male	N.S.	4.0	Daily	4 times at varying intervals	—	Same as above	Berdyshev (1977)
29	Male	N.S.	2.0	Daily	60 days	1	Same as above, but subjects were evaluated on the basis of different types of challenge (stress) situations, i.e., physical exercise, etc.	Berdyshev (1977)

(continued)

177

TABLE II (*Continued*)

Number of subjects	Sex	Age (years)	Dose (ml)	Dosing frequency	Dosing duration	Number of courses[b]	Remarks	References
1000	Both	Adults	4.0	Daily	30 days	5	*Eleutherococcus* was administered to factory workers (metallurgical work, mining, etc.) in a city of the Polar Region (mean daily temperature −5°C) to determine subjective effects in the workers (favorable), number of lost work days (40% reduction in lost days), and general sickness (50% reduction in general sickness), over a period of 1 year. Blood profiles were determined in the subjects, as well as cardiograms (no details given). There were no adverse effects attributable to the *Eleutherococcus* extract mentioned in the article.	Gagarin (1977)

[a] Unless otherwise stated, the extract tested was a standard *E. senticosus* root extract prepared with ethanol and adjusted to a 33% ethanol (w/v) concentration in the final product.

[b] Unless otherwise stated, each dosing period was interrupted by a 2-week period of extract not being administered.

[c] Vestibular stability was low in this treatment group.

[d] Vestibular stability was high in this treatment group.

[e] These subjects were members of the group of 31 immediatedly above, who were given three additional courses of *Eleutherococcus* extract.

[f] A 3.0% solution of a 2% ethanol extract was used.

[g] N.S., not stated in article.

(orally), for periods up to 60 consecutive days. In multiple dosing regimens, there is usually a 2- or 3-week interval between courses. The cited reports represent studies involving as many as five courses of extract administration.

No side effects were reported in any of the studies involving more than 2100 subjects receiving the *Eleutherococcus senticosus* extract. However, in two of the studies, it was recommended that the extract not be given to subjects having blood pressure in excess of 180/90 mm Hg (Dalinger, 1966b; Lapchik, 1967).

B. USE IN SUBJECTS OTHER THAN NORMAL AND/OR STRESSED

A 33% ethanol extract of *Eleutherococcus senticosus* roots has been tested in more than 2200 human subjects having various ailments. Both male and female subjects were involved, ranging in age from 19 to more than 60 years. Oral dosing was always employed, with doses ranging from 0.5 to 6.0 ml, given one to three times daily. Courses of extract administration have been as long as 35 days in these studies, with as many as eight courses being employed (each course is interrupted by 2 to 3 weeks of no extract being administered) (Table III).

One study (Strokina and Mikho, 1968), involving 20 patients with various neuroses (both sexes), ranging in age from 23 to 55 years, utilized 0.06 g of powdered extract of *Eleutherococcus senticosus* twice daily for 3 weeks, rather than the usual ethanol extract. The studies cited (Table III) included patients with atherosclerosis, acute pyelonephritis, diabetes (several types), hypertension, hypotension, acute craniocerebral trauma, various types of neuroses (ambulatory and hospitalized patients were studied), rheumatic heart disease, chronic bronchitis, and cancer. A summary of pertinent features of these studies, involving more than 2200 patients, is presented in Table III.

Very few side effects were reported from the administration of *Eleutherococcus senticosus* extract in these studies. In two studies involving atherosclerotic patients, some incidence of insomnia, shifts in heart rhythm, tachycardia, extrasystole, and hypertonia was reported (P. P. Golikov, 1966a,b). Another study involving 55 patients with rheumatic heart disease (Mikunis *et al.*, 1966a), showed that 2 of the patients (at high dose levels of the extract) reported headaches, pericardial pain, palpitations, and elevated blood pressure. Another study (Koshkarev and Kovinskii, 1966), involving 11 patients diagnosed as hypochondriacs, reported that the *E. senticosus* extract was well tolerated at dose levels of 2.5–3.0 ml three times daily for 60 days, but patients often showed

TABLE III

SUMMARY OF REPORTS ON ADMINISTRATION OF *ELEUTHEROCOCCUS SENTICOSUS* LIQUID EXTRACT ORALLY TO OTHER THAN NORMAL SUBJECTS

Number of patients	Sex	Age (years)	Route	Dose (ml)	Frequency	Duration	Course	Remarks	References
45	Both	N.S.[a]	p.o.	1.5–2.0	3 times daily before meals	25–35 days	6–8[b]	Atherosclerotic patients studied at end of treatment most of the patients had shown an improvement in general feelings and condition; pains in the heart and chest had disappeared; blood pressure was reduced;	Golikov (1967)
19	Both	N.S.	p.o.	1.5–2.0	3 times daily before meals	25–35 days	1	reduction in serum prothrombin and cholesterol levels; and ECG data were improved. Treatment was less effective in patients with high blood pressure or functional disorders of the central nervous system. *Eleutherococcus sen-*	

180

| 54 | Males | 50 (13); | p.o. | 1.5–2.0 | 3 times | 25–35 | 6–8 | Studies carried out in atherosclerotic patients, most having the disease at least 5 years. Blood cholesterol, prothrombin, and blood protein level were measured. Methyl valerate and/or nitroglycerin were also used in conjunction with *E. senticosus* extract. Virtually all of the patients tolerated the extract well. Noted improvement in well-being within first weeks of therapy was noted in some patients, and a majority of patients were notably improved | A. P. Golikov (1966a,b) |
| 10 | Females | 51–59 (47); 60 (4) | | | | | | *ticosus* extract was well tolerated by the patients. | |

(continued)

TABLE III (*Continued*)

Number of patients	Sex	Age (years)	Route	Dose (ml)	Frequency	Duration	Course	Remarks	References
								after the first course of *E. senticosus* therapy (25–35 days), including favorable shifts in protein–lipid metabolism. Side effects observed with the use of *E. senticosus* extract, based on results in some patients, were insomnia, shifts in heart rhythm, tachycardia.	
47	Both	N.S.	p.o.	N.S.	N.S.	N.S.	N.S.	Patients with acute pyelonephritis were given *E. senticosus* extract, which was well tolerated. Patients demonstrated an increased secretion of phenol red and a higher concentration ca-	Voino-Yasenetskii (1966)

N.S.	Both	N.S.	p.o.	2.5	3 times	10–14	N.S.	pacity of the kidneys as compared with those not receiving the extract. Subjects with normal blood sugar, but with gastritis, cholecystitis, and autonomic vascular dystonia.	Kolmakova and Kutolina (1966)
N.S.	Both	N.S.	p.o.	2.5	3 times daily	10–14	N.S.	The *E. senticosus* had no marked hypoglycemic effect in diabetic patients. A slight effect was observed only in a mild form of diabetes.	Kolmakova and Kutolina (1966)
								The *E. senticosus* extract was tolerated well by all patients and exerted a general tonic effect.	Kolmakova and Kutolina (1966)
25	Both	N.S.	p.o.	N.S.	N.S.	N.S.	N.S.	Claimed "successful" in treating various forms of diabetes mellitus (abstract).	Mischenko (1962)

(continued)

183

TABLE III (*Continued*)

Number of patients	Sex	Age (years)	Route	Dose (ml)	Frequency	Duration	Course	Remarks	References
N.S.	N.S.	N.S.	p.o.	N.S.	N.S.	N.S.	N.S.	Patients with hypertension showed a tendency to normalization of arterial blood pressure after taking *E. senticosus* extract.	Mischenko (1962)
N.S.	N.S.	N.S.	p.o.	N.S.	N.S.	N.S.	N.S.	Patients with hypotension showed a tendency to normalization of arterial blood pressure after taking *E. senticosus* extract.	Mischenko (1962)
N.S.	Both	Young healthy subjects	p.o.	N.S.	N.S.	N.S.	N.S.	An extract of *E. senticosus* reduced alimentary hyperglycemia. No side effects were reported.	Brekhman and Oleinikova (1963)
32	N.S.	19–59	p.o.	1.5	3 times daily before meals	3 weeks	1	Administration of *E. senticosus* extract to patients with acute craniocerebral	Sandler and Sandler (1972)

184

(continued)

| 58 | Both | 19–46 | p.o. | 2.5 | Twice daily (a.m. and p.m.) | 4 weeks | N.S. | trauma (light, medium, and heavy types) resulted in the WBC returning to normal, promoted decrease in number of lymphocytes and eosinophils, had no effect on total serum protein, but normalized the various globulin fractions. No side effects due to administration of *E. senticosus* extract were reported.

The *E. senticosus* extract was administered to ambulatory neurotic patients whose main complaints were decreased working abilities, extreme exhaustion, irritability, insomnia, and general state of unrest. | Strokina (1967) |

TABLE III (*Continued*)

Number of patients	Sex	Age (years)	Route	Dose (ml)	Frequency	Duration	Course	Remarks	References
								General increase in sense of well-being and production of sound sleep were the two most beneficial effects. No side effects were reported.	Strokina and Mikho (1968)
36	Both	20–45	p.o.	2.5	Twice daily	4 weeks	N.S.	Ambulatory patients with neurasthenia were given *E. senticosus* extract. Duration of disease in treatment group was 2 months to 10 years. Beneficial effects were noted, with no side effects reported.	
20	Both	23–55	p.o.	0.06 g	Twice daily	3 weeks	N.S.	Neurotic patients in sanatorium were treated with powdered (not extract) roots of *E. senticosus*. Benefits reported with no side effects.	Strokina (1966a,b)

40	Both	28–55	p.o.	1.0	Twice daily before meals	4 weeks	N.S.	Ambulatory neurotic patients were treated with the standard *E. senticosus* liquid extract. Benefits reported with no side effects.	Strokina (1966a,b)
55	26 males 29 females	20 (11); 21–30 (18); 31–40 (12); 41–50 (9); 50 (5)	p.o.	1.0	Three times daily	4 weeks	1	Patients with rheumatic heart lesions were given *E. senticosus* extract. Forty-one of the 55 patients improved, 13 had no change; and 1 worsened. Improvement was judged mainly on results from a series of laboratory tests, including immune reactivity. Side effects were noted in only two of the 55 patients, and included headache, pericardial pain, palpitations, and elevated blood pressure.	Mikunis *et al.* (1966a)

(continued)

TABLE III (*Continued*)

Number of patients	Sex	Age (years)	Route	Dose (ml)	Frequency	Duration	Course	Remarks	References
55	Male	N.S.	p.o.	1.0	Three times daily	4 weeks	1	Patients with rheumatic heart disease showed increased titers of streptococcal antibodies (anti-streptolysin in 23 and anti-hyaluronidase in 12), normalizing effect on blood protein spectrum and reduced blood coagulation properties.	Mikunis *et al.* (1966b)
20	Both	23–55	p.o.	0.5	Once daily	3 weeks	1	Sanatorium patients with neuroses received 0.5 ml extract/day mixed with food. Patients were significantly benefited over controls. No side effects reported.	Strokina and Mikho (1966)
70	Both	17–35	p.o.	1.5	Three times daily	7 days	1	Patients taking *E. senticosus* extract had various	Sandler (1972)

42	Both	20–50	p.o.	2.5	Twice daily	4 weeks	1	stages of acute craniocerebral trauma following rheoencephalography. Hypotension was decreased. No side effects were reported. Neurotic patients were given the E. senticosus extract. No side effects were reported.	Strokina (1966a)
100	Male	N.S.	p.o.	1.5	Three times daily	N.S.	N.S.	Miners with chronic bronchitis received E. senticosus extract plus other therapy. No side effects were reported	Lyubomudrov et al. (1970)
?	Male	N.S.	p.o.	N.S.	N.S.	N.S.	N.S.	Patients with pneumoconiosis and chronic bronchitis were given the E. senticosus extract plus other forms of therapy. No side effects were reported.	Lyobomudrov et al. (1970)

(continued)

TABLE III (*Continued*)

Number of patients	Sex	Age (years)	Route	Dose (ml)	Frequency	Duration	Course	Remarks	References
48	Male	N.S.	p.o.	1.5–2.0	N.S.	N.S.	N.S.	The *E. senticosus* extract was given to miners with pneumoconiosis and was reported to increase lung capacity. No side effects were mentioned.	Lyobomudrov *et al.* (1970)
8	N.S.	N.S.	p.o.	1.5	Three times daily	1 week	1	After treatment with *E. senticosus* extract, 7 of 8 hypertensive patients showed a reduction in high blood pressure. No side effects were reported.	Lyobomudrov *et al.* (1970)
5⁰	N.S.	N.S.	p.o.	1.5	Three times daily	1 week	1	After treatment with *E. senticosus* extract, 4 of 5 hypotensive patients experienced a normalization of blood pressure. No side effects were reported.	Lyobomudrov *et al.* (1970)
22	N.S.	N.S.	p.o.	1.5	Three times daily	1 week	1	The *E. senticosus* extract was administered to	Lyobomudrov *et al.* (1970)

							patients with vibratory illness. No side effects were reported.	Lyobomudrov *et al.* (1970)
42	Male	N.S.	p.o.	N.S.	N.S.	N.S.	N.S.	Miners ill with pneumoconiosis were treated with *E. senticosus* extract and followed with electrocardiograms. The extract decreased changes in heart rhythm activity; sinus bradycardia or arrhythmia, and inhibited introduction of stimulation on the right leg of the Bundle of His. No side effects were reported.	Lyobomudrov *et al.* (1970)
66	N.S.	N.S.	p.o.	1.5	Three times daily	3 weeks	N.S.	The *E. senticosus* extract was administered to patients with vibrating illness in conjunction with diphacil and vitamins B-1 and B-2. No side effects were reported.	Lyobomudrov *et al.* (1970)

(continued)

TABLE III (*Continued*)

Number of patients	Sex	Age (years)	Route	Dose (ml)	Frequency	Duration	Course	Remarks	References
13	N.S.	N.S.	p.o.	1.5	Three times daily	2 weeks	1	The *E. senticosus* extract was administered to patients with chronic saturine intoxication. No side effects were reported.	Lyobomudrov *et al.* (1970)
28	Male	N.S.	p.o.	2.5	Three times daily	5 days	1	The *E. senticosus* extract was given to patients with stage 1 pneumoconiosis who were institutionalized. Respiration was improved with no side effects being reported.	Lyobomudrov *et al.* (1970)
32	N.S.	N.S.	p.o.	2.0	Three times daily	N.S.	1	The *E. senticosus* extract was given to patients with malignant tumors who were undergoing X-ray therapy, in an attempt to diminish the leukopenia attendant with	Sabubova and Titova (1966)

192

								this treatment (abstract).	
11	N.S.	N.S.	p.o.	Variable	Three times daily	60 days	1	The *E. senticosus* extract was given to hypochondriac patients. It was found to cause insomnia, irritability, melancholy and anxiety in some patients. Doses of 2.5–3.0 ml were optimal (minimal side effects), and doses of 4.5–6.0 ml seemed to cause most of the side effects.	Koshkarev and Kovinskii (1966)
1,200	Both	32	p.o.	N.S.	N.S.	2 months		Extract was administered to stressed bus drivers and factory workers to determine the effect on incidence of flu and days work lost to all disorders. Some patients were hypertensive. No side effects were described in the	Galanova (1977)

(continued)

TABLE III (*Continued*)

Number of patients	Sex	Age (years)	Route	Dose (ml)	Frequency	Duration	Course	Remarks	References
								article due to *E. senticosus* extract administration.	
64	Both	9–15	p.o.	1 drop for each year of age	Once daily	6 weeks		Administration to children with abating forms of pulmonary tuberculosis. Duration of illness was 2–4 years. Initial 6 weeks of dosing was interrupted for 2 weeks and then continued for 3½ months total. Improvement in physical exercises was used as a criterion of effectiveness in restoration of good health. No side effects were reported in the treated patients.	Sobkovich (1970)

[a]N.S., not stated.
[b]3–4 months between courses of treatment.
[c]Patients also received methyl valerate and/or nitroglycerin.

insomnia, irritability, melancholy, and anxiety at dose levels of 4.5–6.0 ml.

All other studies cited (Table III) have reported that *Eleutherococcus senticosus* 33% ethanol extract was well tolerated, or that side effects were not observed.

VI. BIOLOGICAL ASSESSMENT, *IN VIVO*

Many studies have been published relating to the biological assessment of *Eleutherococcus senticosus* root ethanol extract in animals. Most of these studies involve experiments designed to demonstrate the "adaptogenic" or "normalizing" effect of *E. senticosus* to a variety of adverse conditions (stress, immobilization, chemical challenge, etc.), or to elucidate the mechanism for these effects. None of these reports suggest toxic effects for *E. senticosus* in a variety of laboratory animals, when administered by several different routes, and at several different dose levels. A brief summary is presented relative to these reports. *Eleutherococcus senticosus* extract (p.o., 1.0 ml/kg·day) given to rabbits with experimentally decreased serum cholinesterase (CHE) levels had no effect on the CHE in the serum (Fruentova, 1965).

An extract of *Eleutherococcus senticosus* roots (p.o., 0.1 or 1.0 ml/kg·day for 12 to 14 days) showed normalizing effects on experimentally induced hypothermia, as well as sedative action, in rats and mice (Abramova *et al.*, 1972a). The extract has also been reported to increase the resistance of rats (p.o., 1.0 ml/kg·day for 21 to 23 days) to the toxic effects of alloxan, but showed little effect on the hyperglycemia induced by the alloxan (Bezdetko, 1966). Along similar lines, the extract (route and dose not indicated) caused a slight reduction in blood sugar levels in rabbits with epinephrine-induced hyperglycemia (Brekhman and Oleinikova, 1963; Saratikov and Pichurina, 1965). Administration of the extract to rabbits (p.o., 1.0 ml/kg for 30 days) caused an increase in appetite and weight gain, as well as an increase in blood sugar level and inorganic blood phosphorous levels and liver glycogen (Bykhovtsova, 1970b).

A freeze-dried aqueous extract of *Eleutherococcus senticosus* roots administered orally to mice induced a dose-dependent hypoglycemia that was significantly different from controls following 3 days of daily administration. A group of mice receiving 80 mg/kg of the extract showed a 40% decrease in blood glucose relative to controls; the 160 mg/kg group showed a 35% decrease, and the 320 mg/kg group showed a decrease in blood glucose of 60% relative to controls. In separate experiments examining the temporal aspects of hypoglycemic action, *E. senticosus* at 80

mg/kg orally produced a statistically significant drop (16%) in blood
glucose levels after 3 days of administration, while producing a 23%
decrease relative to controls on the fourth day. At the 160 mg/kg dose, 3
days of administration similarly induced a significant (22%) decrease in
blood glucose levels, while inducing a 25% decrease in blood glucose
levels on the fourth day (Medon et al., 1981).

The mechanism of this hypoglycemic action of Eleutherococcus senticosus
remains unclear. Several purported adaptogens (E. senticosus, Panax ginseng, Rhaponticum carthamoides, and Rhodiola rosea) have been reported to
reduce blood glucose levels partially by enhanced resynthesis of glycogen
and high-energy phosphate compounds (Brekhman, 1968; Feoktstova,
1966; Sal'nik, 1966; Saratikov, 1966). Panax ginseng has also been reported to increase blood insulin levels, possibly by enhancing the release
of insulin from the pancreas (Medon et al., 1981).

Eleutherococcus senticosus extract (p.o., 10.0 ml/kg, diluted 1 : 10, 1 : 20
or 1 : 40) is reported to decrease the toxicity of diethylglycolic acid in
mice, but does not reduce the severity of electroshock-induced convulsions in the test animals (Kolla and Ovodenko, 1966). Administration to
frogs (ventral lymph sac, 0.1 ml per animal of a 10% root decoction),
followed in 1 hr by injection of varying toxic doses of cardiac glycosides
(gitalin, neodigalen), protected the frogs against lethal doses of these
compounds (Golotkin and Bojko, 1963).

Administration of the extracts to male and female rabbits (s.c., 0.05–
1.0 ml/kg or i.v., 0.05–0.2 ml/kg) produced a stimulating effect on the
central nervous system (CNS), based on a comparison of electroen-
cephalograms of treated versus control animals (Marina, 1966a). Similar
conclusions were drawn following injection of the extract into the ventral
lymph sac of frogs (Marina, 1966b).

The extract has been reported to decrease barbiturate sleeping time in
mice and in rabbits (i.p., 2.5 ml/kg) previously dosed with 2,4-di-
nitrophenol (Kuntsman, 1967; Brekhman, 1963a; El'kin, 1972a). Similarly, Medon and co-workers (1983) have recently reported that extracts
of Eleutherococcus senticosus administered i.p. to mice significantly extended hexobarbital sleeping time and hexobarbital metabolism was significantly decreased in vitro. Hepatic mixed function oxidases were also
decreased in E. senticosus-treated animals challenged with parathion (Ferguson et al., 1983). A concomitant normalization of increased blood
sugar levels accompanied this effect (Pegel, 1964b). Similarly, the extract
(route and dose not indicated) is stated to prevent development of hypoglycemia induced by the s.c. administration of insulin in rabbits (Saratikov and Pichurina, 1965).

Eleutherococcus senticosus extract is claimed to have gonadotropic activity in immature male mice (i.p., 0.25 ml/kg) (Brekhman, 1960; Kuntsman, 1966). A dilution (1 : 1) of the extract administered to immature male mice (i.p., 5.0 ml/kg·day for 30 days) has also been reported to increase the weight of the prostate gland and seminal vesicles by 118 and 70%, respectively (Kuntsman, 1966; Dardymov, 1972). The extract also showed oestrogenic activity in immature female mice (Brekhman, 1963a). An anabolic effect has been reported in immature rats by the *E. senticosus* extract (i.p., 1.0 ml/kg), which was stated to be equivalent to the same effect produced by 6.0 mg/kg of testosterone (i.m.) (Dardymov and Kirillov, 1968).

Parenteral administration of the *Eleutherococcus senticosus* extract for 15 days prior to induced infection increased resistance of mice and rabbits to listeriosis (Cherkashin, 1966, 1968). However, administration of the extract simultaneously with the infection increased the severity of the disease (Cherkashin, 1966). The extract was shown (s.c. or p.o., 1.0 ml/day per animal) to stimulate specific antiviral immunity in guinea pigs and in mice (Fedorov *et al.*, 1966). Daily administration of *E. senticosus* extract to white mice and rabbits (i.p., dose not indicated) for 5 days led to depression of phagocytosis, a decrease in complement titre, and an increased tissue permeability (Trypan blue test) (Krasnozhenov, 1970). In rabbits immunized with typhoid vaccine, *E. senticosus* extract (route not stated, 1.0 ml/kg·day for 20 days) showed a regulatory effect on complement titre and lysozyme activity in the blood (Polozhentseva, 1976) Leucocytosis induced by injection of *Bacillus mesentericus* vaccine was not affected by *E. senticosus* extract given to rabbits (Pegel, 1964b).

Attempts to elucidate the mechanism for the adaptogenic effect on the lymphatic system of *Eleutherococcus senticosus* prompted studies to determine the effect of the extract (i.p., 1.0 ml/kg) on the weights of the thymus and spleen following cortisone-initiated decreased weight of these organs in rats (Kirillov, 1964, 1965; Kirillov and Dardymov, 1966a,b). Following 8 days of *E. senticosus* administration, the spleen and thymus weight decreases due to cortisone were prevented (Kirillov, 1964, 1965; Kirillov and Dardymov, 1966a,b; Kirillov *et al.*, 1966).

The extract of *Eleutherococcus senticosus* showed little or no effect (i.p., dose not indicated) in reversing hypothermia induced in rabbits by injection of ethanol or picrotoxin (Pegel, 1964a). It has been shown that *E. senticosus* (p.o., 0.5 ml/kg) given to rabbits with experimental hypothermia (ice packs) produced stabilized total protein, increased albumin, and decreased β-globulin serum levels, while α- and γ-globulin concentrations were unaffected (Belonosov and Jakovleva, 1965). The extract is claimed to prevent development of leukocytosis induced by the s.c. administration of milk or the i.v. administration of *Shigella flexneri* vaccine (Saratikov and Pichurina, 1965; Zotova, 1966). However, the extract (route and dose not indicated) prevented leukopenia induced in rabbits by the s.c. administration of turpentine oil (Zotova, 1966).

An extract of *Eleutherococcus senticosus* (s.c., 1.0–2.0 ml/kg or i.v., 0.05–0.5 ml/kg) is reported to improve blood supply to the brain of anesthetized cats (Zyryanova, 1966) and to cause dilation of retinal arteries in curarized animals (routes, species, and doses not indicated) (Zyryanova, 1965). The extract has also been studied for its ability to alter vascular permeability in rats (P. P. Golikov, 1966a,b) and is claimed to lower blood pressure (doses and routes not indicated) in rabbits (degree of blood pressure lowering not indicated) (Kuntsman, 1967). Administration of *E. senticosus* extract (s.c., 5.0 ml/kg) decreased the duration of ether anesthesia in male and female mice (Rusin, 1967).

The total eleutherosides from *Eleutherococcus senticosus* roots (i.p., 15.0 mg/kg), administered 1 hr before stress (15 min swimming), delayed the inhibition of RNA polymerase and accelerated its restoration during rest (Bezdetko *et al.*, 1973). The extract also increased tryptophan pyrrolase activity and enhanced enzyme induction in normal and adrenalectomized rats (Brekhman *et al.*, 1971). The total eleutherosides (5.0 ml/kg, i.p.) from *E. senticosus* roots also partially reversed the decrease in rat muscle ATP, glycogen, creatine phosphate, and lactic and pyruvic acid, following 2 hr of stress (swimming) (Brekhman and Dardymov, 1971). During the experiments it was also demonstrated that the total eleutherosides increased work capacity of treated mice (p.o., 5.0 ml/kg) (Gordeicheva *et al.*, 1975; Brekhman, 1963b; Brekhman and Dardymov, 1971). Prolonged prophylactic administration i.p. at 2.0 ml/kg·day (Brekhman and Kirillov, 1966) and p.o. at 2.5 mg/kg (Brekhman and Dardymov, 1969b) to rats subjected to prolonged stress (immobilization) reduced the increase in weight of adrenals and decreased their ascorbic acid and cholesterol content. Gastric ulcer formation due to the stress in these experiments was also prevented by the extract (Brekhman and Kirillov, 1966), as well as in other studies (Akimov and Sudaryshina, 1966). Prevention of the alarm phase of stress in rats (i.p., 1.0 ml/kg), induced by (1) immobilization, (2) tourniquet application on limbs, and/or (3) unilateral adrenalectomy has also been reported following *E. senticosus* administration (Brekhman and Kirillov, 1969).

Irradiated (X-ray) mice treated with *Eleutherococcus senticosus* extract are reported to survive five times longer than controls (Brekhman *et al.*, 1970). A dealcoholized *E. senticosus* extract administered to rats (s.c., 1.0 ml/kg) forced to swim (stress) resulted in an increased total lipid content of the liver of treated animals, in addition to increased nonesterified fatty acids and phospholipids of the blood (Dambueva and Sal'nik, 1966). The iodine number (unsaturated lipids) of muscle lipids was also increased during these experiments. Administration of *E. senticosus* extract to rats (i.p., dose not indicated) has been claimed to result in a more economical utilization of glycogen and high-energy phorphorus compounds and more intense metabolism of lactic and pyruvic acids during stress induced by a 2-hr swim (Dardymov, 1971). The extract (i.p., dose not indicated) is also claimed to restore the inhibition of ^{32}P incorporation into mRNA of rat liver nuclei during swimming-induced stress by a factor of two (Dardymov *et al.*, 1972a). The *E. senticosus* extract (i.p., 0.5 ml/kg) reversed hypertrophy of the adrenals and thyroid in normal as well as in stressed animals (Kirillov and Dardymov, 1966b).

Cholesterol biosynthesis in rabbit liver and adrenals is reported to be inhibited following administration of *Eleutherococcus senticosus* extract (i.p., 1.0 ml per rabbit) (A. P. Golikov, 1966c, 1967; Golikov *et al.*, 1966).

Nitrogen metabolism in normal and in stressed rats (s.c., 1.0 ml/kg) has been reported to be normalized by *Eleutherococcus senticosus* extract (Feoktstova, 1966; Revina and Sal'nik, 1966; Sal'nik, 1966). In male rats, a dealcoholized extract of *E. senticosus* (i.p., 0.2 ml per rat) is reported to have no effect on adrenal ascorbic acid and/or cholesterol levels in normal animals (Kirillov, 1964, 1965). However, i.p. injection of a 33% ethanol solution (which is used in preparing the standard *E. senticosus* extract) resulted in decreased ascorbic acid and cholesterol levels in the adrenals (Brekhman and Kirillov, 1966; Kirillov, 1964, 1965). There was no statistically significant change in the weights of adrenals following administration of *E. senticosus* extract (route and dose not indicated) for 70 days in male rats (V. A. Kirillov and Semashkevich, 1966). However, a pronounced hypertrophy of the adrenal medulla was produced, reaching a maximum in 38 days and persisting for 70 days. In another study, stressed (swimming or immobilization) male adult rats administered *E. senticosus* extract (i.p., 1.0 ml/kg) showed a suppressed weight increase of the adrenals on the first day of treatment, followed by hypertrophy on subsequent days (Kirillov *et al.*, 1966). Similar results have been reported for the thyroid gland (Kirillov *et al.*, 1966). Body weight and weights of the thymus, spleen, seminal vesicles, and prostate were found to be higher in *E. senticosus*-treated animals under stress (Kirillov *et al.*, 1966).

A 10% solution of *Eleutherococcus senticosus* extract was given i.p. to mice (dose not indicated), and the mice were reported to have an increased work ability of 25% following this treatment (Gordeicheva *et al.*, 1975; Kuntsman, 1967).

The *Eleutherococcus senticosus* extract administered to dogs (s.c., 1.0 ml/kg·day for 14 days) having experimentally induced acute brain ischemia, following ligation of the carotid artery or the thoracic brain artery, showed a modified pathologic course of the induced disease (Leonova,

1966). Arterial blood pressure due to this induced condition was also decreased. Rabbits (s.c., 1.0 ml/kg) administered E. senticosus extract showed an increased accumulation of ^{131}I in the thyroid (Saratikov and Cherdyntsev, 1966). In stressed rats (s.c., dose not indicated), the extract is reported to decrease corticosteroid levels (Stolyarova, 1968). Mice, each having a 10.0 g weight attached to its tail and maintained on a vertical perch (stress), showed increased endurance following administration of the E. senticosus extract (s.c., dose not indicated) (Zotova, 1966). Increased duration of swimming time to exhaustion has also been reported in mice following administration of E. senticosus extract (i.p., 1.0 ml/kg) (Brekhman, 1960).

Parenteral administration of Eleutherococcus senticosus extract is reported to elicit an antiedema and antiinflammatory response in mice and rabbits (Ekkert, 1972; P. P. Golikov, 1966a). Adrenalectomized mice (i.v., 0.1 ml of a 10% extract), dosed immediately following injection with formalin and 2–4 hr later, showed a marked antiedema effect attributed to the E. senticosus extract during the fall of the year, but a minimal effect was found in the spring (P. P. Golikov, 1966b). No explanation for this phenomenon was given.

Mice and rats pretreated with Eleutherococcus senticosus extract (route not indicated, 10.0 ml/kg·day for 6 days), followed by induction of cerebrocranial trauma, resulted in an increased survival rate of treated animals over controls (Kaplan, 1965). An extract of E. senticosus (s.c., 1.0 ml/kg·day for 5 days) normalized erythema induced in rabbits by the s.c. injection of cobalt nitrate (Pichurina, 1965). Rabbits subjected to experimental blood loss (amounting to 20% of blood volume), showed an accelerated regeneration of serum protein following E. senticosus administration (i.p., 0.5 ml/kg) (Dzhioev and Prasol, 1966). Similar results were obtained when rats were treated with 9,10-dimethyl-1,2-benzanthracene, followed by E. senticosus extract (i.p., 1.0 ml/kg·day for 3 months) (Dzhioev and Prasol, 1966).

Total eleutherosides from Eleutherococcus senticosus roots (i.p., 5.0 mg/kg·day for 3 days) are reported to stimulate liver regeneration in partially hepatectomized male mice, as evidenced by autoradiographic studies (Li, 1973). This effect was claimed to be due to increasing the number of cells undergoing mitosis, increasing DNA synthesis and shortening of the lag phase before regeneration is initiated (Li, 1973).

Eleutherococcus senticosus extract (p.o., 1.0 ml/kg·day) increased biogenic amine levels in the brain, adrenal gland, and urine of rats following dosing of the animals (Abramova et al., 1972b).

Oral administration to dogs (10.0 ml/kg) increased conditioned responses to stimuli (Kucharenko, 1963).

In young rabbits (p.o., 1.0 ml/kg) given *Eleutherococcus senticosus* extract, weight gain was accelerated sharply, but this effect became stabilized after the animals were 3 months old (Bihkovcova *et al.*, 1966). This effect has been reported to be due partially to an anabolic effect of the extract (rats) (Dardymov and Kirillov, 1965).

Eleutherococcus senticosus extract (dose not indicated) fed to chickens (30 days old) resulted in increased protein concentration and aspartate aminotransferase levels, as well as decreased alanine aminotransferase levels in pectoral muscle (Ponomareva, 1973). Egg yield and weight has also been reported to increase by 20.4 and 24.0%, respectively, in hens given 0.2 ml/day of the *E. senticosus* extract in their ration (Zorikov and Lyapustina, 1974). Increased hatchability of eggs from *E. senticosus*-fed hens has also been claimed (Zorikov *et al.*, 1974). Administration of a 1 : 10 decoction of *E. senticosus* roots i.g. (intragastric) strongly increased food consumption and appetite in rats and decreased alimentary glycosuria (Shulyateva *et al.*, 1966).

Eleutherococcus senticosus liquid extract (20.0 ml/day per animal for 5 days) and powered root (20.0 g/day per animal for 20 days) were administered to cows and bulls to determine the effect on their reproductive capacities (Lapustina, 1967). Blood studies (hemoglobin, total plasma protein, albumin and globulin plasma levels and protein coefficients) were carried out before and during the dosing periods. In the bulls, semen volume measurements and sperm counts were made. Effects on estrus were also determined. It was concluded that the administration of the *E. senticosus* extract or powdered root had no adverse effects on any of the parameters measured, and that reproductive capacity of the bulls was improved and semen production was increased by 28% (Maxsimov, 1967). A related study has also been published, using a larger number of bulls, with similar results (Bykhovtsova, 1970b).

It has been postulated that at least part of the mechanism for the "adaptogenic" effect of *Eleutherococcus senticosus* can be attributed to its antioxidant effect (Starikova, 1970), which in turn inactivates free radicals (Tkhor *et al.*, 1966).

A 10% liquid extract of *Eleutherococcus senticosus* root (5.0 ml/kg, route not specified) was shown to have a slight antidiuretic effect in saline-loaded mice (Dardymov *et al.*, 1965); however, under stress, the animals showed a diuretic effect following administration of the extract (Dardymov *et al.*, 1965). The thyrotropic effect of 6-methylthiouracil was inhibited following administration of *E. senticosus* extract (i.p., rats), or evidenced by a decreased weight of the thyroid glands (Dardymov *et al.*, 1965).

Eleutherococcus senticosus root extract, administered to rabbits (1.0

ml/kg, p.o.) daily for 30 days, showed a 17% increase in blood glucose levels (Bykhovtsova, 1970b). In bulls with experimental atonia (i.d., atropine), *E. senticosus* root extract (0.01 ml/kg) showed no effect on glucose metabolism; however, blood calcium and potassium levels were increased (Agadzhanyan *et al.*, 1972).

Without providing details, it was claimed that *Eleutherococcus senticosus* extract administered to normal and to experimentally hemorrhaged rabbits (to induce anemia) favorably offset hematopoesis in both instances, based on hemoglobin, RBC, WBC, and reticulocyte and thrombocyte determinations (Bykovtsova, 1970a).

In rats (both sexes) having hydrocortisone-induced aseptic inflammation of the adrenals, *Eleutherococcus senticosus* root extract (i.p., 2.0 and/or 4.0 ml/kg) has little effect (Jekkert, 1972).

The extract (i.p., 5.0 ml/kg) inhibited the conditioned avoidance response in mice when administered daily for 36 days (Ilyuchonok and Chaplygina, 1972).

VII. BIOLOGICAL ASSESSMENT, *IN VITRO*

Very few *in vitro* studies have been reported for *Eleutherococcus senticosus* extracts. A mixture of eleutherosides (1.0×10^{-6} g/ml) from the roots of this plant is reported to stimulate the formation of four blastomers by isolated frog egg cells, as well as to stimulate the incorporation of labeled phenylalanine into protein by cells during the final phase of formation of the four blastomers (Voropaev, 1971). Similarly, eleutherosides (1.0×10^{-7} g/ml) are reported to accelerate the rate of growth and protein formation in frog, loach and sea urchin embryos (Voropaev, 1972). Puromycin, a protein synthesis inhibitor, blocks this effect (Voropaev, 1972).

The total eleutherosides had no effect on oxygen utilization by rat liver mitochondria with succinate, glutamate, or tetramethylphenyl substrates (Dardymov *et al.*, 1972a), but did increase oxygen uptake in whole rat liver homogenates. The eleutherosides are reported to potentiate the effect of insulin on glucose consumption *in vitro*, using the rat diaphragm (Dardymov and Khasina, 1972b).

It is claimed that the adaptogenic effect of *Eleutherococcus senticosus* is achieved by regulation of energy, nucleic acid, and protein metabolism in the tissues (Dardymov and Khasina, 1976a). Under stress, a β-lipoprotein (β-LP) glucocorticoid complex is generated in the blood; the complex inhibits permeation of cell membranes by sugars and also inhibits hexokinase activity *in vivo* and *in vitro* (Dardymov and Khasina, 1972a,b,

1976a). Under the influence of *E. senticosus* extracts, increased formation of glucose-6-phosphate (Gl-6-P) decreases the competition between the different pathways of its utilization (Dardymov and Khasina, 1976b). In animal tissues deficient in ATP, Gl-6-P is oxidized in the pentose phosphate shunt, which yields substrates for the biosynthesis of nucleic acids and proteins (Dardymov and Khasina, 1976b). The *E. senticosus* glycosides are said to stimulate Gl-6-P dehydrogenase *in vitro* (Dardymov and Khasina, 1976b).

In support of reports of the glucocorticoidlike effects *in vivo* for *Eleutherococcus senticosus*, Pearce *et al.* (1982) have shown that extracts of this plant bind to classical progestin, mineralocorticoid, and glucocorticoid receptors *in vitro*, and to estrogen receptors as well.

A recent report (Wacker and Eilmes, 1978) has shown that when *Eleutherococcus senticosus* liquid extract (100 μg, after removal of ethanol) and a suspension of vasicular stomatitis virus are simultaneously introduced into mouse fibroblast cultures, virus growth is not inhibited. However, when the extract is introduced into the mouse fibroblast culture before contact with the virus, the cells become resistant to the virus. The duration of this effect, however, is only ~ 6 hr.

VIII. BIOLOGICAL ASSESSMENT OF SUBSTANCES PRESENT IN *ELEUTHEROCOCCUS* ROOTS

None of the pure compounds from *Eleutherococcus senticosus* roots have been evaluated for biological effects in humans. The following remarks concern either *in vitro* or *in vivo* biological assessments.

A. ELEUTHEROSIDE A (DAUCOSTEROL, β-SITOSTEROL GLUCOSIDE) (EA)

This ubiquitous sterol glucoside has not been studied extensively, but all indications are that it is devoid of marked biological effects, as would be predicted from its structure. *In vitro*, it is inactive as a cytotoxic agent when evaluated in the NCI 9KB cell culture (Farnsworth, 1983). It is also devoid of inhibitory effects, *in vitro*, against *Mycobacterium tuberculosis* (Mitscher *et al.*, 1975), and has no stimulant or relaxant effects on the isolated uterus *in vitro* (DasGupta and Pandey, 1970).

At 5.0 mg/kg (p.o.), EA is devoid of analgesic activity in mice, as evidenced by failure to inhibit the acetic acid writhing effect (Tanaka *et al.*, 1977).

In male mice, EA (i.p., dose not indicated) increased the time to com-

204   N. R. FARNSWORTH ET AL

plete fatigue in a variety of stress tests (Brekhman and Dardymov, 1969b).

B. ELEUTHEROSIDE B (SYRINGIN) (EB)

At concentrations up to 0.1 mg/ml, EB has no effect on early embryogenesis in the sea urchin (Anisimov *et al.*, 1973); at 1.0×10^{-7} g/ml, this compound stimulates hexokinase activity (Dardymov and Khasina, 1972b) and also suppresses the inhibitory action of diabetic β-lipoprotein (from alloxan-treated rabbits) *in vitro*. In male mice (i.p., dose not indicated), EB increased the time to fatigue in a variety of stress tests (Brekhman and Dardymov, 1969b). Also in mice (i.p., 5.0 mg/kg), EB had no effect on conditioned-avoidance response (Ilyuchonok and Chaplygina, 1972). The compound (i.p., 0.5 mg/kg) showed an antistress effect in rats (immobilized to induce stress), based on a comparison of the weights of the adrenals, thymus, spleen, and thyroid, as well as ascorbic acid and cholesterol levels in the adrenals, in EB-treated animals versus controls (Brekhman and Dardymov, 1969b). EB showed an androgenic response in immature mice (i.p., 25.0–50.0 mg/kg) (Dardymov, 1972), as well as an increase in RNA content of the seminal vesicles and prostate of the treated animals (Dardymov, 1972).

EB (5.0 mg/kg, s.c.) is reported to inhibit the formation in the blood of rats of a glucose inhibitor in stress animals (Dardymov and Khasina, 1977).

C. ELEUTHEROSIDE B₁ (ISOFRAXIDIN-7-O-α-l-GLUCOSIDE) (EB-l)

EB-1 (i.p., 25.0–50.0 mg/kg) in immature mice showed an androgenic effect, as well as an increased RNA content of the seminal vesicles and prostate (Dardymov, 1972). The compound showed a weak effect on the antialarm reaction in stressed animals (Brekhman and Dardymov, 1969b).

D. ELEUTHEROSIDE B₄ [(−)-SESAMIN] (EB-4)

(−)-Sesamin is a lignan that is well known for its synergistic effect on the insecticidal activity of pyrethrin extracts. The chemistry and pharmacology of this compound have been reviewed (Budowski, 1964). It has some degree of activity against *Mycobacterium tuberculosis in vitro* (Ramaswamy and Sirsi, 1957, 1958; Gangadharan *et al.*, 1953) and inhibits the growth of silkworm larvae (Kamikado *et al.*, 1975). Patch tests conducted on 13 patients having a contact allergy to sesame oil (*Sesamum indicum*), which contains high concentrations of (+)-sesamin, showed 12/13 to be positive to (+)-sesamin (Neering *et al.*, 1975).

E. ELEUTHEROSIDE C (METHYL-β-GALACTOSIDE) (EC)

EC increased the time to complete fatigue in stressed male mice (i.p., dose not stated) in a variety of tests (Brekhman and Dardymov, 1969a,b). The compound (i.p., dose not indicated), showed antistress (immobilization of animals) effects in rats based on a comparison of weights of the adrenals, thymus, spleen, and thyroid glands, as well as ascorbic acid and cholesterol levels of the adrenals in stress versus control animals. EC showed no effects in rats against the alarm phase of stress (Brekhman and Dardymov, 1969a,b).

F. ELEUTHEROSIDE D AND E [(−)-SYRINGARESINOL DI-O-β-D-GLUCOSIDE)] (EDE)[2]

In stressed (immobilization) rats, EDE (i.p., 20.0 mg/kg) is reported to have a counteracting effect, based on comparison of weights of adrenals, thymus, spleen, and thyroid glands as well as ascorbic acid and cholesterol levels of the adrenals, in treated animals versus controls (Brekhman and Dardymov, 1969b). Treatment of male mice with EDE (i.p., dose not stated) increased the time required for complete fatigue in a number of stress situations (Brekhmann and Dardymov, 1969b).

EDE showed androgenic effects (i.p., 20.0 mg/kg) in immature male mice, as well as increasing the RNA content of seminal vesicles and the prostate (Dardymov, 1972).

G. ELEUTHEROSIDE F (STRUCTURE UNKNOWN) (EF)

Male mice (i.p., dose not indicated) showed an increase in the time to fatigue in a variety of stress tests when given EF (Brekhman and Dardymov, 1969a).

H. ISOFRAXIDIN (IF)

This rare, nonglycosidic, trace component of *Eleutherococcus senticosus* roots has been reported to induce choleretic effects in rats (i.v. or intra-

---

[2]Eleutherosides D and E are chemically equivalent, showing differences only in their crystal form.

duodenal, 10.0 mg/kg) (Danielak *et al.*, 1973; Nieschulz and Schmersahl, 1968).

I. POLYSACCHARIDES PES-A AND PES-B

Polysaccharides PES-A and PES-B, isolated from *Eleutherococcus senticosus* roots, have been shown to be immunopotentiating agents. They were effective in potentiating the antibody response against sheep red blood cells and also in stimulating the phagocytosis of peritoneal macrophages of mice. By *in vitro* lymphocyte transformation test they were shown to have the ability of promoting the response of spleen cells to lipopolysaccharide but not to concanavalin A. By *in vivo* test they showed no significant effect on the response of spleen cells to lipopolysaccharide or to concanavalin A at therapeutic doses. These two compounds could partially diminish intoxication to thioacetamide, phytohemagglutinin, and X-rays in mice and also inhibited the growth of transplanted sarcoma 180 tumor cells in mice to about 50% (Xu *et al.*, 1982; Zhu *et al.*, 1982).

The possible mechanism(s) for the adaptogenic effect of *Eleutherococcus senticosus* root extracts and certain of the eleutherosides has been reviewed (Dardymov, 1976c).

IX. MISCELLANEOUS LITERATURE

Methods for the macroscopic and microscopic identification of *Eleutherococcus senticosus* have been published (Suprunov, 1972), as well as data concerning thin-layer chromatographic patterns for extracts of the roots (Suprunov, 1972; Wagner and Wurmboeck, 1977), which are useful for establishing the authenticity of the crude drug.

Extracts of the roots (as well as stems and leaves) of *Eleutherococcus senticosus* are reported to inhibit germination of radish (*Raphanus sativus* cultivar Rubin) seeds (Lapchik, 1967).

X. SUMMARY AND FUTURE RESEARCH SUGGESTIONS

Eleutherococcus senticosus (ES) is said to be an "adaptogen." Such a substance is defined as having three characteristics. The first aspect of an adaptogen relates to its lack of toxicity. It must be an innocuous substance and cause minimal disorders in the physiological functions of an organism. The results of animal studies performed on ES clearly demonstrate the innocuous nature of ES extracts. A large variety of species have been exposed to ES extracts, including mice, rats, rabbits, dogs, minks, deer, lambs, and piglets. In one study that used powdered ES root, the acute oral LD_{50} in mice was more than 30 g/kg. Teratogenicity has also been studied in at least three species—rats, rabbits, and minks—with no adverse effects observed. On the basis of animal studies, ES extracts are nontoxic.

However, the only sure test for human toxicity or safety requires that a substance be administered to humans. Numerous studies in humans involving more than 6000 subjects have been performed to observe the effects of ES extracts on human performance. Serious toxicity was not reported in any of the studies.

The second aspect of an adaptogen is its nonspecific action. Numerous animal models have been utilized to evaluate ES action on the performance of animals under various conditions of stress, such as heat and cold, excessive exercise, swimming to exhaustion, restricted movement, and others. The ability of ES extracts to modulate such artificially induced stress in a variety of conditions in a number of animal species is good evidence of the nonspecific nature of the adaptogenic action of ES extracts.

Studies of the effects of ES extracts on stress in humans have also been performed, monitoring a variety of conditions, including physical exertion, visual acuity, color discrimination, hearing, motion sickness, and mental alertness related to motor activity. The ability to improve human subject performance under this wide variety of difficult situations provides excellent proof of the wide range of activities associated with ES administration.

The third characteristic of an adaptogen is its normalizing action irrespective of the direction of the changes from physiological norms caused by the presence of a pathological condition. There are numerous reports of the ability of ES extracts to inhibit or modulate a disease process, including atherosclerosis, pyelonephritis, hypertension, rheumatic heart lesions, pneumoconiosis, hypotension, and arrhythmias. The activity of ES extracts on all of the major organ systems to correct a pathological condition is evidence of the nonspecificity of its action. The return to normal of both hypertensive and hypotensive patients following ES treatment is an example of the ability of this extract to normalize pathological conditions regardless of the direction of change.

Evidence to support the adaptogenic nature of ES extracts in both animal models and humans is extensive. Pharmacological explanations of the mechanism of activity of these actions, however, are not well characterized, in spite of numerous investigations. Many pharmacologists remain unconvinced of the existence of a true "adaptogenic activity." It is difficult to rationalize the existence of an agent that acts as an adaptogen using current pharmacological concepts.

The technology available to scientists several years ago suggested that the central nervous system was the prime controller of all bodily functions. The concept of neurons leading to far reaches of the body to provide the input to the central nervous system for responding to all the situations encountered by the body was most logical according to current knowledge available at the time. Elucidation of the autonomic nervous system and the endocrine system were steps in the recognition that many of the bodily control mechanisms were not necessarily under the direct control of the central nervous system.

Further research has now revealed the presence of several substances called autocoids. These substances are usually synthesized in local tissue sites and are probably responsible for mediating only local types of tissue responses. Examples are prostaglandins, which are involved in blood coagulation and smooth muscle contraction; kinins, which are involved in perception of pain and vascular tone; and leucotrienes, which are involved in smooth muscle contraction and chemotaxis. It is possible to envision control mechanisms for these autocoids that respond to demands of local tissues to deal with conditions not conducive to homeostasis and induced by stress or disease states. Existing models have demonstrated the ability of ES to modify physiological function without any conclusive evidence of the biochemical site of its action. A reason for this may be the failure of the models to consider the detailed physiological mechanisms which respond to the adaptogens. It is possible that some control mechanism for the autocoids not yet elucidated is responsible for such actions: a control mechanism that coordinates the local effect of autocoids to the condition of the site at which it is located, not dependent on the overall status of the body.

It is our belief that research in the area of adaptogens has only just begun. The extensive studies on *Eleutherococcus senticosus* and *Panax* species have contributed much to a beginning of an understanding of the adaptogenic response. Further studies centered around novel *in vivo* and *in vitro* systems to elucidate the adaptogenic mechanism should now evolve, so that more and better drugs will be developed to benefit mankind.

ACKNOWLEDGMENTS

The authors acknowledge Dr. P. J. Medon and Dr. E. B. Thompson for helpful discussions and are grateful to Dorothy Guilty for typing so many revisions of the manuscript.

George Matwyshyn was especially helpful in providing most of the translation assistance of papers written in the Russian language.

REFERENCES

Abramova, Z. I., Chernyi, Z. K., Natalenko, V. P., and Gutman, A. M. (1972a). *Lek. Sredstva Dal'nego Vostoka* **11**, 102.
Abramova, Z. I., Chernyi, Z. K., Natalenko, V. P., and Gutman, A. M. (1972b). *Lek. Sredtsva Dal'nego Vostoka* **11**, 106.
Afanas'ev, B. G., Zhestovskii, V, A., Mazunov, K. V., and Maevskii, K. L. (1973). *Vopr. Pitan.* **1**, 3.
Agadzhanyan, G. M., Vartanyan, L. V., and Manasyan, A. B. (1972). *Biol. Zh. Arm.* **25**(4), 96; *Biol. Abstr.* **56**, 27580 (1972).
Akimov, A. A., and Sudaryshina, N. (1966). *Mater. Itogovoi Nauchn. Konf. Inst. Onkol., 1966; Biol. Abstr.* **48**, 122030 (1966).
Amirov, R. 0., and Abdulova, E. B. (1966). *Uch. Zap. Azerb. Inst. Usoversh. Vrachel* **1**, 3.
Anisimov, M. M., Fronert, E. B., Frolova, G. M., and Suprunov, N. I. (1973). *Izv. Akad. Nauk Tadzh. SSR, Otd. Biol. Nauk* **4**, 590; *Biol. Abstr.* **60**, 43822 (1973).
Baburin, E. F. (1966a). *Lek. Sredstva Dal'nego Vostoka* **7**, 173.
Baburin, E. F. (1966b). *Lek. Sredstva Dal'nego Vostoka* **7**, 179.
Baburin, E. F. (1976a). *In* "Processes of Adaptation and Biologically Active Substances" (I. I. Brekhman *et al.*, eds.). Vladivostok.
Baburin, E. F. (1976b). *In* "Processes of Adaptation and Biologically Active Substances" (I. I. Brekhman *et al.*, eds.), p. 91. Vladivostok.
Baburin, E. F., Porkalov, P. V., and Polonskij, V. V. (1970). *Lek. Sredstva Dal'nego Vostoka* **10**, 67.
Baburin, E. F., Tarasov, I. K., and Alexeev, V. N. (1972). *Lek. Sredstva Dal'nego Vostoka* **11**, 120.
Baranov, A. I. (1982). *J. Ethnopharmacol.* **6**, 339.
Belonosov, I. S., and Jakovleva, E. G. (1965). *Mater. Sci. Sess., 22nd, 1964, Minist. Health RSFSR Khaborovsk Reg. Med. Inst., Khabarovsk, USSR* p. 210.
Belonosov, I. S., Jakovlena, E. G., and Jakovlev, Y. I. (1965). *Mater. Sci. Sess., 22nd, 1964, Minist. Health, RSFSR, Khabarovsk Reg. Med. Inst., Khabarovsk, USSR* p. 211.
Berdyshev, V. V. (1970). *Lek. Sredstv. Dal'nego Vostoka* **10**, 64.
Berdyshev, V. V. (1977). *Adaptation Adaptogens* **2**, 119.
Bezdetko, G. N. (1966). *Lek. Sredstva Dal'nego Vostoka* **7**, 81.
Bezdetko, G. N., Brekhman, I. I., Dardymov, I. V., Zil'ber, M. L., and Rogozkin, V. A. (1973). *Vopr. Med. Khim.* **19**(3), 245.
Bihkovcova, T. L., Polozenceva, M. I., and Shestak, J. A. (1966). *Lek. Sredtsva Dal'nego Vostoka* **7**, 41.
Blokhin, B. N. (1966a). *Lek. Sredstva Dal'nego Vostoka* **7**, 191.
Blokhin, B. N. (1966b). *Mater. Conf. Tomsk. Med. Inst., C.N.I.L., 3rd,* Vol. 3, p. 134.
Brandis, S. A., and Pilovitskaya, V. N. (1966a). *Lek. Sredstva Dal'nego Vostoka* **7**, 141.
Brandis, S. A., and Pilovitskaya, V. N. (1966b). *Lek. Sredtsva Dal'nego Vostoka* **7**, 155.
Brekhman, I. I. (1960). *In* "Eleutherococcus Root—New Stimulating and Tonic Remedy," V. I. Lenin Red Banner Military Institute of Physical Culture and Sport, Leningrad.
Brekhman, I. I. (1963a). *Lek. Sredtsva Dal'nego Vostoka* **5**, 219.
Brekhman, I. I. (1963b). *Biochem. Pharmacol.* **12S**, 50.
Brekhman, I. I. (1964). *Proc. Int. Pharmacol. Meet., 2nd, 1963* Vol. 7, No. 2, p. 97.

Brekhman, I. I. (1966). "Man and Biologically Active Substances." Acad. Sci. U.S.S.R., Leningrad.

Brekhman, I. I. (1968). "*Eleutherococcus*," 1st ed. Nauka Publ. House, Leningrad; *Biol. Abstr.* **51**, 25876 (1970).

Brekhman, I. I. (1970). "*Eleutherococcus*. Clinical Data." U.S.S.R. Foreign Trade Publication 28524/2, Moscow.

Brekhman, I. I. (1977). "Adaptation and Adaptogens." Acad. Sci. U.S.S.R., Far Eastern Center for Science, Vladivostok, USSR.

Brekhman, I. I., and Dardymov, I. V. (1969a). *Lloydia* **32**, 46.

Brekhman, I. I., and Dardymov, I. V. (1969b). *Annu. Rev. Pharmacol.* **9**, 410.

Brekhman, I. I., and Dardymov, I. V. (1971). *Sb. Rab. Inst. Tsitol., Akad. Nauk SSSR* **14**, 82.

Brekhman, I. I., and Kirillov, O. I. (1966). *Lek. Sredstv. Dal'nego Vostoka* **7**, 9.

Brekhman, I. I., and Kirillov, O. I. (1969). *Life Sci.* **8**, 113.

Brekhman, I. I., and Oleinikova, T. P. (1963). *Lek. Sredstva Dal'nego Vostoka* **5**, 249.

Brekhman, I. I., Oskotsky, L. I., and Khakham, A. R. (1970). *Izv. Akad. Nauk SSSR, Ser. Biol.* **6**, 33.

Brekhman, I. I., Berdyshev, G. D., and Golotin, V. G. (1971). *Izv. Akad. Nauk SSSR, Ser. Biol.* **1**, 31.

Budowski, P. (1964). *J. Am. Oil Chem. Soc.* **41**, 280.

Bykhovtsova, T. L. (1970a). *Izv. Akad. Nauk SSSR, Ser. Biol.* **5**, 713. *Biol. Abstr.* **53**, 56701 (1970).

Bykhovtsova, T. L. (1970b). *Izv. Akad. Nauk SSSR, Ser. Biol.* **6**, 915.

Cherkashin, G. V. (1966). *Cent. Nerv. Syst. Stimulants, 1966* p. 91; *Biol. Abstr.* **48**, 113118 (1966).

Cherkashin, G. V. (1968). *Izv. Sib. Otd. Akad. Nauk SSSR, Ser. Biol.-Med. Nauk* **1**, 116; *Biol. Abstr.* **50**, 99491 (1968).

Curtze, A. (1975). *Dtsch. Apoth.* **27**, 501.

Dalinger, O. I. (1966a). *Cent. Nerv. Syst. Stimulants, 1966* p. 106; *Biol. Abstr.* **48**, 69624 (1966).

Dalinger, O. I. (1966b). *Cent. Nerv. Syst. Stimulants, 1966* p. 112; *Biol. Abstr.* **48**, 106779 (1966).

Dambueva, E. A., and Sal'nik, B. Yu (1966). *Cent. Nerv. Syst. Stimulants, 1966* p. 51; *Biol. Abstr.* **48**, 69561 (1966).

Danielak, R., Popwska, E., and Borkowski, B. (1973). *Pol. J. Pharmacol. Pharm.* **25**, 271.

Dardymov, I. V. (1963). *Lek. Sredstva Dal'nega Vostoka* **8**, 245.

Dardymov, I. V. (1971). *Sb. Rab. Inst. Tsitol., Akad. Nauk SSSR* **11**, 76; *Chem. Abstr.* **82**, 54331w (1971).

Dardymov, I. V. (1972). *Lek. Sredtsva Dal'nego Vostoka* **11**, 60.

Dardymov, I. V. (1976a). *Khim. Zhizn.* **(3)**, 66; *Chem. Abstr.* **85**, 15p (1976).

Dardymov, I. V. (1976b). "Ginseng and *Eleutherococcus*, the Mechanism of their Biological Actions," 1st ed. Nauka Publ. House, Moscow.

Dardymov, I. V. (1976c). *In* "Processes of Adaptation and Biologically Active Substances" (I. I. Brekhman *et al.*, eds.), p. 113. Vladivostok.

Dardymov, I. V., and Khasina, E. I. (1972a). *Lek. Sredstva Dal'nego Vostoka* **11**, 52.

Dardymov, I. V., and Khasina, E. I. (1972b). *Lek. Sredtsva Dal'nego Vostoka* **11**, 56.

Dardymov, I. V., and Khasina, E. I. (1976a). *In* "Processes of Adaptation and Biologically Active Substances" (I. I. Brekhman, *et al.*, eds.), p. 69. Vladivostok.

Dardymov, I. V., and Khasina, E. I. (1976b). *Proc. Southeast Asian Pac. Reg. Meet. Pharmacol., 1976* Abstract 77.

Dardymov, I. V., and Khasina, V. I. (1977). *Adaptation Adaptogens* **2**, 108.

Dardymov, I. V,, and Kirillov, O. I. (1965). *Mater. Sci. Sess., 12th, 1964, Minist. Health RSFSR Khaborovsk Reg. Med. Inst., Khabarovsk, USSR* p. 215.
Dardymov, I. V., and Kirillov, O. I. (1968). *Lek. Sredtsva Dal'nego Vostoka* 7, 43.
Dardymov. I. V., Kirillov, O. I., and Urgenc, I. L. (1965). *Mater. Sci. Sess., 22nd, 1964, Minist. Health RSFSR Khaborovsk Reg. Med. Inst., Khabarovsk, USSR.* p. 216.
Dardymov, I. V., Berdyshev, V. V., Golikob, P. P., and Fedorets, B. A. (1966). *Lek. Sredtsva Dal'nego Vostoka* 7, 133.
Dardymov. I. V., Bezdetko, G. N., and Brekhman, I. I. (1972a). *Vopr. Med. Khim.* 18(3), 267; *Chem. Abstr.* 77, 97282u (1972).
Dardymov, I. V., Suprunov, N. I., and Sokolenko, L. A. (1972b). *Lek. Sredtsva Dal'nego Vostoka* 11, 66.
DasGupta, B., and Pandey, V. B. (1970). *Experientia* 26, 745.
Dzhioev, F. K. (1965). *Vopr. Onkol.* 11(9), 51.
Dzhioev, F. K., and Prasol, S. D. (1966). *Lek. Sredtsva Dal'nego Vostoka* 7, 69.
Egorov, Y. N., and Baburin, E. F. (1966). *Lek. Sredtsva Dal'nego Vostoka* 7, 167.
Ekkert, L. G. (1972). *Lek. Sredtsva Dal'nego Vostoka* 11, 98.
El'kin, A. I. (1972a). *Lek. Sredtsva Dal'nego Vostoka* 11, 91.
El'kin, A. I. (1972b). *Lek. Sredtsva Dal'engo Vostoka* 11, 94.
Elyakov, G. P., and Ovodov, V. S. (1972). *Khim. Prir. Soedin.* 8, 697.
Farnsworth, N. R. (1983). University of Illinois Health Sciences Center, Chicago, Illinois. Unpublished data obtained from the National Cancer Institute, Bethesda, Maryland.
Fedorov, Yu. V., Vasil'eva, O. A., and Vasil'ev, N. V. (1966). *Cent. Nerv. Syst. Stimulants, 1966* p. 99; *Chem. Abstr.* 69, 104731e (1966).
Feoktstova, G. I. (1966). *Cent. Nerv. Syst. Stimulants, 1966* p. 55; *Biol. Abstr.* 48, 106706 (1966).
Ferguson, F. W., Medon, P. J., Briley, T. C., and Watson, C. F. (1983). *Toxicologist* 3(1), 51.
Frolova, G. M., and Ovodov, Y. S. (1971). *Khim. Prir. Soedin.* 7, 618.
Frolova, G. M., Ovodov, Y. S., and Suprunov, N. I. (1971). *Khim. Prir. Soedin.* 7, 614.
Fruentova, T. A. (1965). *Mater. Sci. Sess., 22nd, 1964, Minist. Health RSFSR Khabarovsk Reg. Med. Ins., Kharbarovsk USSR* p. 225.
Gagarin, I. A. (1977). *Adaptation Adaptogens* 2, 128.
Galanova, L. K. (1977). *Adaptation Adaptogens* 2, 126.
Gangadharan, P. R. J., Natarajan, S., Wadhwani, T. K., Biri, K. V., Narayanamurity, N. L., and Iyer, B. H. (1953). *J. Indian Inst. Sci., Sect. A* 35, 69.
Gol'berg, E. D., Shubina, T. S., and Shternberg, I. B. (1971). *Antibiotiki (Moscow)* 16(2), 113.
Golikov, A. P. (1966a). *Velnno-Med. Zh.* No. 9, p. 24.
Golikov, A. P. (1966b). *Lek. Sredtsva Dal'nego Vostoka* 6, 213.
Golikov, A. P. (1966c). *Lek. Sredtsva Dal'nego Vostoka* 7, 63.
Golikov, A. P. (1967). *Kazan. Med. Zh.* 5, 76.
Golikov, A. P., Reshetnev, V. G., and Golikov, P. P. (1966). *Lek. Sredtsva Dal'nego Vostoka* 7, 67.
Golikov, P. P. (1963). *Lek. Sredtsva Dal'nego Vostoka* 5, 233.
Golikov, P. P. (1966a). *Lek. Sredtsva Dal'nego Vostoka* 7, 17.
Golikov, P. P. (1966b). *Lek. Sredtsva Dal'nego Vostoka* 7, 295.
Golotkin, G. F., and Bojko, S. N. (1963). *Mater. Study Ginseng Other Med. Plants Far East* 5, 257.
Golotkin, V. G., Brekhman, I. I., Dobryakov, P. G., and Li, S. E. (1972). *Lek. Sredtsva Dal'nego Vostoka* 11, 37.
Gordeicheva, N. V. et al. (1975). *Kosm. Biol. Aviakosm. Med.* 9(5), 6; *Chem. Abstr.* 83, 188498j (1975).

Gorovi, P. G. (1963). *Soobshch. Dal'nevost. Fil. Sib. Otd. Akad. Nauk SSR* **16,** 97; *Biol. Abstr.* **45,** 40829 (1963).

Gorshkov, G. I., and Antrushin, M. C. (1967). *Eleutherococcus Livest. Rear., 1967* p. 66.

Grinkevich, N. I., and Gribovskaya, I. F. (1968). *Rast. Resur.* 4(4), 506; *Chem. Abstr.* **70,** 118015g (1968).

Grinkevich, N. I., Koval'skii, V. V., and Gribovskaya, I. F. (1967). *Biol. Rol. Medi, Simp.* p. 333; *Chem. Abstr.* **79,** 84086w (1967).

Hacker, B., and Medon, P. J. (1984). *J. Pharm. Sci.* **73,** 270.

Ilyuchonok, R. Y., and Chaplygina, S. R. (1972). *Lek. Sredtsva Dal'nego Vostoka* **11,** 83.

Jekkert, L. G. (1972). *Lek. Sredtsva Dal'nego Vostoka* **11,** 98.

Judin, A. M. (1967). *Eleutherococcus Livest. Rear., 1967,* p. 78.

Kamikado, T., Chang, C.-F., Murakoshi, S., Sakurai, A., and Tamura, S. (1975). *Agric. Biol. Chem.* **39,** 833.

Kaplan, E. Ya. (1965). *Lek. Sredtsva Dal'nego Vostoka* **7,** 77; *Biol. Abstr.* **48,** 107025 (1965).

Karimov, M. A. (1969). *Adravokhr, Kaz.* **11,** 50; *Biol. Abstr.* **54,** 61495 (1969).

Kirillov, O. I. (1964). *Soobshch. Dal'neovst Fil. Sib. Otd. Akad. Nauk SSSR* **23,** 3; *Biol. Abstr.* **47,** 32626 (1964).

Kirillov, O. I. (1965). *Mater. Sci. Sess., 12th, 1964, Minist. Health RSFSR Khaborovsk Reg. Med. Ins., Khabarovsk, USSR* p. 228.

Kirillov, O. I., and Dardymov. I. V. (1966a). *Lek. Sredstv. Dal'nego Vostoka* **7,** 55.

Kirillov, O. I., and Dardymov. I. V. (1966b). *Mater. Conf. Tomsk Med. Inst. C.N.I.L., 3rd,* **3,** 124.

Kirillov, O. I., Yurgens, I. L., and Galkin, V. V. (1966). *Lek. Sredstv. Dal'nego Vostoka* **7,** 13.

Kirillov, V. A., and Semashkevish, G. M. (1966). *Tr. Blagoveshch. Gos. Med. Inst.* **7,** 133; *Chem. Abstr.* **67,** 10233a (1966).

Kolla, V. F., and Ovodenko, L. A. (1966). *Lek. Sredtsva Dal'nego Vostoka* **7,** 33.

Kolmakova, L. F., and Kutolina, N. I. (1966). *Cent. Nerv. Syst. Stimulants, 1966* p. 131; *Biol. Abstr.* **48,** 69715 (1966).

Komarova, V. L. (1967). "*Eleutherococcus* in Livestock Rearing." Acad. Sci. USSR, Siberian Division, Mining-Siberian Forest Station, Akad. Nauk SSSR, Sib. Otd.

Koshkarev, K. I., and Kovinskii, K. P. (1966). *Cent. Nerv. Syst. Stimulants, 1966* p. 128; *Biol. Abstr.* **48,** 107098 (1966).

Krasnozhenov, E. P. (1970). *Izv. Sib. Otd. Akad. Nauk SSSR, Ser. Biol. Nauk* No. 2, p. 139; *Chem. Abstr.* **74,** 86141j (1970).

Kucharenko, T. M. (1963). *Mater. Study Ginseng Other Med. Plants Far East* **5,** 229.

Kuntsman, I. Ya. (1966). *Lek. Sredstva Dal'nego Vostoka* **7,** 129; *Biol. Abstr.* **48,** 111611 (1966).

Kuntsman, I. Ya. (1967). *Lek. Sredstva Dal'nego Vostoka* **11,** 121; *Biol. Abstr.* **48,** 107099 (1967).

Lapchik, V. F. (1967). *Visn. Kiiv. Univ., Ser. Biol.* **9,** 131; *Biol. Abstr.* **50,** 49717 (1967).

Lapchik, V. F., and Ovodov, Y. S. (1969). *Visn. Kiiv. Univ., Ser. Biol.* **11,** 105; *Biol. Abstr.* **52,** 132288 (1969).

Lapchik, V. F., and Ovodov, Y. S. (1970). *Rast. Resur.* 6(2), 228; *Chem. Abstr.* **74,** 1080r (1970).

Lapchik, V. F., Frolova, G. M., and Ovodov, Y. S. (1969). *Rast. Resur.* 5(3), 455; *Chem. Abstr.* **72,** 19118c (1969).

Lapustina, I. A. (1967). *Eleutherococcus Livest. Rear., 1967* p. 88.

Lazarev, N. V. (1962). *Vopr. Onkol.* 8(11), 20.

Lazarev, N. V. (1963). *Mezhdunar. Protivorakovyi Kongr., 7th,* **3,** 383. *Biol. Abstr.* **47,** 73026 (1963).

Lazarev, N. V. (1965a). *Curr. Probl. Oncol.* 54; *Biol. Abstr.* **48,** 18723 (1965).

Lazarev, N. V. (1965b). *Vopr. Onkol.* **11**(12), 48.
Lazarev, N. V., and Brekhman, I. I. (1967). *Med. Sci. Serv.* No. 9, p. 9.
Lee, S.-W., Kozukue, N., Bae, H.-W., and Lee, J.-H. (1978). *Hanguk Sikp'um Kwahakhoe Chi* **10**, 245; *Chem. Abstr.* **90**, 3173c (1979).
Lee, S.-W., Kozukue, N., Bae, H.-W., and Yoon, T.-H. (1979). *Hanguk Sikp'um Kwahakhoe Chi* **11**, 273; *Chem. Abstr.* **92**, 169117d (1980).
Leonova, E. V. (1966). *Aktual. Probl. Teor. Klin. Med.* p. 262; *Chem. Abstr.* **87**, 95994k (1966).
Leshchev, L. S. (1966). *Vopr. Onkol.* **12**(5), 60.
Li, S. E. (1973). *Lek. Sredstva Dal'nego Vostoka* **11**, 70.
Lui, J. H.-C., and Staba, E. J. (1980). *J. Nat. Prod.* **43**, 340.
Lyobomudrov, V. E., Basamygina, L. Y., Bikezine, V. G., Mukhina, M. S., Mikhailova, T. I., Osadchuk, V. S., Shidlovsky, E. F., Bondarenko, G. A., and Demkovich, O. A. (1970). *Vrach, Delo* No. 1, p. 102.
Malyugina, L. L. (1964). *Acta Unio Int. Cancrum* **20**, 199.
Malyugina, L. L. (1966). *Vopr. Onkol.* **12**(7), 53.
Malyugina, L. L. (1969). *Vopr. Onkol.* **15**(4), 87.
Marina, T. F. (1966a) *Cent. Nerv. Syst. Stimulants, 1966* p. 24; *Chem. Abstr.* **66**, 93852e (1966).
Marina, T. F. (1966b). *Cent. Nerv. Syst. Stimulants, 1966* p. 31; *Chem. Abstr.* **66**, 1039122r (1966).
Maxsimov, J. L. (1967). *Eleutherococcus Livest. Rear. 1967* p. 96.
Medon, P. J., Thompson, E. B., and Farnsworth, N. R. (1981). *Acta Pharmacol. Sin.* **2**, 281.
Medon, P. J., Ferguson, P. W., and Watson, C. F. (1983). *Abstr. Annu. Meet. Am. Pharm. Assoc.*
Medvedev, M. A. (1966). *Lek. Sredstva Dal'nego Vostoka* **7**, 179.
Mikhailova, L. I., and Fruentov, N. K. (1972). *Lek. Sredtsva Dal'nego Vostoka* **11**, 86.
Mikunis, R. I., Serkova, V. K., and Shirkova, T. A. (1966a). *Lek. Sredtstv. Dal'nego Vostoka,* **7**, 221.
Mikunis, R. I., Serkova, V. K., and Shirkova, T. A. (1966b). *Lek. Sredtstva Dal'nego Vostoka* **7**, 227.
Mironova, A. I. (1963). *Vopr. Onkol.* **9**(1), 42.
Mishchenko, E. D. (1962). *Simp. Eleutherokokku Zhen'shenyu, XX Sess. Kom. Po Izuch. Zhen'shenya Drugikh Lakarstv. Rast. Dal'n. Vost., 20th, Vladivostok* p. 54; *Chem. Abstr.* **59**, 14483f (1962).
Mitscher, L. A., Bathala, M. S., Clark, G. W., and Beal, J. L. (1975). *Lloydia* **38**, 117.
Monakhov, B. V. (1965). *Vopr. Onkol.* **11**(12), 60.
Monakhov, B. V. (1967a). *Vopr. Onkol.* **13**(3), 71.
Monakhov, B. V. (1967b). *Vopr. Onkol.* **13**(8), 94.
Moskalik, K. G. (1970a). *Vopr. Onkol.* **16**(7), 74.
Moskalik, K. G. (1970b). *Patol Fiziol. Eksp. Ter.* **14**(5), 73; *Biol. Abstr.* **53**, 17410 (1970).
Moskalik, K. G. (1973). *Onkologiya (Kiev)* **4**, 170; *Chem. Abstr.* **81**, 72621y (1973).
Neering, H., Vitanyi, B. E. J., Malten, K. E., van Ketel, W. G., and van Dijk, E. (1975). *Acta Dermatol.* **55**, 31.
Nieschulz, O., and Schmersahl, P. (1968). *Arzneim-Forsch.* **18**, 1330.
Oleinichenko, V. F. (1966). *Stimul. Tsentr. Nerv. Sist.* p. 124; *Biol. Abstr.* **48**, 111816 (1966).
Ovodov, Y. S., and Shibaeva, V. I. (1969). *Khim. Prir. Soedin.* **5**, 589.
Ovodov, Y. S., Frolova, G. M., Elyakova, L. A., and Elyakov, G. B. (1965a). *Izv. Akad. Nauk SSSR, Ser. Khim.* p. 2065; *Chem. Abstr.* **64**, 6737 (1966).
Ovodov, Y. S., Ovodova, R. G., Solov'eva, T. F., Elyakov, G. B., and Kochetkov, N. K. (1965b). *Khim. Prir. Soedin.* **1**, 3.

Ovodov, Y. S., Frolova, G. M., Nefedova, M. Y., and Elyakov, G. B. (1967). *Khim. Prir. Soedin.* **3**, 63.

Ovodov, Y. S., Frolova, G. M., Dzizenko, A. K., and Litvinenko, V. I. (1969). *Izv. Akad. Nauk SSSR, Ser. Khim.* p. 1370.

Padkin, V. V., and Baburin, E. F. (1966). *Lek. Sredstva Dal'nego Vostoka* **7**, 185.

Pearce, P. T., Zois, I., Wynne, K. N., and Funder, J. W. (1982). *Endocrinol. Jpn.* **29**, 567.

Pegel, N. B. (1964a). *Mater. Teor. Klin. Med.* **3**, 130; *Biol. Abstr.* **48**, 89309 (1964).

Pegel, N. B. (1964b). *Mater. Teor. Klin. Med.* **3**, 137; *Biol. Abstr.* **49**, 83736 (1964).

Pichurina, E. A. (1965). *Mater. Teor. Klin. Med.* **5**, 79; *Biol. Abstr.* **49**, 13012 (1965).

Polozhentseva, M. I. (1976). *In* "Processes of Adaptation and Biologically Active Substances" (I. I. Brekhman *et al.*, eds.), p. 87. Vladivostok.

Ponomareva, M. F. (1973). *Byull. Vses. Nauchno-Issled. Inst. Fiziol., Biokhim. Pitan. S-kh. Zhivotn.* **7**(2), 56; *Chem. Abstr.* **82**, 3118y (1973).

Ramaswamy, A. S., and Sirsi, M. (1957). *Naturwissenschaften* **44**, 380.

Ramaswamy, A. S., and Sirsi, M. (1958). *Chemother., Proc. Symp., Lucknow, 1958* p. 46; *Chem. Abstr.* **54**, 4910g (1958).

Revina, T. A., and Sal'nik, B. Yu. (1966). *Cent. Nerv. Syst. Stimulants, 1966* p. 59; *Chem. Abstr.* **67**, 2001h (1966).

Ronichevskaya, G. M. (1967). *Vopr. Onkol.* **13**(3), 67; *Biol. Abstr.* **49**, 84607 (1967).

Rozin, M. A. (1966). *In Nonspecific Prophylaxis with Drugs and the Therapy of Cancer,* p. 21. Meditsina Publ. House, Leningrad; *Biol. Abstr.* **48**, 106964 (1966).

Rusin, V. Ya. (1967), *Lek. Sredstva Dal'nego Vostoka* **7**, 27.

Sabubova, V. A., and Titova, A. A. (1966). *Kazan. Med. Zh.* **6**, 38; *Biol. Abstr.* **50**, 93138 (1966).

Sal'nik, B. Yu. (1966). *Cent. Nerv. Syst. Stimulants, 1966* p. 44; *Chem. Abstr.* **67**, 1999c (1966).

Sandberg, F. (1973). *Planta Med.* **24**, 392.

Sandler, B. I. (1972). *Lek. Sredstva Dal'nego Vostoka* **11**, 109.

Sandler, B. I., and Sandler, T. V. (1972). *Lek. Sredstva Dal'nego Vostoka* **11**, 114.

Saratikov, A. S. (1966). *Cent. Nerv. Syst. Stimulants, 1966* p. 3; *Chem. Abstr.* **70**, 10069p (1966).

Saratikov, A. S., and Cherdyntsev, S. G. (1966). *Cent. Nerv. Syst. Stimulants, 1966* p. 62; *Chem. Abstr.* **66**, 93853f (1966).

Saratikov, A. S., and Pichurina, R. A. (1965). *Izv. Sib. Otd. Akad. Nauk SSSR, Ser. Biol-Med. Nauk* **1**, 113; *Biol. Abstr.* **47**, 62277 (1965).

Shulyateva, L. D., Gulyaev, V. G., and El'chenkov, E. N. (1966). *Lek. Sredtsva Dal'nego Vostoka* **7**, 37.

Sobkovich, L. N. (1970). *Lek. Sredtsva Dal'nego Vostoka* **10**, 82.

Soejarto, D. D., and Farnsworth, N. R. (1978). *Bot. Mus. Leafl., Harv. Univ.* **26**, 339.

Solov'eva, T. F., Prudnikova, T. I., and Ovodov, Y. S. (1968). *Rast. Resur.* **4**(4), 497; *Biol. Abstr.* **50**, 117110 (1968).

Starikova, M. P. (1970). *Zhivotnovodstvo* **32**(1), 78; *Chem. Abstr.* **73**, 22877a (1970).

Stolyarova, N. V. (1968). *Fiziol Zh. SSSR im. I.M. Sechenova* **54**(7), 838; *Chem. Abstr.* **69**, 84796z (1968).

Strokina, T. I. (1966a). *Mater. Conf. Tomsk. Med. Inst. C.N.I.L. 3rd,* **3**, 317.

Strokina, T. I. (1966b). *Lek. Sredstva Dal'nego Vostoka* **7**, 201.

Strokina, T. I. (1967). *Zh. Nevropatol. Psikhiatr. im. S.S. Korsakova* **67**, 903.

Strokina, T. I., and Mikho, T. B. (1966). *Lek. Sredstva Dal'nego Vostoka* **7**, 195.

Strokina, T. I., and Mikho, T. B. (1968). *Sb. Nauchn. Tr., Vladivost. Med. Inst.* **4**, 19; *Biol. Abstr.* **51**, 14332.

Stukov, A. N. (1965). *Vopr. Onkol.* **11**(12), 64.

Stukov, A. N. (1966). *Vopr. Onkol.* **12**(7), 57.
Stukov, A. N. (1967). *Vopr. Onkol.* **13**(1), 94.
Suprunov, N. I. (1967). *Eleutherococcus Livest. Rear., 1967* p. 57.
Suprunov, N. I. (1970). *Khim. Prir. Soedin.* **6**, 486.
Suprunov, N. I. (1972). *Lek. Sredstva Dal'nego Vostoka* **11**, 221.
Suprunov, N. I., and Dzizenko, S. N. (1971). *Khim. Prir. Soedin.* **7**, 524.
Suprunov, N. I., and Samojlov, T. P. (1967). "*Eleutherococcus* in Livestock Rearing." Acad. Sci., SSSR, Sib. Div., Vladivostok.
Tanaka, S., Ikeshiro, Y., Tabata, M., and Konoshima, M. (1977). *Arzneim.-Forsch.* **27**, 2039.
Tkhor, L. F., Taranenko, G. A., and Kozlov, Yu. P. (1966). *Tr. Mosk. O-va. Ispyt. Prir., Otd. Biol.* **16**, 73; *Biol. Abstr.* **48**, 94207 (1966).
Tsirlina, E. V. (1965). *Vopr. Onkol.* **11**(10), 70.
Voino-Yasenetskii, A. M. (1966). *Urol. Nefrol.* **31**(6), 21.
Voropaev, V. M. (1971). *Sb. Rab. Inst. Tsitol., Akad. Nauk SSSR* No. 14, p. 65; *Chem. Abstr.* **78**, 661b (1973).
Voropaev, V. M. (1972). *Lek. Sredstva Dal'nego Vostoka* **11**, 74.
Wacker, A., and Eilmes, H. G. (1978). *Kassenarzt* **18**(4), 1.
Wagner, H., and Wurmboeck, A. (1977). *Dtsch. Apoth.-Ztg.* **117**, 743.
Wagner, H., Heur, Y. H., Obermeier, A., Tittel, G., and Bladt, S. (1982). *Planta Med.* **44**, 193.
Xu, R.-S., Feng, S.-C., and Fan, Z.-Y. (1980). *Planta Med.* **39**, 278.
Xu, R.-S., Feng, S.-C., Fan, Z.-Y., Ye, C.-J., Zhai, S.-K., and Shen, M.-L. (1982). *Chem. Nat. Prod. Proc. Sino-Am. Symp., 1980* pp. 271–274.
Yaremenko, K. V. (1966). *Lek. Sredstva Dal'nego Vostoka* **7**, 109.
Yaremenko, K. V., and Moskalik, K. G. (1967). *Vopr. Onkol.* **13**(9), 65.
Yaremenko, K. V., and Moskalik, K. G. (1971). *Vopr. Onkol.* **17**(2), 66.
Zhu, C., Tu, G.-R., and Shen, M.-L. (1982). *Yao Hsueh T'ung Pao* **17**, 178.
Zorikov, P. S., and Burii, T. P. (1974). *Deposit. Doc. VINITI* **734–774**, 18; *Chem. Abstr.* **86**, 11762a (1974).
Zorikov, P. S., and Lyapustina, T. A. (1974). *Deposit. Doc. VINITI* **732–74**, 58; *Chem. Abstr.* **86**, 11973f (1974).
Zorikov, P. S., Lyapustina, T. A., and Frolov, Yu. D. (1974). *Deposit. Doc. VINITI* **732–74**, 34; *Chem. Abstr.* **86**, 119751e (1974).
Zotova, M. I. (1966). *Cent. Nerv. Syst. Stimulants, 1966* p. 67; *Biol. Abstr.* **48**, 107086 (1966).
Zyryanova, T. M. (1965). *Mater. Toeor. Klin. Med.* **5**, 69; *Biol. Abstr.* **48**, 69689 (1965).
Zyryanova, T. M. (1966). *Cent. Nerv. Syst. Stimulants, 1966* p. 37; *Biol. Abstr.* **48**, 106881 (1966).

6

Chemistry and Pharmacology of *Panax*

SHOJI SHIBATA
Meiji College of Pharmacy
Tokyo, Japan

OSAMU TANAKA
Institute of Pharmaceutical Sciences
School of Medicine, Hiroshima University
Hiroshima, Japan

JUNZO SHOJI
School of Pharmaceutical Sciences
Showa University
Tokyo, Japan

HIROSHI SAITO
Faculty of Pharmaceutical Sciences
University of Tokyo
Tokyo, Japan

217

ECONOMIC AND MEDICINAL PLANT RESEARCH
VOLUME 1

I. INTRODUCTION

Ginseng or Korean ginseng, the root of the araliaceous plant *Panax ginseng* C. A. Meyer, since ancient times has been used by east Asians as a medicine in the general belief that it is a panacea or promotes longevity. Originally, it grew in northeastern China, Korea, and eastern Siberia, but now the wild plant is very rare, if still extant. Almost all the ginseng roots in the market at present are the products of cultivation in northeastern China, Korea, and Japan.

The earliest description of ginseng appeared in the oldest Chinese pharmacopoeia, *Shen-Nung Pen T'sao Ching,* which is believed to have been written in the first century during the late Han dynasty and revised later by Tao Hung-Ching (AD 452–536):

> It is used for repairing five viscera, quieting the spirit, curbing the emotion, stopping agitation, removing noxious influence, brightening the eyes, enlightening the mind, and increasing the wisdom. [English translation by Dr. (Mrs.) Shin-Yin Hu.]

Several prescriptions including ginseng as a major component were recorded in the oldest medical text, *Shang Han Lun,* which was said to be written by Chang Chung-Ching (AD 142–220) in the second century, recompiled by Wang Su-ho (AD 210–285), and revised later by Lin Yi in the Sung Dynasty (AD 960–1280).

The general idea of the efficacy of ginseng in Chinese medicine is to restore Yang. By the philosophical concepts of traditional Chinese medicine, Yin and Yang must be equilibrium in the viscera to make a healthy balance.

Ginseng is currently produced by cultivation. It takes 4–6 years from seeding to harvest, and must grow in soil that has not grown ginseng in 10–15 years, and under shade to avoid direct sunshine. Therefore, mass production of ginseng to meet heavy demand is still not easy.

Ginseng is processed in two forms, white ginseng and red ginseng. The former is the dried root, whose peripheral skin is frequently peeled off, and the latter is the steamed root, which shows a caramellike colour and resists invasion of fungi and worms. The white and red lateral roots are also available in the market.

Some other congeners of ginseng are also used as medicines. American ginseng, the root of *Panax quinquefolium* L; which initially grew wild in the northeastern United States, is now cultivated for export to the Hong Kong market. Chinese people use it in almost the same manner as Korean ginseng, but with slightly different purposes.

Sanchi ginseng, the root of *Panax notoginseng* Burk. (*P. wangianus* Sun), which is cultivated in southwestern China, Yunnan and Kwangsi, and part of Vietnam, has been used in Chinese medicine for haemorrhages and bruises, and partly for restoring Yang, as with Korean ginseng.

Japanese Chikusetsu ginseng is the rhizome of wild growing *Panax pseudoginseng* Wall. subsp. *japonicus* Hara (= *P. japonicus* C. A. Meyer). It has been used as a medicine in Japan in place of Korean ginseng. Recommended by the *Shang-Han-Lun* for the syndrome of stiffness at the pit of stomach, Chikusetsu ginseng sometimes has been used in place of Korean ginseng.

Panax pseudoginseng subsp. *himalaicus* Hara is grows wild in Nepal and the East Himalayan district, and some other species of *Panax*, such as *P. pseudoginseng* var. *major* (Bark) Li [*P. japonicus* C. A. Meyer var. *major* (Burk) C. Y. Wu *et* K. M. Feng], which is also used in China as a drug, are found in western China (Table I).

In traditional Chinese medicine, ginseng has sometimes been replaced by the root of *Adenophora triphilla* DC. subsp. *aperticampanulata* Kitamura (*sha-shen*) or *A. remotiflora* Miq. (Campanulaceae) or by the root of *Codonopsis tangshen* Oliv. (Campanulaceae) (*tangshen*).

TABLE I

ORIGINAL GINSENG AND ITS CONGENERS

Scientific name	Organ used	Common name
Panax ginseng C. A. Meyer	Root	Ginseng (Korean ginseng)
P. quinquefolium L.	Root	American ginseng
P. pseudoginseng Wall. var. *notoginseng* (Burk.) Hoo et Tseng [*P. notoginseng* (Burk.) F. H. Chen]	Root	San-chi ginseng
P. pseudoginseng (Will.) subsp. *japonicus* Hara (*P. japonicus* C. A. Meyer)	Rhizome	Chikusetsu ginseng
P. pseudoginseng subsp. *himalaicus* Hara	Rhizome	Himalayan ginseng
P. pseudoginseng var. *major* [*P. japonicus* C. A. Meyer var. *major* (Burk.) C. Y. Wu et K. M. Feng] [*P. major* (Burk.) Ting]	Rhizome	Zhuzishen

II. HISTORY OF MODERN SCIENTIFIC RESEARCH ON GINSENG

The research on the principle of action in a modern scientific sense was first made by Garriques (1854) on American ginseng. He isolated an amorphous compound, $C_{32}H_{56}O_{14}$, and named it panaquilon. In 1889 Davydow (1889) obtained the same substance from the ginseng collected from Ussuri region in Far East Siberia. From Korean ginseng, Asahina and Taguchi (1906) reported the isolation of a saponin (m.p. 190°C), which was hydrolyzed into a sapogenin, m.p. > 270°C, and glucose. From 1915 to 1920, Kondo *et al.* (Kondo and Tanaka, 1915; Kondo and Yamaguchi, 1918; Kondo and Amano, 1920) studied the chemical principles of ginseng and isolated a saponin, m.p. 220°C, from which a crystaline sapogenin, $C_{27}H_{48}O_3$ (m.p. 242.5°C) and glucose were obtained.

In 1930 Kotake isolated a nonhaemolytic saponin named panaxin and a prosapogenin, α-panaxin, $C_{38}H_{66}O_{12}$, from which the chlorine-containing sapogenin $C_{30}H_{53}O_3Cl$ was formed on hydrolysis with concentrated HCl. At that time, no reliable evidence for the chemical structure of such a complex compound was obtained.

After some interruption of chemical investigations on ginseng, pharmacological studies on ginseng extracts were reported by Brekhman (1957) in the U.S.S.R. and Petkov (1959) in Bulgaria in the end of the

1950s. These works have drawn general interest of Western people to ginseng and stimulated the chemical investigations on its active principles.

Hörhammer, Wagner, and Lay (1961) studied the sapogenins of ginseng obtained by the hydrolysis of 50% ethanolic extracts with 7% H_2SO_4 and determined the presence of oleanolic acid. Wagner-Jauregg and Roth (1962) isolated a sapogenin, $C_{29}H_{50}O_3$, named panaxol, and suggested the presence of two or three hydroxyls and an ether linkage in the molecule. Elyakov *et al.* (1962) reported the isolation from ginseng of saponins designated panaxosides A–F, from which a sapogenin, panaxogenin B, was obtained. Lin (1961), a former associate of Kotake, later isolated from ginseng a sapogenin, ginsengenin, $C_{30}H_{50}O_3$, along with stigmasterol.

Fujita *et al.* (1962) and Shibata *et al.* (1962a, 1963a,b) reported the first of their series of works on ginseng saponins. The white and red ginseng prepared in Japan were extracted with methanol to obtain a fraction of faint yellowish crude saponin. On treatment of the saponin under mildly acidic conditions, a prosapogenin and glucose were obtained, and the former was hydrolyzed again to give a sapogenin and glucose.

III. CHEMICAL STRUCTURES OF SAPOGENINS OF GINSENG

The sapogenin panaxadiol (**1**) was investigated by the following chemical reactions and by mass spectrometry. Panaxadiol, $C_{30}H_{52}O_3$, $[\alpha]_D^{18.8}$ + 1.0 (in chloroform), gave a red-violet Liebermann–Burchard reaction, suggesting a triterpenoid structure. However, it does not belong to the ordinary oleanane type, because of negative tetranitromethane reaction and absence of ultraviolet (UV) absorption at 210 nm. Oxidation of panaxadiol monoacetate with CrO_3 yielded a monoketo compound, panaxanolone acetate (**2**) whose carbonyl group was eliminated by the Wolff–Kishner reaction to give **3**. On treatment with hydrochloric acid in acetic acid followed by the catalytic reduction, **3** was converted into isotirucallenyl acetate (**5**). Thus panaxadiol must be a triterpenoid with a tetracyclic structure. A hindered hydroxyl group in panaxadiol was suggested to be at C-12 based on the formation of an α,β-unsaturated ketonic compound (**6**) from **2** by the action of H_2SO_4 in acetic acid.

The presence of a trimethyltetrahydropyrane ring in the panaxadiol molecule was proved by the mass ionic fragments with m/z 127 (**7**) and m/z 341 (**8**). The molecular formula, $C_{30}H_{52}O_3$, of panaxadiol was also supported by the molecular ion peak (M^+) at 460.

The stereochemistry at C-12, C-13, C-17, and C-20 was established by [¹H]NMR spectral analysis (Shibata *et al.,* 1962b, 1963c).

Panaxadiol (**1**) was soon proven to be an artifact produced by the action of acid during the process of hydrolysis of native saponins of

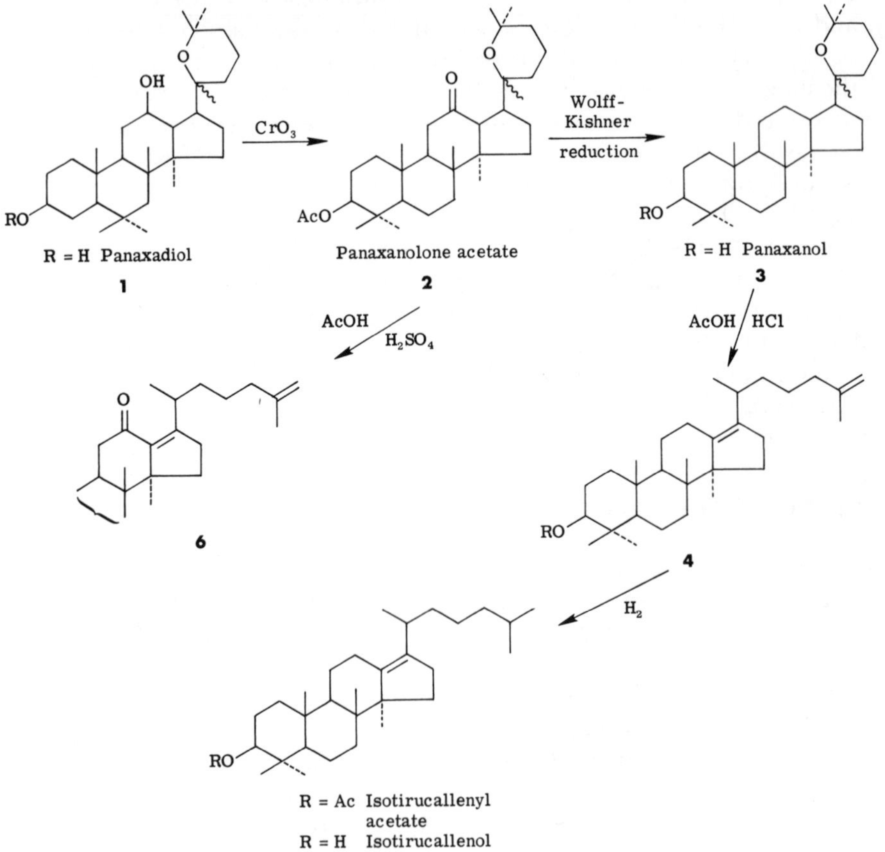

R = H Panaxadiol
1

Panaxanolone acetate
2

R = H Panaxanol
3

R = Ac Isotirucallenyl
                acetate
R = H Isotirucallenol
5

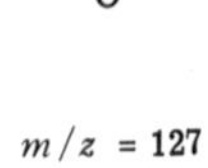

6

4

$m/z = 127$
7

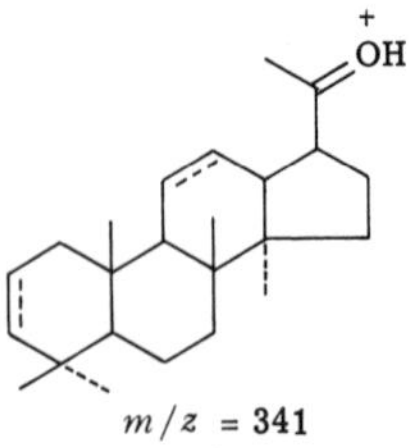

$m/z = 341$
8

Ginseng (Shibata *et al.*, 1963d, 1966a). On treatment with 0.7% H_2SO_4, the neutral saponin of ginseng yielded prosapogenin, which was converted into panaxadiol (**1**) on refluxing with 7% HCl, and into Kotake's chlorine-containing sapogenin, $C_{30}H_{53}O_3Cl$, on treatment with concentrated HCl at room temperature.

Dehydrochlorination of the chlorine-containing sapogenin with potassium *tert*-butoxide yielded protopanaxadiol, which possesses three hydroxyls (Ohsawa *et al.*, 1972). By infrared (IR) spectral analysis, two of three hydroxyls of protopanaxadiol were shown to be hydrogen-bonded and, one of them participates to form the tetrahydropyrane ring of panaxadiol on heating with dilute HCl.

On the other hand, dihydroprotopanaxadiol (**9**) was obtained by the acid hydrolysis of the hydrogenated saponin of ginseng.

These facts revealed that a tertiary hydroxyl attached to C-20 participates in ring closure with a double bond at C-24 or C-25 on acid treatment of protopanaxadiol.

The stereochemistry of protopanaxadiol was elucidated by the following reactions (Tanaka *et al.*, 1964, 1966). The chromic acid oxidation of the 3- and 12-hydroxyls of **9** afforded a 3,12-diketo compound (**10**), which on $LiAlH_4$ reduction yielded a 3β,12α-dihydroxy derivative (= 12-*epi*-dihydroprotopanaxadiol) (**11**). The 3-monoacetate (**12**) of this compound was oxidized with chromic acid to yield a 12-keto compound (**13**), which was submitted to the modified Wolff–Kishner reduction to form dammaranediol I monoacetate (**14**) (Mills and Werner, 1955; Mills, 1956). This series of reactions proved that protopanaxadiol is a deriva-

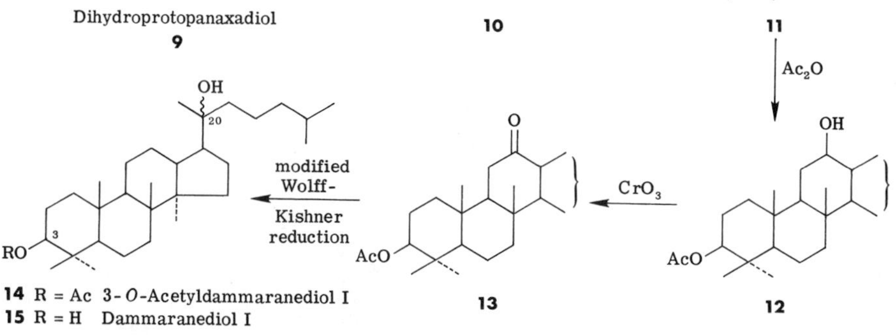

Dihydroprotopanaxadiol
9

10

11

modified
Wolff–
Kishner
reduction

13

12

14 R = Ac 3-*O*-Acetyldammaranediol I
15 R = H Dammaranediol I

tive of a dammarane-type tetracyclic triterpene, having two secondary β-hydroxyls at C-3 and C-12 and one tertiary hydroxyl at C-20.

On the other hand, the stereochemical correlation between dammarenediol II and betulafolienetriol, which was isolated from the leaves of white birch, was established by Fischer and Seiler (1959, 1961a,b).

Dammarenediols I and II were tetracyclic triterpenes isolated by Mills

(1956) from dammar resin, whose stereochemistry at C-20 had not been established at that point. The establishment of the stereochemistry at C-20 of panaxadiol was achieved by the following reactions and found to correlate to R- ($-$) - linalool (**24**) (M. Nagai *et al.*, 1966, 1967, 1971, 1972; Tanaka *et al.*, 1967). The determination of the stereochemistry at C-20 of panaxadiol as R by the above reaction led the establishment of stereochemistry at C-20 of dammarenediol I as R and dammarenediol II as S. At almost the same time, Biellmann (1966) assigned S configuration to C-20 of dipterocarpol (= 3-keto-3-deoxydammarenediol II). On treatment of betulafolienetriol [3-*epi*-12β-hydroxydammarenediol II (20-S)] with 60% ethanolic H_2SO_4,3-*epi*-12β-hydroxydammarenediol I (20R) was partly produced to form an equilibrium mixture in which the R form was dominant. Therefore, the hydrolysis of ginseng saponins with acidic

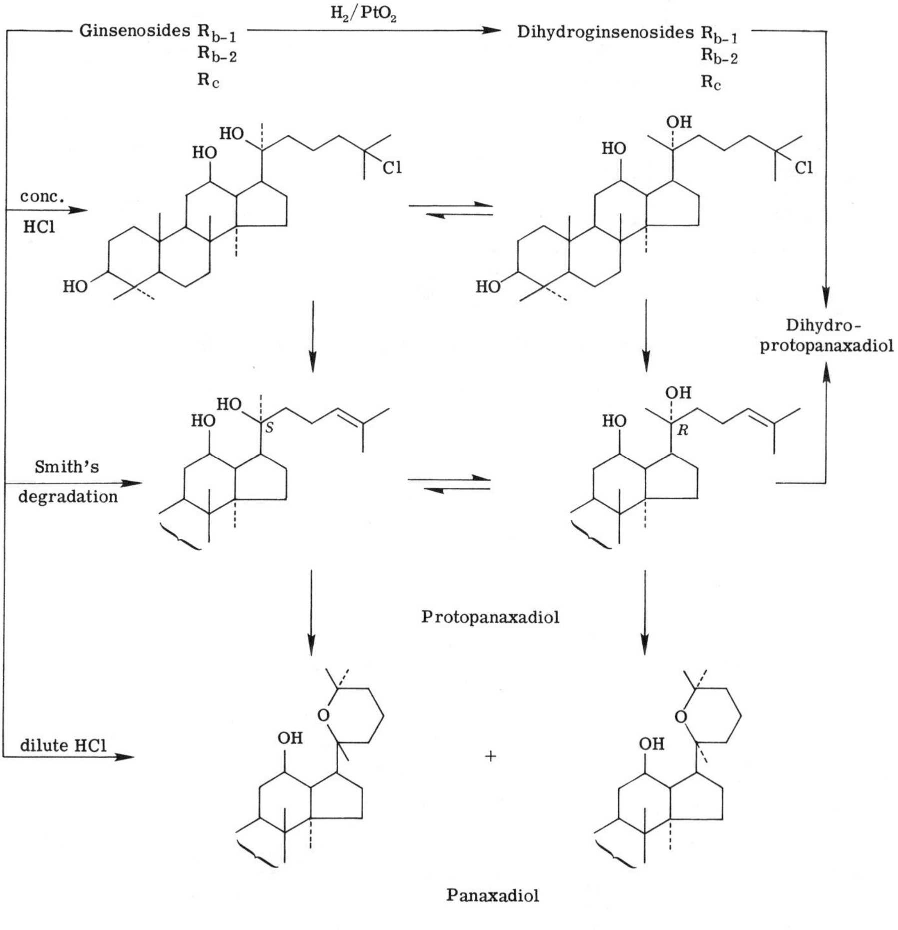

Ginsenosides R_{b-1} $\xrightarrow{\text{H}_2/\text{PtO}_2}$ Dihydroginsenosides R_{b-1}
R_{b-2} R_{b-2}
R_c R_c

conc. HCl

Dihydro-protopanaxadiol

Smith's degradation

Protopanaxadiol

dilute HCl

+

Panaxadiol

reagent would result in forming an equilibrium mixture of protopanaza-diols having S and R configurations at C-20.

To avoid this confusion, ginseng saponins, ginsenosides R_{b-1}, R_{b-2} and R_c, were treated under the Smith–de Mayo procedure with metaperio-date, followed by reduction with $NaBH_4$ and hydrolysis with $2N$ H_2SO_4 at pH 1.8–2.0 at room temperature, to yield 20-*epi*-protopanaxadiol [= 3-*epi*-betulafolienetriol (20S)], which must be a genuine sapogenin of ginseng.

IV. ISOLATION OF GINSENG SAPONINS

By thin-layer chromatography (TLC), the presence of more than 13 saponins in the butanol-soluble fraction of methanolic extracts of gin-seng was comparatively studied with those of American, Sanchi, and Japanese Chikusetsu ginseng. The ginseng saponins were designated as ginsenosides R_x (x = o, a-1, a-2, b-1, b-2, b-3, c, d, e, f, 20 gluco-f, g-1, g-2, h-1, . . .) by the sequence of R_f value of the spots on the TLC from the bottom to the top[14]. The refined TLC and the isolation procedure of ginsenosides R_x presented by Sanada *et al.* (1974a,b, 1978) are illus-trated as follows:

The physical data for ginsenosides R_x are shown in Table II.

On acid hydrolysis or via the Smith–de Mayo degradation, ginsenoside R_o afforded oleanolic acid as the sapogenin, while ginsenosides R_{a-1}–R_d yielded panaxadiol or (20S)-protopanaxadiol, and ginsenosides R_e–R_{h-1} yielded panaxatriol or (20S)-protopanaxatriol as the sapogenins.

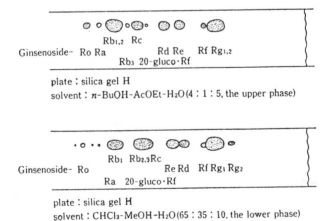

FIG. I Thin-layer chromatograms of Ginseng saponins. From Oura *et al.* (1981), with permission.

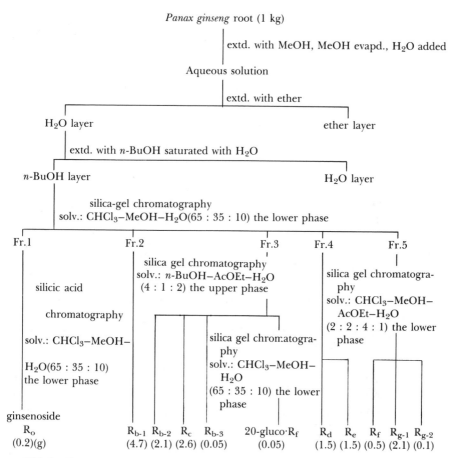

Panax ginseng root (1 kg)

extd. with MeOH, MeOH evapd., H₂O added

Aqueous solution

extd. with ether

H₂O layer | ether layer

extd. with *n*-BuOH saturated with H₂O

n-BuOH layer | H₂O layer

silica-gel chromatography
solv.: CHCl₃–MeOH–H₂O(65 : 35 : 10) the lower phase

Fr.1 | Fr.2 | Fr.3 | Fr.4 | Fr.5

silicic acid chromatography solv.: CHCl₃–MeOH–H₂O(65 : 35 : 10) the lower phase

silica gel chromatography solv.: *n*-BuOH–AcOEt–H₂O (4 : 1 : 2) the upper phase

silica gel chromatography solv.: CHCl₃–MeOH–AcOEt–H₂O (2 : 2 : 4 : 1) the lower phase

silica gel chromatography solv.: CHCl₃–MeOH–H₂O (65 : 35 : 10) the lower phase

ginsenoside Rₒ (0.2)(g)

Rᵦ₋₁ (4.7) | Rᵦ₋₂ (2.1) | R_c (2.6) | Rᵦ₋₃ (0.05) | 20-gluco·R_f (0.05) | R_d (1.5) | R_e (1.5) | R_f (0.5) | R_g₋₁ (2.1) | R_g₋₂ (0.1)

FIG. 2 Chromatographic separation of ginseng saponins. From Oura *et al.* (1981), with permission.

A. GINSENOSIDE Rₒ

Ginsenoside Rₒ is a minor component of ginseng saponins and is the only oleanane-type compound among them. It has been identified with chikusetsusaponin V isolated from Japanese Chikusetsuginseng. The structure was elucidated by the reactions illustrated below (Kondo *et al.*, 1970).

B. GINSENOSIDE R_g₋₁

Y. Nagai *et al.* (1971) reported the determination of ginsenoside R_g₋₁, one of the main components of ginseng saponins, as follows.

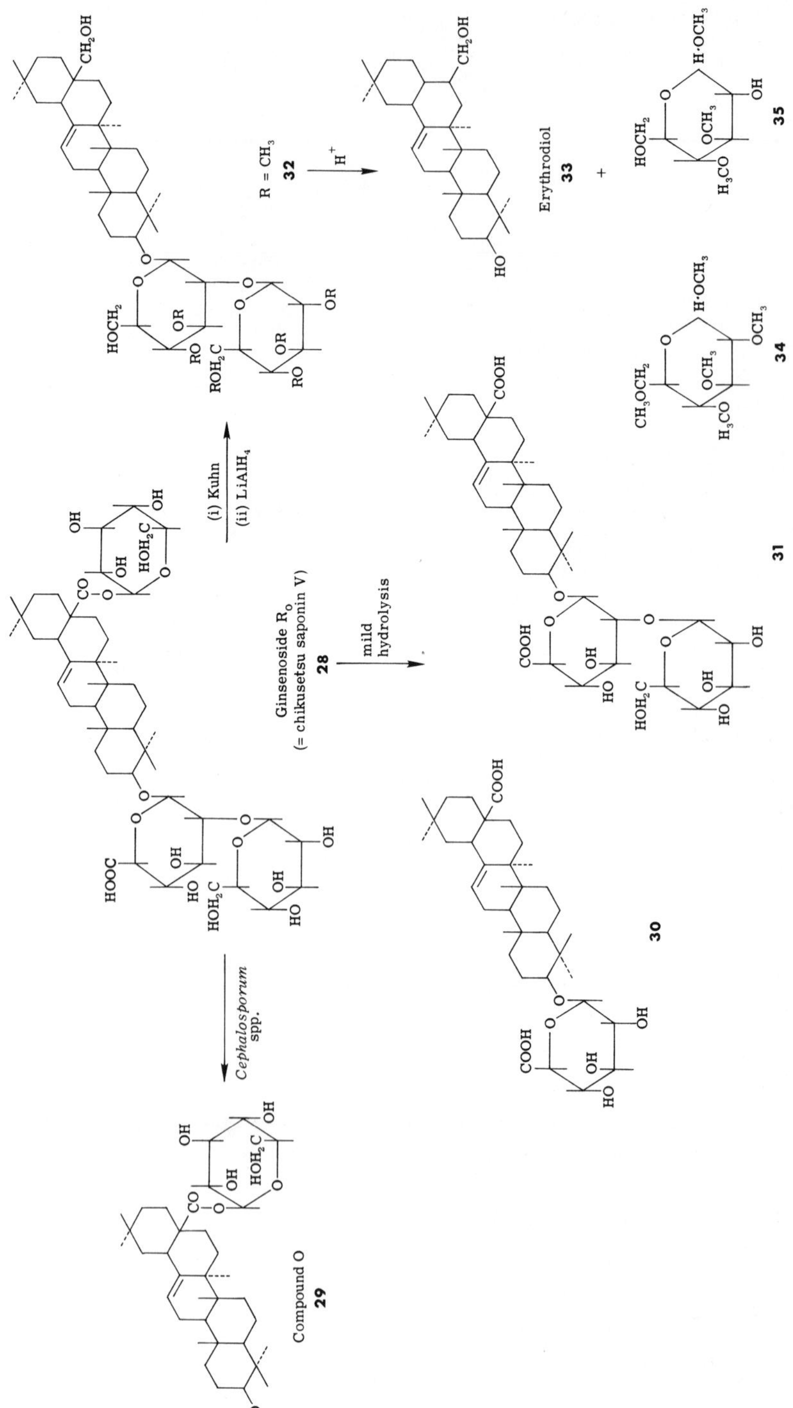

Ginsenoside R_0
(= chikusetsu saponin V)
28
(i) Kuhn
(ii) $LiAlH_4$
$R = CH_3$
32
H^+
Erythrodiol
33
34
35
mild hydrolysis
31
30
Cephalosporum spp.
Compound O
29

TABLE II

PHYSICAL PROPERTIES OF GINSENG SAPONINS (GINSENOSIDES R_x)[a]

Ginsenoside	Properties	m.p.(°C)	$[\alpha]_D$ (c in MeOH)	Formula	IR (KBr) cm^{-1}
Ro	Colourless needles (MeOH)	239–241	+15.33 (0.91)	$C_{48}H_{76}O_{19}$	3400 (OH) 1740 (COOR) 1728 (COOH)
Rb_1	White powder (EtOH–BuOH)	(197–198)	+12.42 (0.91)	$C_{54}H_{92}O_{23}$	3400 (OH) 1620 (C = C)
Rb_2	White powder (EtOH–BuOH)	(200–203)	+ 3.05 (0.98)	$C_{53}H_{90}O_{22}$	3400 (OH) 1620 (C = C)
Rb_3	White powder (iso-PrOH)	(193–195)	+19.39 (0.98)	$C_{53}H_{90}O_{22}$	3420 (OH) 1620 (C = C)
Rc	White powder (EtOH–BuOH)	(199–201)	+ 1.93 (1.03)	$C_{53}H_{90}O_{22}$	3420 (OH) 1620 (C = C)
20-gluco-Rf	White powder (iso-PrOH)	(182–184)	+21.00 (1.01)	$C_{48}H_{82}O_{19}$	3420 (OH) 1620 (C = C)
Rd	White powder (EtOH–AcOEt)	(206–209)	+19.39 (1.03)	$C_{48}H_{82}O_{18}$	3400 (OH) 1620 (C = C)
Re	Colourless needles (50% MeOH)	201–203	− 1.00 (1.00)	$C_{48}H_{82}O_{18}$	3380 (OH) 1620 (C = C)
Rf	White powder (acetone)	(197–198)	+ 6.99 (1.00)	$C_{42}H_{72}O_{14}$	3380 (OH) 1620 (C = C)
Rg_1[b]	White powder (BuOH–MeCOEt)	194–196.5	+32.0 (pyr.)	$C_{42}H_{72}O_{14}$	3400 (OH) 1620 (C = C)
Rg_2	Colourless needles (EtOH)	187–189	−13.0 (1.03)	$C_{42}H_{72}O_{13}$	3400 (OH) 1620 (C = C)

[a]From Oura et al. (1981), with permission.
[b]From Y. Nagai et al. (1971).

229

A crude saponin fraction (8.0 g) obtained from ginseng powder (2.0 kg) by the extraction with methanol was chromatographed over a silica-gel column by gradient elution using chloroform and chloroform–methanol (95 : 5–90 : 10) to isolate crude ginsenoside R_{g-1}, which was acetylated to decaacetate for purification. Ginsenoside R_{g-1}, $C_{42}H_{72}O_{14}$, was obtained in a pure form by deacetylation of the decaacetate using 5% methanolic KOH. On acid hydrolysis of ginsenoside R_{g-1}, panaxatriol [6α-hydroxy-(20R)-panaxadiol] and 2 moles of D-glucose were yielded, and on Smith–de Mayo degradation, (20S)-protopanaxatriol [= 6α-hydroxy-(20S)-protopanaxadiol] was afforded as the genuine sapogenin.

The β-linkage of D-glucose moieties was proved by the [¹H]NMR signals of anomeric protons of decamethyl ether of ginsenoside R_{g-1}.

On methanolysis of ginsenoside R_{g-1}, decamethyl ether, methyl 2,3,4,6-tetramethylglucopyranoside was obtained as a single carbohydrate product. This result revealed that D-glucosyl moieties attached individually to two out of four hydroxyls of (20-S)-protopanaxatriol.

The decamethyl ether of dihydroginsenoside R_{g-1} was hydrolyzed with concentrated HCl to obtain di-O-methyl-20-dihydroprotopanaxatriol, which was submitted to Jones oxidation to prepare a di-O-methylketonic compound, whose IR spectrum revealed the presence of 12-O-methyl-20-hydroxy structure. The di-O-methyl-ketonic compound showed an optical rotatory dispersion (ORD) curve having (−) Cotton effect, which resembled that of the 6-keto compound derived from panaxatriol and differed from that of 3β-acetoxy-(20R)-dammaran-12-one. Moreover, the NMR chemical shift and coupling pattern of the di-O-methyl-ketonic compound resembled those of the 6-ketonic derivative, leading to the conclusion that ginsenoside R_{g-1} is 6,20-di-O-β-D-glucosyl-(20S)-protopanaxatriol.

On the other hand, Elyakov et al. (1964) proposed the presence of six saponins named panaxoside A–F in ginseng, and the resemblance of panaxoside A and ginsenoside R_{g-1} was suggested by the comparison of their physical properties, R_f values from TLC, and Debye–Scheller's X-ray patterns. However, Elyakov et al. (1965, 1968) proposed a different structure for panaxoside A than that of ginsenoside R_{g-1}. In 1978, after X-ray crystallography of the prosapogenin of panaxoside A, Elyakov withdrew his earlier structure of panaxoside A and agreed on structure of ginsenoside R_{g-1} (Iljin et al., 1978).

C. PARTIAL DEGLYCOSYLATION OF GINSENOSIDE R_x

On treatment with 50% acetic acid, the saponins that possess glycosyl group(s) at C-20 of (20S)-protopanaxadiol or -triol afford a mixture of (20S)- and (20R) prosapogenins in detaching sugar group(s) from there under simultaneous partial epimerization (Nagai *et al.*, 1972; Shibata *et al.*, 1966b). Thus by this procedure ginsenosides 20-gluco-R_f and R_e yielded ginsenosides R_f and R_{g-2}, respectively, in liberating 1 mole each of D-glucose, whereas ginsenosides R_f and R_{g-2} gave no sugar fragment, indicating the absence of glycosyl moiety at C-20. Under the same condition, arabinopyranosyl linkage was not cleaved as indicated in the case of ginsenoside R_{b-2}, giving prosapogenin (protopanaxadiol 3-*O*-β-D-glucopyranosyl (1 → 2)-β-D-glucopyranoside) and a biose consisting of D-glucose and L-arabinose, whereas ginsenoside R_c yielded D-glucose and L-arabinose, the latter of which must be a furanose.

D. METHYLATION OF GINSENOSIDES R_x

Methylation of ginsenosides was performed for determination of position of glycosylation and the linkage of sugar moieties. Hakomori's method was generally applied for this purpose, except for ginsenoside R_o, which possesses an ester linkage of sugar moiety and for which Kuhn's method was adopted.

O-Methylginsenosides
R_{b-1}, R_{b-2}, R_c, and R_d

H_2/PtO_2

Dihydro-*O*-methylginsenosides
R_{b-1}, R_{b-2}, R_c, and R_d

CrO_3

Monoketo-dihydro-*O*-methylginsenosides
R_{b-1}, R_{b-2}, R_c, and R_d
IR ν_{max} (Nujol) cm^{-1}:OH (nil),
1710–1714 (6-membered C=O)

7% HCl-MeOH

3-Hydroxy-12-oxo-dammarane (13,17)
UV λ_{max}^{EtOH} nm (log ϵ): 265 (3.39)
UV $\lambda_{max}^{cyclohexane}$ nm (log ϵ): 257 (3.42)

The methylation of ginsenosides R_x (except R_e) showed the presence of a resistant hydroxyl group in the sapogenins by the IR spectral analysis giving free OH absorption at 3400–3438 cm^{-1} for the methylated derivatives of R_{b-1}, R_{b-2}, R_{b-3}, R_c, 20-gluco-R_f, and R_d, and at 3360–3380 cm^{-1} for those derived from R_f and R_{g-2}. It has been proved by the chemical reactions that the resistant hydroxyl is located at C-12, or C-20, if any, as follows:

E. THE STRUCTURES OF GINSENOSIDES R_x

The sugar moieties of the ginsenosides that have sapogenin (20S)-protopanaxadiol are located at the C-3 OH and C-20 OH, while those of the ginsenosides that have sapogenin (20S)-protopanaxatriol are attached at the C-6 OH and C-20 OH (R_e, R_{g-1}, 20-gluco-R_f) or C-6 OH (R_f, R_{g-2}, R_{h-1}) only.

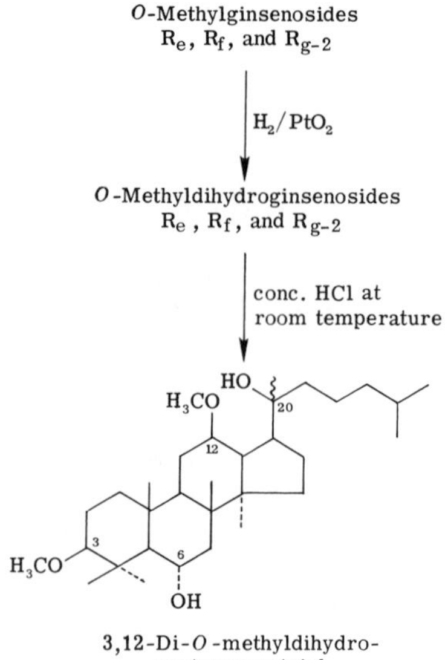

O-Methylginsenosides
R_e, R_f, and R_{g-2}

H_2/PtO_2

O-Methyldihydroginsenosides
R_e, R_f, and R_{g-2}

conc. HCl at
room temperature

3,12-Di-O-methyldihydro-
protopanaxatriol

The structural determination of sugar moieties was performed by the methanolysis of O-methylated ginsenoside R_x. From O-methylated ginsenoside R_{b-1}, methyl 2,3,4,6-tetra-O-methylglucopyranoside, methyl 3,4,6-tri-O-methylglucopyranoside, and methyl 2,3,4,-tri-O-methylglucopyranoside were obtained, indicating that glucopyranosyl (1 → 2)-glucosidic and glucopyranosyl (1 → 6)-glucosidic moieties are attached individually to the sapogenin, (20S)-protopanaxadiol. In this way, the sugar moieties of all other ginsenosides were determined.

The configurations of all the monosaccharide components of the sugar moieties of ginsenosides R_x were determined by the Klyne rule (1950) and by [^{1}H]NMR chemical shifts and coupling constants of anomeric protons (Table III). Thus the structures of all the saponins, ginsenosides R_x, so far isolated from Ginseng have been established as formulated in Table IV.

Kaku *et al.* (1977) reported the isolation of ginsenoside R_{g-3}, (20-*R*) protopanaxadiol 3-*O*-β-D-glucopyranosyl (1 → 2)-β-D-glucopyranoside, from ginseng, but this compound might be an artifact, since all other ginsenosides R_x possess (20*S*) configuration.

F. THE SAPONINS OF AMERICAN GINSENG AND SAN-CHI GINSENG

The presence of ginsenosides R_x in American ginseng, the root of *Panax quinquefolium* L., and in San-Chi ginseng, the root of *P. pseudoginseng* var. *notoginseng* Burk, was demonstrated by Ando *et al.* (1971) and Otsuka *et al.* (1977) with thin-layer chromatography (TLC) and droplet countercurrent chromatography (DCC) in comparison with those of Korean ginseng.

Kim and Staba (1973) first isolated several saponins from American ginseng and named panquilins A, B, C, D, E-1, E-2, E-3, G-1, and G-2, which were later identified with individually corresponding ginsenosides R_x (Fig. 3).

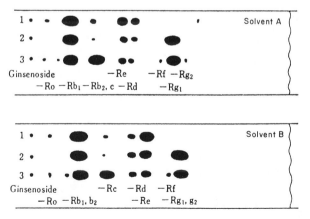

FIG. 3 Thin-layer chromatograms of saponins of American ginseng and San-chi ginseng. 1, American ginseng; 2, San-chi ginseng; 3, ginseng. Plate, Kieselgel H. Solvents: A, $CHCl_3$–MeOH–H_2O (13 : 7 : 2) lower phase; B, BuOH–AcOEt–H_2O (4 : 1 : 5) upper phase. Colour reagent, 10% H_2SO_4. From Oura *et al.* (1981), with permission.

TABLE III

DETERMINATION OF STEREOCHEMISTRY OF ANOMERIC PROTONS OF THE SUGAR PORTIONS OF GINSENOSIDES R_x

The anomeric configurations of rhamnose in ginsenosides R_e and R_{g-2} and arabinose in ginsenoside R_c determined by the Klyne rule:

$[M]_D$ (ginsenoside R_e) − $[M]_D$ (ginsenoside R_{g-1}) = −240.3°

  Methyl α-L-rhamnopyranoside, $[M]_D$ −111°
  Methyl β-L-rhamnopyranoside, $[M]_D$ +170°

$[M]_D$ (ginsenoside R_c) − $[M]_D$ (ginsenoside R_d) = −162.8°

  Methyl α-L-arabinofuranoside, $[M]_D$ −226°
  Methyl β-L-arabinofuranoside, $[M]_D$ +208°

The anomeric configurations of glucose, xylose, and arabinopyranose in ginsenosides R_x determined by [1H]NMR:

O-Methylated ginsenoside	[1H]NMR chemical shifts (δ) and coupling constants (Hz) of anomeric protons		
R_o	4.40 (1H, d, J = 7)	4.55 (1H, d, J = 7)	5.30–5.45 (1H, overlapped C = CH)
R_{b-1}	4.17 (1H, d, J = 7)	4.26 (1H, d, J = 7)	4.38 (1H, d, J = 7) 4.57 (1H, d, J = 7)
R_{b-2}	4.23 (1H, d, J = 7)	4.30 (1H, d, J = 7)	4.40 (1H, d, J = 7) 4.65 (1H, d, J = 7)
R_{b-3}	4.25 (1H, d, J = 7)	4.29 (1H, d, J = 7)	4.45 (1H, d, J = 7) 4.66 (1H, d, J = 7)
R_c	4.27 (1H, d, J = 8)	4.47 (1H, d, J = 7)	4.67 (1H, d, J = 7) 5.00 (1H, s,)
20-Gluco-R_f	4.34 (1H, d, J = 7)	4.48 (1H, d, J = 7)	4.65 (1H, d, J = 7)
R_d	4.24 (1H, d, J = 7)	4.45 (1H, d, J = 7)	4.65 (1H, d, J = 7)
R_e	4.40 (1H, d, J = 7)	4.60 (1H, d, J = 7)	5.31 (1H, br. s)
R_f	4.39 (1H, d, J = 7)	4.77 (1H, d, J = 7)	
R_{g-2}	4.60 (1H, d, J = 7)	5.31 (1H, br. s)	

TABLE IV

SAPONINS AND SAPOGENINS OF *PANAX GINSENG* ROOT

Compound		R^1	R^2	R^3
(20*S*)-Protopanaxadiol		H	H	H
(20*S*)-Protopanaxatriol		H	OH	H
Ginsenoside	R_{a-1}	Glc—^{2}Glc-	H	Xyl—^{4}Ara (p)—^{6}Glc-
	R_{a-2}	Glc—^{2}Glc-	H	Xyl—^{2}Ara (f)—^{6}Glc-
	R_{b-1}	Glc—^{2}Glc-	H	Glc—^{6}Glc
	R_{b-2}	Glc—^{2}Glc-	H	Ara (p)—^{6}Glc-
	R_{b-3}	Glc—^{2}Glc	H	Xyl—^{6}Glc-
	R_c	Glc—^{2}Glc	H	Ara (f)—^{6}Glc-
	R_d	Glc—^{2}Glc-	H	Glc-
	R_e	H	Rha—^{2}Glc-O-	Glc-
	R_f	H	Glc—^{2}Glc-O-	H
	R_{g-1}	H	Glc-O-	Glc-
	R_{g-2}	H	Rha—^{2}Glc-O-	H
	20-Gluco-R_f	H	Glc—^{2}Glc-O-	Glc
	R_{h-1}	H	Glc-O-	H

From American Ginseng cultivated in Japan, Sanada and Shoji (1978b) isolated ginsenosides R_o (0.07%), R_{b-1} (1.57%), R_{b-2} (0.02%), R_c (0.22%), R_d (0.77%), and R_e (0.89%) by column chromatography of the butanol-soluble fraction (6.2%).

The presence of ginsenoside R_{g-2} and some other nonidentified saponins whose R_f values were close to R_{b-1} were shown by TLC.

Besso *et al.* (1982b) identified from American ginseng the ginsenosides R_{b-3}, R_{g-1}, R_{g-2}, and F_2, and pseudoginsenoside-F_{11}, gypenoside XVII, and quinquenoside R_1, in addition to the above saponins.

From San-chi ginseng, Zhou *et al.* (1981) and Besso *et al.* (1981) isolated butanol-soluble crude saponins in higher yield (8.1%), from which ginsenoside R_{b-1} (1.62%), R_d (0.32%), R_e (0.51%), and R_{g-1} (2.07%) were separated. Zhou *et al.* (1981) and Matsuura *et al.* (1983) reported the isolation of ginsenosides R_{g-2} and R_{h-1}, gyenoside XVII, and some new minor saponins from Sanchi ginseng and determined their structures as follows:

Notoginsenoside R_1 (0.16%): 20S)-Protopanaxatriol 6-O-[β-D-xylopyranosyl (1 → 2)-β-D-glucopyranosido]-20-O-β-D-glucopyranoside
Notoginsenoside R_2 (0.04%): 20S)-Protopanaxatriol 6-O-β-D-xylopyranosyl (1 → 2)-β-D-glucopyranoside
Notoginsenoside R_3 (0.007%): 20S)-Protopanaxatriol 6-O-β-D-glucopyranosido-20-O-β-D-glucopyranosyl (1 → 6)-β-D-gluopyranoside
Notoginsenoside R_4 (0.028%): 20S)-Protopanaxadiol-3-O-[β-D-glucopyranosyl (1 → 2)-β-D -glucopyranosido]-20-O-β-D-xylopyranosyl (1 → 6)-β-D-glucopyranosyl (1 → 6)-β-D-glucopyranoside
Notoginsenoside R_5 (0.02%)

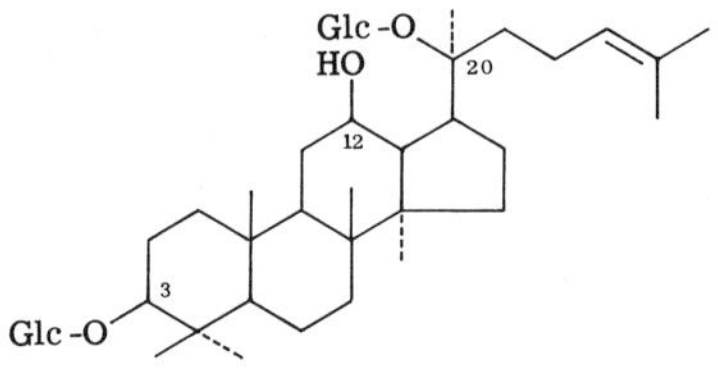

Ginsenoside F_2

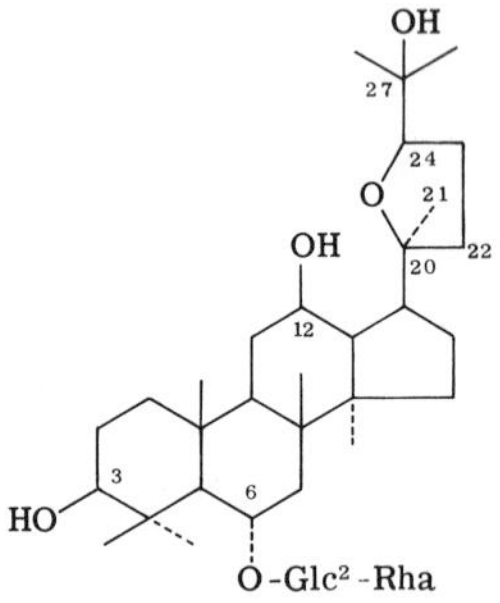

Pseudoginsenoside F_{11}

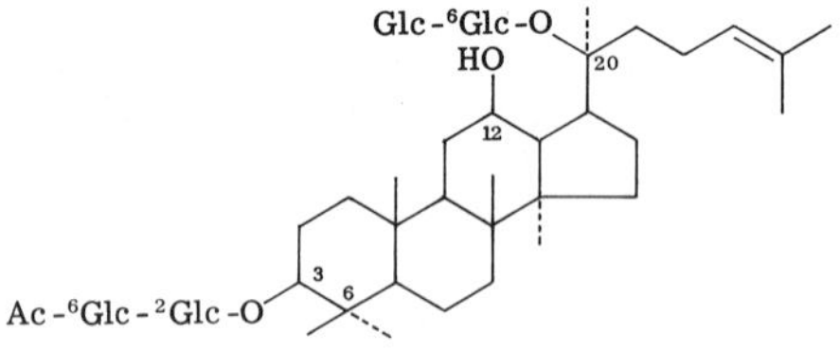

Quinquenoside R_1

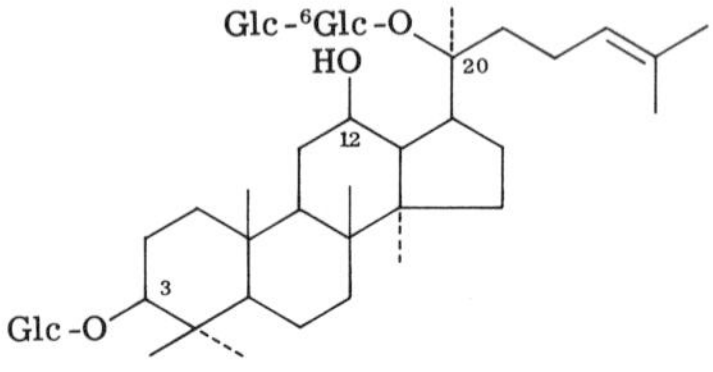

Gypenoside XVII

Gypenoside XVII is (20S) protopanaxadiol-3-O-β-D-glucopyrano-sido-20-O-[β-D-glucopyranosyl (1 → 6)-β-D-glucopyranoside], which was first isolated by Takemoto *et al.* (1983) from the cucurbitaceous plant *Gynostemma pentaphyllum*, along with number of dammarane-type oligoglycosides (Table V).

G. THE SAPONINS OF JAPANESE CHIKUSETSU GINSENG

The Chikusetsu ginseng is the rhizome of *Panax pseudoginseng* subsp. *japonicus* (= *P. japonicus*), which grows wild in Japan. The aerial part of this plant is very similar to *P. ginseng*, but the ground part is mainly rhizone with lateral roots. In Japan, Chikusetsu ginseng has been used as a substitute for Korean ginseng in the prescriptions of traditional medi-

TABLE V

SAPONINS AND SAPOGENINS OF SAN-CHI GINSENG

Compound		R^1	R^2	R^3
(20S)-Protopanaxadiol		H	H	H
(20S)-Protopanaxatriol		H	OH	H
Ginsenoside	$R_{b\text{-}1}$	Glc—^{2}Glc-	H	Glc—^{6}Glc-
	R_d	Glc—^{2}Glc-	H	Glc-
	R_e	H	Rha—^{2}Glc-O-	Glc-
	$R_{g\text{-}1}$	H	Glc-O-	Glc-
	$R_{g\text{-}2}$	H	Rha—^{2}Glc-O-	H
	$R_{h\text{-}1}$	H	Glc-O-	H
Notoginsenoside	R_1	H	Xyl—^{2}Glc-O-	Glc-
	R_2	H	Xyl—Glc-O-	H
	R_3	H	Glc-O-	Glc—^{6}Glc-
	R_4	Glc—^{2}Glc-	H	Xyl—^{6}Glc—^{6}Glc-
Gypenoside XVII		Glc-	H	Glc—^{6}Glc-

cine. The areas of efficacy of Chikusetsu ginseng are stomachic, expectorant, and antipyretic, without restoration of Yang.

The description of ginseng in *Shang-Han-Lun,* the oldest Chinese medical text, suggested, in contrast to *Shin Nung Pen T'sao Ching,* that ginseng is used to resolve the syndrome of stiffness at the pit of the stomach, and mentioned nothing about the restoration of Yang. Yoshimasu Todo (1701–1773), an outstanding medical scholar in the middle Edo Era in Japan, emphasized this point in his book *Yakucho.* From his description of ginseng, it is obvious that he employed clinically Japanese Chikusetsu ginseng in place of Korean ginseng without noting their botanical nonidentity.

The first modern scientific investigation of Chikusetsu ginseng was made by Murayama and Itagaki (1923) and Murayama and Tanaka (1927), who isolated from Chikusetsu ginseng a saponin, which Aoyama (1929, 1930), Kotake (1930), Kotake *et al.* (1930); Kotake and Kimoto, (1932), and Kitasato and Sone (1932) later studied. Oleanolic acid was isolated as the sapogenin.

238   S. SHIBATA *ET AL.*

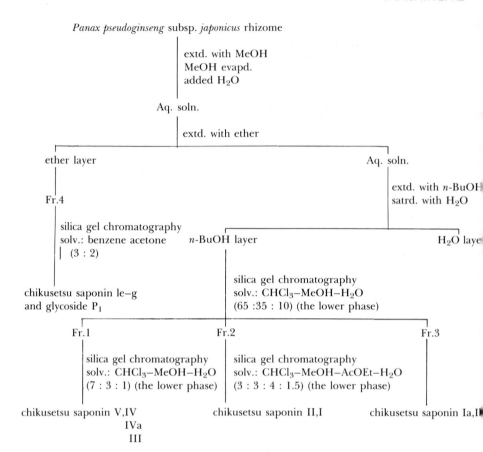

FIG. 4   Separation of Chikusetsu saponins. From Oura *et al.* (1981), with permission.

In 1962, Fujita *et al.* demonstrated the presence of panaxadiol as well as oleanolic acid in the hydrolysate of saponins of Chikusetsu ginseng. From the saponins of aereal part of Japanese *Panax pseudoginseng,* panaxatriol was isolated as well as the sapogenins already mentioned.

Since 1968, Kondo *et al.* (1962, 1969, 1970) developed the studies of the saponins of Chikusetsu ginseng further. The procedure adopted for the separation of Chikusetsu saponins is shown in Figs. 4 and 5.

The total yield of crude saponins of Chikusetsu ginseng is ~ 24%, from which Chikusetsu saponins V (= ginsenoside R_o) (5.35%), IV (0.43%), III (1.17%), and other minor saponins were isolated.

In contrast to Korean ginseng, in Chikusetsu ginseng oleanolic acid-type saponins are dominant to dammarane-type saponins (Table VI).

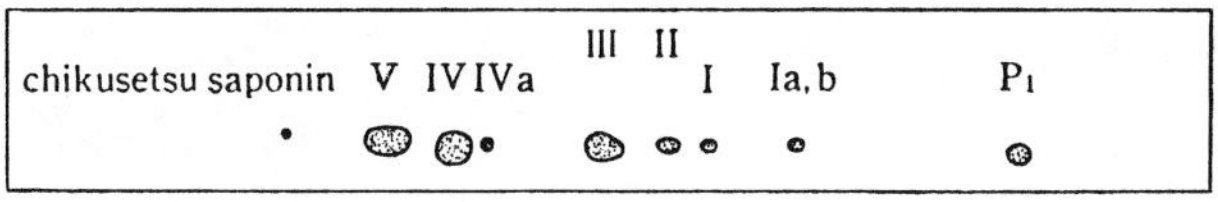

FIG. 5 Thin-layer chromatogram of saponins of Chikusetsu ginseng. Plate, silica gel II. Solvent, $CHCl_3$–MeOH–H (65 : 35 : 10 the lower phase). From Oura *et al.* (1981), with permission.

TABLE VI

SAPONINS AND SAPOGENINS OF CHIKUSETSU GINSENG

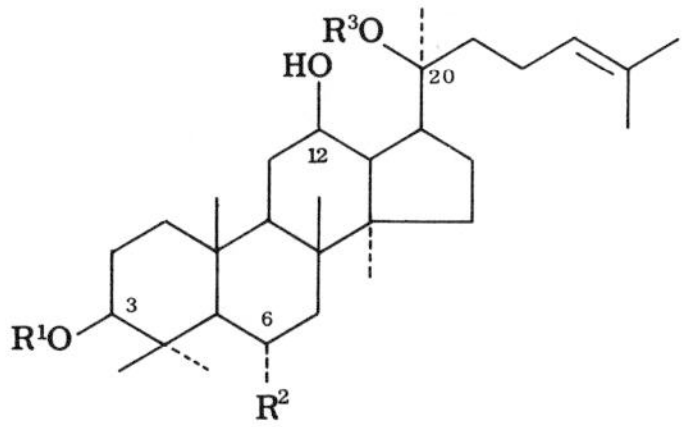

Compound	R^1	R^2	R^3
(20S)-Protopanaxadiol	H	H	H
(20S)-Protopanaxatriol	H	OH	H
Chikusetsu saponin Ia	Xyl—^{6}Glc-	H	H
Chikusetsu saponin III	Xyl—$^6_2$Glc- | Glc	H	H
Chikusetsu saponin I (= ginsenoside $R_{g\text{-}2}$)	H	Rha—^{2}Glc-O-	H

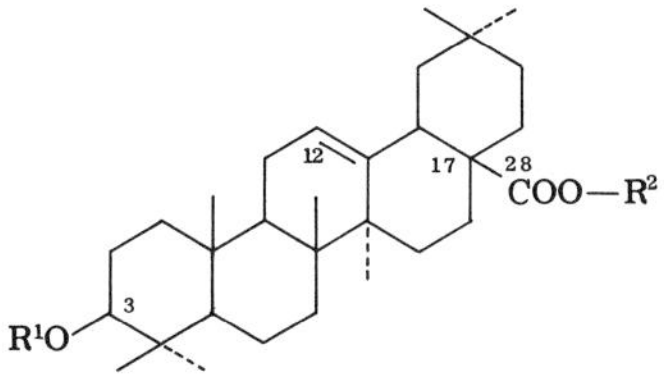

Compound	R^1	R^2
Chikusetsu saponin I$_b$	Ara (f)—$^4_6$GlcA- / Glc	H
Chikusetsu saponin IVa	GlcA-	Glc-
Chikusetsu saponin IV	Ara (f)—^{4}GlcA-	Glc-
Chikusetsu saponin V	Glc—^{2}GlcA	Glc-

TABLE VII

PHYSICAL PROPERTIES OF SAPONINS OF JAPANESE CHIKUSETSU GINSENG[a]

Chikusetsu saponin	Properties	m.p.(°C)	$[\alpha]_D$ (c in MeOH)	Formula	IR cm^{-1}
I	Colorless needles (EtOH)	189–191	−13.0 (1.03)	$C_{42}H_{72}O_{13}$	3400 (OH) 1630 (C = C) (KBr)
Ia	Colorless needles (CHCl$_3$–MeOH–AcOEt–H$_2$O, 3:3:4:1.5)	194	− 3.51 (1.42)	$C_{41}H_{70}O_{12}$	3400 (OH) 1630 (C = C) (KBr)
Ib	Colorless prisms (CH$_3$COCH$_3$)	177	−17.98 (1.02)	$C_{47}H_{74}O_{18}$	3400 (OH) 1730 (COOH, COOR) (KBr)
III	Colorless needles (MeOH–AcOEt satd. H$_2$O)	196–197	+ 1.5 (1.69)	$C_{47}H_{80}O_{17}$	3200–3500 (OH) 1630 (C = C) (Nujol)
IV	Colorless prisms (BuOH satd. with H$_2$O–AcOEt)	>235 (decomp.)	− 9.7 (0.92 in pyridine)	$C_{47}H_{74}O_{18}$	3200–3500 (OH) 1740 (COOR) (Nujol)
IVa	Colorless needles (BuOH satd. with H$_2$O–AcOEt)	221	+21.07 (0.85)	$C_{42}H_{66}O_{14}$	3400 (OH) 1730 (COOR) (KBr)
V	White powder (MeOH–AcOEt)	(240–241)	+ 2.85 (2.01)	$C_{48}H_{76}O_{19}$	3300 (OH) 1742 (COOR) 1725 (COOH) (Nujol)
Glycoside P$_1$	White powder (CHCl$_3$–MeOH, 1:1)	(162)	−45.37 (1.41 in CHCl$_3$)	$C_{50}H_{88}O_7$ $C_{51}H_{90}O_7$	3360 (OH) 1733 (COOR) (CCl$_4$)

[a]From Oura et al. (1981), with permission.

The physical properties of saponins of Chikusetsu ginseng are tabulated in Table VII.

H. THE SAPONINS OF HIMALAYAN GINSENG

The rhizome of *Panax pseudoginseng* subsp. *himalaicus,* collected by Tanaka at Khosa (1800 m) in western Bhutan, was extracted by Kondo *et al.* (1973, 1975) with methanol and fractionated with butanol to isolate saponins. The TLC of the saponin fractions of Himalayan ginseng and that of Chikusetsu ginseng were compared to suggest that Himalayan ginseng is intermediate in saponin production between Korean ginseng and Japanese Chikusetsu ginseng.

I. THE SAPONINS IN "RED GINSENG"

As mentioned previously, ginseng in the market is classified into white ginseng and red ginseng, which are the results of different ways of processing. White ginseng is prepared by drying after removal of peel or without peeling, while red ginseng is processed by steaming without peeling.

Kitagawa *et al.* (1981) reported that red ginseng contains panaxatriol, $(20R)$ ginsenoside R_{g-2}, ginsenoside R_{g-3}, $(20S)$- and $(20R)$-ginsenoside R_{h-2}, and all the saponins so far isolated from white ginseng. Besso *et al.* (1981) found in the red ginseng extracts some new additional saponins: quinquenoside R_1, notoginsenoside R_1, and ginsenosides R_{S-1} and R_{S-2} (see Fig. 6, Table VIII, and below).

Ginsenoside R_{S-1}: Ara (p)-^{6}Glc— = R
Ginsenoside R_{S-2}: Ara (f)- ^{6}Glc— = R

J. THE SAPONINS OF THE RHIZOMES OF *PANAX PSEUDOGINSENG* GROWING IN YUNNAN, CHINA

There are several *Panax* spp. growing in Yunnan, China, that are morphologically similar to *P. pseudoginseng* subsp. *japonicus* and *P. pseudoginseng* subsp. *himalaicus* in having well-developed rhizomes.

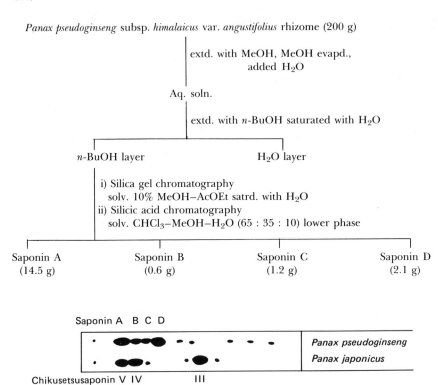

Panax pseudoginseng subsp. *himalaicus* var. *angustifolius* rhizome (200 g)

extd. with MeOH, MeOH evapd.,
added H_2O

Aq. soln.

extd. with *n*-BuOH saturated with H_2O

n-BuOH layer H_2O layer

i) Silica gel chromatography
   solv. 10% MeOH–AcOEt satrd. with H_2O
ii) Silicic acid chromatography
   solv. $CHCl_3$–MeOH–H_2O (65 : 35 : 10) lower phase

| Saponin A | Saponin B | Saponin C | Saponin D |
| (14.5 g) | (0.6 g) | (1.2 g) | (2.1 g) |

Saponin A B C D

Panax pseudoginseng
Panax japonicus

Chikusetsusaponin V IV   III

plate : Silica gel H
solvent : $CHCl_3$–MeOH–H_2O (65 : 35 : 10) lower phase

FIG. 6 Separation of saponins of red ginseng. From Oura *et al.* (1981), with permission.

Morita *et al.* (1983) isolated from Yunnan-chikusetsu ginseng (zhu-jieshen) the chikusetsusaponins IV (3.4%), IV a (2.8%), and V (3.1%), as the oleanolic acid saponins, and ginsenosides R_d (0.04%), R_e (0.12%), R_{g-1} (0.15%), R_{g-2} (0.05%) and notoginsenoside R_2 (0.02%) as the dammarane-type saponins, plus a (24R)-ocotillol-type saponin, pseudoginsenoside F_{11} (0.24%). The presence of ginsenoside R_d and ocotillol-type saponins is characteristic in Yunnan zhujieshen, distingusihing it from Japanese Chikusetsu ginseng and Himalayan ginseng.

From the rhizome of *Panax pseudoginseng* subsp. *japonicus* var. *major* (= *P. japonicus* var. *major*), Morita *et al.* (1982) obtained the main oleanolic acid saponins, chikusetsu saponins IVa (0.19%) and V (0.95%), the dammarane-type saponins ginsenosides R_d (0.67%), 20-gluco-R_f (0.01%), and notoginsenoside R_2 (0.03%), and the (24S)-ocotillol-type saponins majonosides R_1 (0.07%) and R_2 (0.11%).

TABLE VIII

PHYSICAL PROPERTIES OF SAPONINS OF HIMALAYAN GINSENG[a]

Saponin[a]	Properties	m.p.(°C)	$[\alpha]_D$ (c in solv.)	Formula	IR (Nujol) cm^{-1}
A	White powder (MeOH–AcOEt)	241	+14.5 (0.76 MeOH)	$C_{48}H_{70}O_{19}$	3400 (broad OH) 1740 (COOR) 1728 (COOH)
B	White powder (MeOH–AcOEt)	235 (decomp.)	−9.10 (0.66 Pyr.)	$C_{47}H_{74}O_{18}$	3200–3500 (OH) 1750 (sh. COOR) 1720 (COOH)
C	White powder (BuOH satd. with H₂O–AcOEt)	216 (decomp.)	+15.8 (0.62 MeOH)	$C_{42}H_{66}O_{14}$	3400 (OH) 1750 (sh. COOR) 1730 (broad COOH)
D	White powder (EtOH)	189	+15.78 (1.13 MeOH)	$C_{54}H_{92}O_{23}$	3400 (OH) 1620 (C = C)

[a]From Oura et al. (1981), with permission.
[b]Structures:

Saponin A (= Chikusetsu saponin V = Ginsenoside R₀)

Glc–²GlcA–O

Saponin B (= Chikusetsu saponin IV)

Ara(f)–⁴GlcA–O

Saponin C (= Chikusetsu saponin III)

GlcA–O

Saponin D (= Ginsenoside R_b-1)

Glc–²Glc–O

Glc–⁶Glc–O
OH

Majonoside R_1: Glc-^{2}Glc— = R
Majonoside R_2: Xyl-^{2}Glc— = R

V. THE SAPONINS OF THE AERIAL PARTS OF *PANAX* PLANTS

The flowers and leaves of *Panax ginseng* C. A. Meyer have been used in East Asia as a tea drink and as a bath salt, but nothing has been known about the chemical principles of those parts of the plant.

The aerial part of the plant contains chlorophyll, carotenoids, and phenolics other than saponins, so that the methanol extracts are suspended with water and extracted first with ether to remove such lipophilic constituents. The saponin fraction is obtained by shaking the aqueous layer with *n*-butanol, followed by column chromatography over alumina or polyamide in separating it from the phenolic glycosides.

The purification of saponins is achieved by silica-gel column chromatography or high-performance liquid chromatography (HPLC).

A. *PANAX GINSENG*

The flavonoid of the leaves of *Panax ginseng* was studied by Komatsu *et al.* (1966); Komatsu and Tomimori, (1966), and the saponins were preliminarily studied by Shibata *et al.* (1971) with TLC.

Yahara *et al.* (1976a, 1979) isolated from the leaves known saponins (ginsenosides R_{b-1}, R_{b-2}, R_c, R_d, R_e, and R_{g-1} in yields of 0.1, 0.4, 0.2, 1.5, 1.5, and 1.5%, respectively), and new saponins (ginsenosides; F_1, F_2, and F_3 in yields of 0.4, 0.2, and 0.2%, respectively). The structures of the new saponins are as follows.

The leaves of *Panax ginseng* contain the protopanaxatriol-type saponins in higher content than does the root.

From the flower buds of *Panax ginseng,* which are usually picked be-

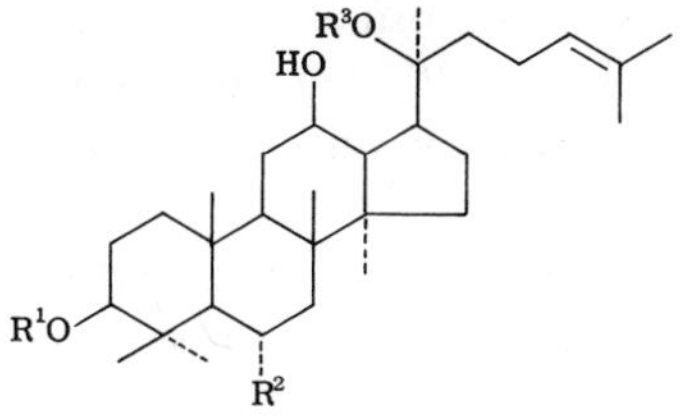

R³O
HO

R¹O

R²

Ginsenoside	R¹	R²	R³
F_1	H	OH	Glc
F_2	Glc	H	Glc
F_3	H	OH	Ara(p)-⁶Glc—

fore blossoming to stimulate the growth of root, the known saponins found were ginsenosides R_{b-1}, R_{b-2}, R_c, R_d, R_e, R_{g-1}, and F_3, in yields of 0.2, 0.2, 0.2, 0.5, 2.8, 0.2 and 0.03%, respectively; a new saponin, ginsenoside M_{7cd}, was also obtained. The high yield of R_e (2.8%) from the flower buds was noted (Yahara *et al.*, 1976b, 1979).

From the fruits, ginsenosides R_{b-2}, R_c, R_d, R_e, and R_{g-1} were isolated in yields of 0.2, 0.1, 0.1, 0.6, and 0.04%, respectively. The high yield of R_e is characteristic in the fruits (Yahara *et al.*, 1976b).

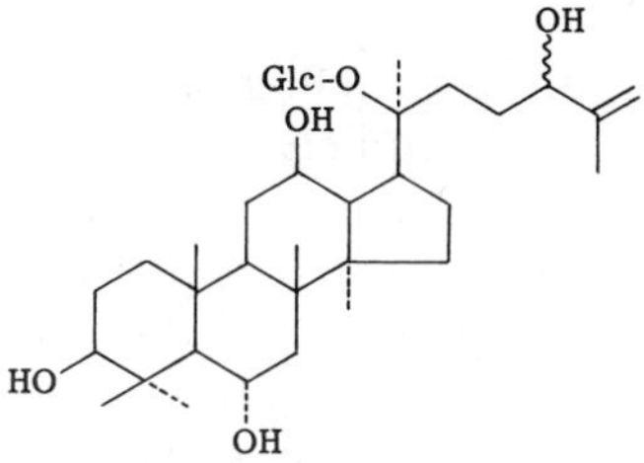

OH
Glc -O
OH

HO

OH

Ginsenoside M_{7cd}

B. *PANAX QUINQUEFOLIUM*

The saponins of the aerial part of American ginseng, *Panax quiniquefolium* L. were studied first by Chen and Staba (1978) and Lui and Staba (1980) and later by Chen *et al.* (1981), who isolated ginsenosides R_{b-3} (0.1%), R_d (0.2%), R_e (0.1%), and R_{g-1} (0.1%) and an ocotillol-type saponin, pseudoginsenoside F_{11} (0.1%).

C. *PANAX PSEUDOGINSENG* SUBSP. *HIMALAICUS*

The rhizomes of this plant, Himalayan ginseng, contain oleanolic acid saponins predominantly resembling Japanese Chikusetsu ginseng, whereas the leaves contain dammarane-type saponins, ginsenosides R_e (0.1%), R_d (0.1%), and R_{b-3} (0.9%), and a new saponin, pseudoginsenoside F_8 (0.1%), besides an ocotillol-type saponin, pseudoginsenoside F_{11} (0.4%).

Ginsenoside R_{b-3}, which is a minor saponin in Korean ginseng, is contained in the leaves of Himalayan ginseng plant in fairly good yield (Tanaka and Yahara, 1978).

Pseudoginsenoside F_8 Pseudoginsenoside F_{11}

D. PANAX NOTOGINSENG

The flower buds of Chinese Sanchi ginseng plant, *Panax notoginseng*, used in China as a tonic, were studied by Taniyasu *et al.* (1982) with the collaboration of Chinese scientists in Yunnan. The results showed that the flower buds contain ginsenosides R_{b-1} (0.4%), R_{b-2} (0.1%), R_c (1.0%), R_d (0.1%), and F_2 (0.1%). Noted that all these saponins belong to protopanaxadiol series compounds, and none to the protopanaxatriol series of saponins, while R_c is found at a high level.

From the leaves of Sanchi ginseng plant, Yang *et al.* (1983) isolated ginsenosides R_{b-1} (0.03%), R_{b-3} (0.71%), R_c (0.39%), notoginsenosides F_a (0.01%), F_c (0.05%), F_d (gypenoside IX) (0.03%), and F_e (0.005%).

Notoginsenoside	R^1	R^2
F_a	Xyl-^{2}Glc-^{2}Glc-	Glc-^{6}Glc-
F_c	Xyl-^{2}Glc-^{2}Glc-	Xyl-^{6}Glc-
F_d (Gypenoside IX)	Glc-	Xyl-^{6}Glc-
F_e	Glc-	Ara(f)-^{6}Glc-

E. *PANAX PSEUDOGINSENG* SUBSP. *JAPONICUS* (JAPANESE CHIKUSETSU GINSENG PLANT)

The leaves of Japanese Chikusetsu ginseng plant (*Panax pseudoginseng* subsp. *japonicus*) gave a pattern of saponin contents differing from those of *P. ginseng* and *P. pseudoginseng* subsp. *himalaicus*.

In the series of studies carried out by Yahara *et al.* (1977, 1978), remarkable differences were found in the distribution of saponins among the Chikusetsu ginseng plants growing wild in different localities in Japanese islands, as shown in the accompanying tabulation.

THE SAPONINS IN THE LEAVES OF *PANAX PSEUDOGINSENG* SUBSP. *JAPONICUS*

Hiroshima and other places on the Pacific side	Tottori (Mt. Daisen)	Niigata
Ginsenosides R_e (0.1%) F_1 (0.01%), F_3 (0.1%) Chikusetsu saponins L_5 (0.7%), L_{9a} (0.1%), L_{9bc} (0.2%), L_{10} (0.2%)	Chikusetsu saponins LT_5, LT_8	Chikusetsu saponin LN_4

Chikusetsu saponin	R^1	R^2
L_5	H	Xyl-^{4}Ara(p)-^{6}Glc-
L_{10}	Glc	H

Chikusetsu saponin L_9

Chikusetsu saponin L_{9bc}

Chikusetsu saponin	R¹	R²
LT₅	Glc	Glc-⁶Glc-
LT₈	Glc	Glc
LN₄	Xyl-⁶Glc-	Ara(p)-⁶Glc-

VI. TISSUE CULTURE OF PANAX GINSENG

The cultivation of *Panax ginseng* takes at least 4–6 years, and special precautions. The plant must be cultivated in soil in which ginseng has not been grown for at least 10–15 years, otherwise the roots are affected by rot. The beds are prepared with rich applications of organic fertilizers under straw thatch to avoid direct sunshine, as the plant grows originally in deep forest. Thus the field cultivation of *Panax ginseng* is very laborious and cost-consuming, so the price of ginseng in the market is very high. Therefore, the tissue culture of *P. ginseng* has been attempted, to produce ginseng or its biologically active principles in the fermenter on a manufacturing scale.

The experiments by Furuya *et al.* (1970, 1973; Furuya and Ishii, 1981) demonstrated that the callus of ginseng is formed by the cultivation of root or petiole of the original plant on Murashige–Skoog agar medium containing 2,4-dichlorophenoxyacetic acid (2,4-D) at 0.1 ppm for 4–5 weeks at 25°C in the dark. The callus was inoculated every 4–5 weeks repeatedly to the new medium in decreasing the content of 2,4-D in the medium to 0.01, 0.001, and 0.0001 ppm, successively. Finally, the habituated callus was obtained, which could grow on the medium without 2,4-D. The habituated callus showed a good growth rate; however, it produced saponins only in low yield.

Addition of kinetin in the medium and illumination with 2500–3000 lux induced partial differentiation of stems, leaves and roots in the habituated callus, and it showed a high growth rate and a high production of saponins.

The differentiated root portion thus induced was cultivated on a medium containing indole butyric acid (IBA) to obtain a good multiplication of root pieces, which showed a high yield of saponins.

The cultivation of callus in a shaking flask and a jar fermenter gave a good result in growth and production of saponins. The crude extracts of the callus were analyzed on HPLC using Shodex OH Pak B-804, reflex index (RI) detector, and CH_3CN/H_2O (85 : 15) as a solvent system to give almost superimposable chromatographical profiles of ginsenosides R_x when compared with those isolated from the commercial ginseng.

Manufacture of differentiated root pieces of *Panax ginseng* in a large-scale fermenter is under investigation.

VII. CHEMICAL ANALYSIS OF *PANAX* SAPONINS

A. PREPARATIVE ISOLATION USING COLUMN CHROMATOGRAPHY

The preparation of *Panax* saponins has been performed with silica-gel column chromatography of the butanol-soluble fraction of methanolic extracts of roots, rhizomes, leaves, and flower buds of *Panax* plants. The procedure has been developed by Sanada *et al.* (1978a) as described in the previous chapter, using chloroform–methanol–water or butanol–ethyl acetate–water as the eluting solvents.

The lateral roots contain saponins in higher percentages than the main roots. The saponins of the protopanaxatriol series are contained in a higher yield in the leaves and the flower buds of *Panax ginseng*, and those of the protopanaxadiol series are rich in the flower buds of San-chi ginseng, which are available in the Chinese drug market.

B. SEPARATION WITH DROPLET COUNTERCURRENT CHROMATOGRAPHY

Droplet countercurrent chromatography (DCC) consists of a few hundred lengths of glass tubing (1.65 mm ID × 40 cm) connected to each other top-to-bottom with thin plastic tubes (0.65 mm ID × 55 cm). The stationary phase fills up the tubing system, and the mobile phase moves upward or downward to form small droplets. A solution of sample mixture to be separated is pumped into this system, mixing with the mobile phase, and the separation is performed by the difference of partition coefficients of the components during the course of moving; each 3–4 ml eluent is separately collected into the glass tubes on a fraction collector, connected at the end of the system. This method has advantages in automatic separation without using any adsorbent, to avoid the loss of materials and the oxidative degradation that might be caused by exposing to the air.

The DDC process takes rather a long time to separate a complex mixture of compounds, but it could be employed conveniently for semi-microscale preparation of *Panax* saponins, for example.

Otsuka *et al.* (1977) reported DCC separation of saponins of ginseng in comparison with those of American, San-chi, Japanese Chikusetsu, and Himalayan Ginseng (Fig. 7).

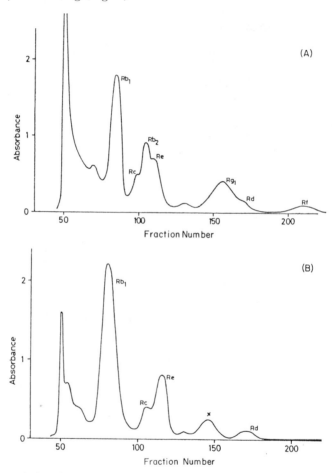

FIG. 7 (A) DCC elution diagram of the butanol-soluble saponins of white Korean ginseng (Nagano). (B) DCC elution diagram of the butanol-soluble saponins of American ginseng cultivated at Nagano; x, an unknown component. (C) DCC elution diagram of the butanol-soluble saponins of San-chi ginseng. (D) DCC elution of the butanol-soluble saponins of Chikusetsu ginseng.

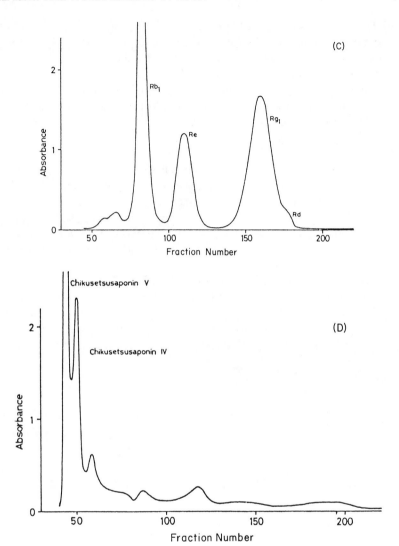

FIG. 7 *Continued*

C. PREPARATIVE SEPARATION WITH HIGH-PERFORMANCE LIQUID CHROMATOGRAPHY

For the preparative isolation of *Panax* saponins, several types of reversed-phase HPLC have been applied using the octadecyl silica gel (ODS) and Waters carbohydrate analysis column eluted with the solvent system of acetonitrile–water (CH_3CN/H_2O). The proportion of CH_3CN

and H_2O for separation of ginsenosides in general was $80:20$ (v/v), while it was varied to $81:19$ for separation of R_{b-1} and R_{b-2}; $82:18$ for R_c; $86:14$ for R_d and R_e; $89:11$ for R_{g-1}; $86:14$ for R_f and R_{g-2}; and $89:11$ for R_{h-1} (Nagasawa et al., 1980a,b).

For the separation of ginsenosides R_x, Sticher and Soldati (1979, 1980) used a μ-Bondapak-C_{18} (Waters) column, the solvent system, CH_3CN/H_2O ($30:70$), and a UV detector at $202–207$ nm. The separation of R_e and R_{g-1} was pursued using the eluting solvent CH_3CN/H_2O ($18:82$).

Zhou et al. (1981) reported that ginsenoside R_e and notoginsenoside R_1 extracted from Sanchi ginseng, which were separated sparingly by silica-gel chromatography, were separated on a TSK gel LS 410 column (Toyo-soda) using methanol water ($55:45$) as the eluting solvent (Fig. 8).

As the ginseng saponins possess no conjugated system in the molecule, the detection with reflex index (RI) is more suitable than UV for HPLC analysis.

A remarkable improvement has been made in HPLC separation of water-soluble compounds including glycosides using a new column, Aquasil (Senshu-Sci. Co. Ltd.), and a solvent system, chloroform/methanol/ethanol/water ($62:16:16:6$). The crude saponin fraction of *Panax* was applied to this system to perform preparative separation of ginsenosides

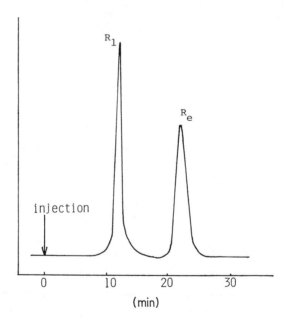

FIG. 8 Separation of ginsenoside R_3 and notoginsenoside R_1 by reversed phase HPLC.

R_x. On evaporation of eluant, some ginseng saponins were obtained in a pure crystalline form. The pretreated Waters 500 column can also be used using the same solvent system for preparative purpose (Kaizuka and Takahashi, 1983).

D. QUANTITATIVE SEPARATION OF *PANAX* SAPONINS

Sanada *et al.* (1978c) reported a quantitative determination of ginsenosides R_x in Korean, American, and San-chi ginseng using TLC on a silica-gel plate 60F$_{254}$ (Merck) and a dual-wavelength TLC zig-zag scanner (λ_S 525 nm, λ_R 760 nm) (Shimadzu) (Table IX). For this method, several developing solvent systems were used for comparison (Fig. 9).

Solvent A: n-BuOH/AcOEt/H$_2$O (4 : 1 : 5 upper layer)
Solvent B: CHCl$_3$/MeOH/H$_2$O (65 : 35 : 10 lower layer)
Solvent C: CHCl$_3$/MeOH/AcOEt/H$_2$O (2 : 2 : 4 : 1 lower layer)
Solvent D: CHCl$_3$/n-BuOH/MeOH/H$_2$O (4 : 8 : 3 : 4 lower layer)

The reversed-phase TLC [HPTLC plate, RP-8 F254; HPTLC plate RP-18 F254 (Merck)] could also be employed for the separation of ginsenosides R_x.

Some studies on gas–liquid chromatographical separation of panax-adiol and panaxatriol as their trimethyl silyl (TMS) derivatives were reported by Sakamoto *et al.* (1975).

By these methods, ginsenosides R_x were only determined as classified into two groups, the protopanaxadiol and protopanaxatriol series.

The quantitative determination of panaxadiol and panaxatriol on TLC was reported by Woo *et al.* (1973).

A two-dimensional TLC of ginsenosides R_x developed with p-anisalde-hyde–H$_2$SO$_4$ and determined with a densitometer was reported by Lee and Marderosian (1981).

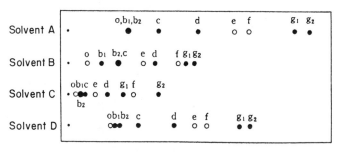

FIG. 9 Thin-layer chromatogram of ginseng saponins. See text for solvents. From Oura *et al.* (1981), with permission.

TABLE IX

CONTENTS OF GINSENOSIDES IN GINSENG AND ITS CONGENERS[a]

Samples	Place harvested	Ginsenoside- (%)										
		R^o	R_{b-1}	R_{b-2}	R_c	R_d	R_e	R_f	R_{g-1}	R_{g-2}
Ginseng	Commercial	0.1	0.4	0.4	0.2	0.1	0.1	0.1	0.1	—
White ginseng	Commercial	0.2	0.4	0.1	0.1	0.1	0.1	0.1	0.2	—
Red ginseng	Nagano	0.2	0.6	0.4	0.5	0.1	0.2	0.1	0.3	—
Ginseng peelings	Nagano	0.1	1.0	0.5	0.5	0.6	0.8	—	0.2	0.1
Lateral root of ginseng	Nagano	0.6	3.3	2.3	2.5	0.9	1.5	0.3	0.9	0.1
American ginseng	Commercial	0.1	1.5	—	—	0.3	0.4	—	0.4	—
American ginseng	Nagano	0.1	1.6	0.1	0.1	0.1	0.4	—	—	—
San-chi ginseng	Commercial	—	1.7	—	—	0.4	0.7	—	2.1	—

[a] Measured by Sanada *et al.* (1978), using a TLC dual-wavelength zig-zag scanner.

The application of HPLC for the quantitative determination of gin-senosides R_x was studied by Sticher and Soldati (1980) using a reversed-phase system on a μ-Bondapak C_{18} column (3.9 mm ID × 30 cm) and the solvent system CH_3CN/H_2O at 30 : 70 and 18 : 82 with UV detection at 203 nm (Table X) Nagasawa *et al.* (1980a,b) reported the HPLC analysis of ginsenoside R_x using a reverse–phase carbohydrate analysis column (Waters) (3.9 mm ID × 30 cm) with a moving phase of CH_3CN/H_2O (80 : 20 or 86 : 14) and RI detection (Fig. 10).

Owing to the high sensitivity of the UV detection (240 nm or 250 nm) of HPLC, Besso *et al.* (1979) used benzoates derived from ginseng saponins for separation on a silica-gel (TSK-gel LS 310 Toyo-Soda) col-umn with the eluant *n*-hexane/CH_2Cl_2/CH_3CN (15 : 3 : 2).

The HPLC on an Aqua-sil column (4.6 mm ID × 30 cm) was most efficiently used for the simultaneous quantitative determination of gin-senosides R_x using the solvent system $CHCl_3/MeOH/H_2O$ (30 : 17 : 2) with RI detection. Under this condition, the separation of saponins on the baseline was performed within 15 min, and the column was tolerable for multiple uses (Fig. 11) (Kaizuka and Takahashi, 1983).

E. IDENTIFICATION OF GINSENG SAPONINS

The general identification procedures for organic compounds, such as mixed melting point determination, elemental analysis, IR spectral mea-surement, and TLC, could not unambiguously be applied to saponins of higher molecular weight.

HPLC on a silica gel as well as on a reversed–phase column can be used for identification of saponins more reliably than these other procedures.

The most unequivocal identification of ginsenosides R_x, including the stereochemistry at C-20 of aglycones as well as the establishment of the position of glycosyl linkages, have been performed by [^{13}C]NMR spec-troscopy. The substantial assignments of dammarane-type triterpenes, including ginseng saponins, were reported by Kasai *et al.* (1977a,b,c), Mizutani *et al.* (1980), Yahara *et al.* (1977, 1978, 1979), Tanaka and Yahara (1978), and Zhou *et al.* (1981) to reveal that the carbon resonance displacements of both sugar and aglycone moieties (glycosylation shifts) upon glycosylation are closely related to the stereochemical linkage of a sugar unit and aglycone alcohol. The [^{13}C]NMR chemical shift values of the carbons underlined in the Tables XI and XII indicate the glycosyla-tion shifts, which show a downfield shift at the glycosylated carbons, C-3 or C-6 and C-20, and a upfield shift at the neighbouring carbons, C-2 or C-7, C-17, and C-21, in comparison with those in the corresponding aglycones. Some anomalous glycosylation shifts for sugar carbon signals

TABLE X

DISTRIBUTION OF GINSENOSIDES IN *PANAX GINSENG*[a]

Location	Content (%)								
	$R_{g\text{-}1}$	R_e	R_f	$R_{g\text{-}2}$	$R_{b\text{-}1}$	R_c	$R_{b\text{-}2}$	R_d	Total
Leaves	1.078	1.524	—	—	0.184	0.736	0.553	1.113	5.188
Leafstalks	0.327	0.141	—	—	—	0.190	—	0.107	0.765
Stem	0.292	0.070	—	—	—	—	0.397	—	0.759
Main root	0.379	0.153	0.092	0.023	0.342	0.190	0.131	0.038	1.348
Lateral roots	0.406	0.668	0.203	0.090	0.850	0.738	0.434	0.143	3.532
Root hairs	0.376	1.512	0.150	0.249	1.351	1.349	0.780	0.381	6.148

[a]Measured by Sticher and Soldti (1980) using reverse-phase HPLC on μ-Bondapak C_{18} column.

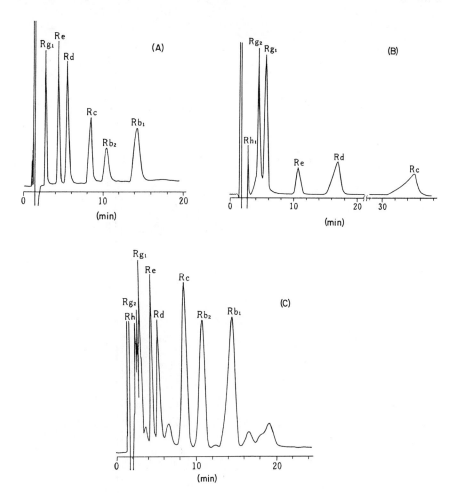

FIG. 10 (A) Chromatogram of ginsenoside R_{g-1}, R_e, R_c, R_d, R_{b-2}, and R_{b-1}. Conditions: column, 30 cm × 3.9 mm; packing, carbohydrate analysis; mobile phase, AcCN/H$_2$O, 80 : 20 (v/v); flow rate, 2 ml/min; RI detector, attenuation 16×. (B) Chromatogram of gin-senosides R_{h-1}, R_{g-2}, R_{g-1}, R_e, R_d, and R_c. Conditions: column, 30 cm × 3.9 mm; packing, carbohydrate analysis; mobile phase, AcCN/H$_2$O, 86 : 14 (v/v); flow rate, 2 ml/min; RI detector, attenuation 16×. (C) Chromatogram of crude saponin from Ginseng Radix. Conditions: column, 30 cm × 3.9 mm; packing, carbohydrate analysis; mobile phase, AcCN/H$_2$O, 80 : 20 (v/v); flow rate, 2 ml/min; RI detector, attenuation 16×. From Nagasawa *et al.* (1980a), with permission.

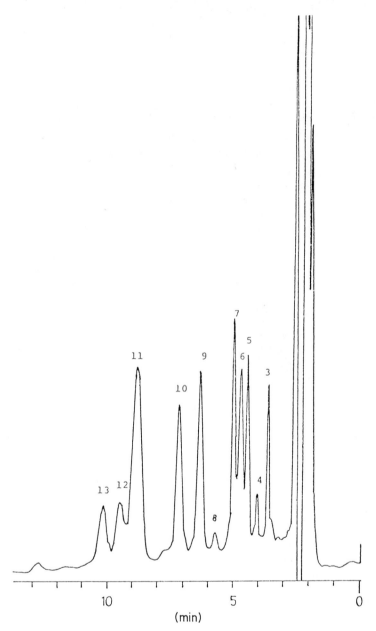

FIG. II HPLC profile of the BuOH fraction of ginseng. Column: Aquasil SS-452N (4.6 mm ID × 30 cm). Eluent: $CHCl_3$–MeOH–H_2O (30 : 17 : 2). Flow rate: 1.0 ml/min. Detector: RI. Ginsenoside R_x : g-1 (3); d (5); e (7); c (9); b-2 (10); b-1 (11).

TABLE

[^{13}C]NMR CHEMICAL SHIFTS OF SAPOGENIN PARTS OF GINSENOSIDES R_X (IN D_5 PYRIDINE)[a]

C	(20S) ppd[c]	R_d	R_c	R_{b-2}	R_{b-1}	(20S) ppt[c]	R_{g-1}	R_e
1	39.5	39.1	39.0	39.4	39.1	39.2	39.5	39.4
2	28.2	26.7	26.6	26.6	26.6	28.0	27.6	27.4
3	77.9	88.0	89.0	89.1	89.3	78.3	78.6	78.7
4	39.5	39.6	39.6	39.6	39.6	40.2	40.1	39.8
5	56.3	56.4	56.3	56.4	56.3	61.7	61.3	60.7
6	18.7	18.5	18.3	18.3	18.6	67.6	77.8	74.6
7	35.2	35.2	35.1	35.1	35.1	47.4	44.9	45.7
8	40.0	40.0	39.9	39.9	39.9	41.1	41.0	41.0
9	50.4	50.2	50.1	50.1	50.1	50.1	49.9	49.4
10	37.3	36.9	36.8	36.8	36.8	39.3	39.5	39.4
11	32.0	30.8	30.7	30.7	30.8	31.9	30.8[b]	30.6
12	70.9	70.2	70.2	70.1	70.1	70.9	70.3	70.3
13	48.5	49.4	49.5	49.4	49.3	48.1	48.9	48.8
14	51.6	51.4	51.4	51.3	51.3	51.6	51.3	51.3
15	31.8	30.8	30.8	30.7	30.8	31.3	30.6[b]	30.6
16	26.8	26.7	26.6	26.4	26.6	26.8	26.4	26.5
17	54.7	51.7	51.6	51.6	51.6	54.6	51.6	51.8
18	16.2[b]	16.3[b]	16.2[b]	16.2[b]	16.2[b]	17.5[b]	17.4[b]	17.4[b]
19	15.8[b]	15.9[b]	15.9[b]	15.9[b]	15.9[b]	17.4[b]	17.4[b]	17.4[b]
20	72.9	83.3	83.1	83.5	83.5	72.9	83.3	83.2
21	26.9	22.4	22.2	22.2	22.6	26.9	22.3	22.4
22	35.8	36.0	36.0	36.3	36.1	35.7	35.9	35.8
23	22.9	23.2	23.1	23.1	23.1	22.9	23.2	23.3
24	126.2	125.9	125.9	124.8	125.8	126.2	125.8	125.7
25	130.6	130.9	130.9	131.0	131.0	130.6	130.9	130.9
26	25.8	25.8	25.7	25.8	25.8	25.8	25.7	25.7
27	17.6[b]	17.8[b]	17.8[b]	17.9[b]	17.9[b]	17.7[b]	17.7[b]	17.7[b]
28	28.6	28.0	28.0	28.0	28.0	31.9	31.6	32.0
29	16.4[b]	16.6[b]	16.5[b]	16.5[b]	16.5[b]	16.4[b]	16.2[b]	17.1[b]
30	17.0[b]	17.3[b]	17.3[b]	17.3[b]	17.3[b]	17.0[b]	17.0[b]	17.1[b]

[a]From Oura et al. (1981), with permission. Underscored numbers represent signals shifted by glycosylation.
[b]Values in any vertical column may be reversed although those given here are preferred.
[c]ppd, Protopanaxadiol; ppt, protopanaxatriol.

TABLE XII

[^{13}C]NMR CHEMICAL SHIFTS OF SUGAR MOIETIES OF GINSENOSIDES R_X (IN D^5 PYRIDINE)[a]

Sugar		R_d	R_c	$R_{b\text{-}2}$	$R_{b\text{-}1}$	Sugar		$R_{g\text{-}1}$	R_e
3-Glc	1	105.0	104.9	105.0	105.0	6-Glc	1	105.7	101.6
	2	83.3	83.1	83.0	82.9		2	75.3	79.1
	3	78.1[b]	77.8[b]	78.1[b]	77.2[b]		3	80.0[c]	78.0[c]
	4	71.6	71.5	71.5	71.5		4	71.6[d]	72.1
	5	78.1[b]	77.8[b]	78.1[b]	78.0[b]		5	79.3[c]	78.0[c]
	6	62.7	62.6	62.7	62.6		6	62.9	62.9
Glc	1	105.9	105.6	105.7	105.6	Rha			101.6
	2	77.0	76.8	76.9	76.7		2		72.1
	3	79.1[b]	78.7[b]	79.0[b]	78.8[b]		3		72.1
	4	71.6	71.5	71.5	71.5		4		73.8
	5	78.1[b]	78.0[b]	78.7[b]	78.0[b]		5		69.3
	6	62.7	62.6	62.7	62.6		6		18.6
20-Glc	1	98.2	97.9	97.9	97.9	12-Glc	1		
	2	75.0	74.9	74.8	74.9		2		
	3	78.1[b]	78.0[b]	78.7[b]	78.0[b]		3		
	4	71.6	71.5	71.5	71.5		4		
	5	78.1[b]	76.3	76.6	76.7		5		
	6	62.7	68.3	69.0	71.5		6		
Glc	1				105.0	20-Glc	1	98.1	98.1
	2				74.9		2	74.9	75.0
	3				78.0[b]		3	78.8[c]	79.1
	4				71.5		4	71.3[d]	71.1
	5				78.0[b]		5	77.8[c]	78.5[c]
	6				62.6		6	62.6	62.7
Ara	1		109.9	104.5					
	2		83.3	72.0					
	3		78.9	73.9					
	4		85.8	68.5					
	5		62.6	65.5					

[a] From Oura et al. (1981), with permission.
[b,c,d] Values in any vertical column may be reversed, although those given here are preferred.

were observed in the spectra of the 1,2-linked biosyl moiety attached to the C-6 OH of (20S)-protopanaxatriol (i.e., ginsenosides R_e, R_f, and $R_{g\text{-}2}$, notoginsenosides R_1 and R_2, etc.).

The anomeric configuration of the rhamnosyl group could not be determined by [^{1}H]NMR, whereas [^{13}C]NMR readily assigned it (Kasai et al., 1979).

Mass spectrometry (MS) is also applicable to the identification of ginsenosides R_x. For the measurement of electron impact mass spectra (EI-

MS), the samples must be transformed into volatile derivatives, such as acetate, trimethylsilyl ether (TMSi), or methyl ether.

Trimethylsilylation of ginsenosides R_x and their saponins (20*S*)-protopanaxadiol and (20*S*)-protopanaxatriol was performed with trimethylsilylimidazole, a potential on-column TMSi-reagent, with which even the hindered *tert*-C-20 OH could readily be blocked (Komori *et al.*, 1974; Kasai *et al.* 1977a).

Komori *et al.* (1974) reported that the glycosyl linkage at the C-20 OH of dammarane saponins is easily cleaved, and the MS of their acetates or TMSi showed no M^+ or fragment ions having an intact O-glycosyl group at C-20, exhibiting ions (A and A') illustrated as follows.

On the other hand, Zhou *et al.* (1981) revealed that the spectra of TMSi of the compound having a free C-20, such as ginsenoside R_f, showed the characteristic ions B (very strong) and C, along with ions A and A', being diagnostic for the absence of a glycosyl linkage at C-20.

Mass fragmentation also provides useful information for the sequence of linkage of sugar unit.

As already mentioned, the molecular weights of ginseng saponins are difficult to determine by EI-MS, whereas by means of field desorption mass spectrometry (FD-MS), M^+ (as its cluster ion) could definitely be observed (Komori *et al.*, 1979).

F. RADIOIMMUNOASSAY OF GINSENOSIDES R_x

Radioimmunoassay is used for the ultramicrodetermination of drugs or their metabolites in the biological fluids, tissues, and organs. It is highly sensitive and specific. It has been developed to apply to the determination of alkaloids and glycosides in the plant tissue. Sankawa *et al.* (1982) studied radioimmunoassay of ginsenoside R_{g-1} to determine its content in ginseng plant tissues and in biological fluids of the animals fed with ginseng or its saponins.

The preparation of ginsenoside R_{g-1} and bovine serum albumin (BSA) conjugate was carried out as follows.

Ginsenoside R_{g-1}–BSA carries 3.4–4.6 molecules of ginsenoside R_{g-1} per molecule of BSA.

Immunization was carried out by the procedure given by Kawashima (1981), and antiserum for ginsenoside R_{g-1} was obtained 5 months from the first immunization.

An ^{125}I-labeled compound, GSR_{g-1}–[^{125}I]tyramine was also prepared. The radioimmunoassay of ginsenoside R_{g-1} was carried out using these materials to determine 250 pg to 10 ng of ginsenoside R_{g-1} in the biological samples.

M+ m/e 764 (676)

Fragment ion B
m/e 674 (586)

$-$ TMSiOH

B_1 m/e 584 (496)

B_2 m/e 494 (406)

B_3 m/e 404

$-$ TMSiOH

Fragment ion A
m/e 681 (593)

Fragment ion C
m/e 199

$-$ TMSiOH

C_1 m/e 109

$-$ TMSiOH

A_1 m/e 591 (413) $\xrightarrow{\text{$-$ TMSiOH}}$ A_2 m/e 501 (413) $\xrightarrow{\text{$-$ TMSiOH}}$ A_3 m/e 411 (323)

$-$ TMSiOH

A_4 m/e 321

(A)

(A′)

(B)

(C)

Ginsenoside R$_{g1}$

1. Ac$_2$O-pyridine
2. OsO$_4$
3. HIO$_4$

Wittig-
Horner

DMSO

1. NH$_2$NH$_2$
2. HNO$_2$

BSA

H$_2$O

GSR$_{g1}$-BSA

1. H$_2$/PtO$_2$
2. NH$_2$NH$_2$
3. HNO$_2$

tyramine

Na ^{125}I

GSR$_{g1}$-[^{125}I]tyramine

VIII. SOME CHEMICAL CONSTITUENTS OTHER THAN SAPONINS IN GINSENG

Ginseng contains ~ 0.05% volatile oil, which Takahashi *et al.* (1961, 1962, 1964, 1966) studied by ether extraction, isolating β-elemene from the low-boiling oil fraction and panaxynol from the higher-boiling fraction. The structure of panaxynol was proved synthetically to be 1,9-*cis*-heptadecadiene-1,6-diyn-3-ol.

Poplawski *et al.* (1980) and Dabrowski *et al.* (1980) isolated two other β-Elemene

$$CH_3(CH_2)_6CH=CH-CH_2(C\equiv C)_2\underset{OH}{CH}-CH=CH_2$$

Panaxynol

$$n\text{-}C_7H_{15}-\underset{\diagdown O\diagup}{CH-CH}-CH_2-(C\equiv C)_2-\underset{OH}{CH}-CH=CH_2$$

Panaxydol

$$CH_3-(CH_2)_7-\underset{OH}{CH}-CH_2-(C\equiv C)_2-\underset{OH}{CH}-CH=CH_2$$

acetylenic compounds, 9,10-epoxy-3-hydroxyheptadeca-1-en-4,6-diyne (panaxydol) and heptadeca-1-en-4,6-diyn-3,9-diol, from aqueous alcoholic extracts of ginseng. Peptides were isolated from white ginseng by Gstirner and Vogt (1966). Ginseng was extracted with methanol/water (1 : 1) and the extracts were subjected to paper electrophoresis under 30 V/cm using AcOH/AcONa buffer at pH 5.0. Five ninhydrin-positive spots were obtained, which were concentrated by two-dimensional high-voltage electrophoresis on Sephadex G-50 (1000 V, 8 hr) to give four low molecular weight peptides. On hydrolysis of the peptides with concentrated HCl/90% HCOOH (1 : 1) at 120°C for 36 hr, amino acid components were afforded, which were analyzed by the Stein–Moore method on Amberlite IR-120 to give aspartic acid, serine, glutamic acid, glycine, alanine, and arginine as the common components. Besides these amino acids, threonine, proline, leucine, isoleucine, lysine, and histidine are contained in fraction 1; threonine, valine, β-aminobutyric acid, β-aminoisobutyric acid, lysine, histidine, hydroxyproline, and two unidentified components in fraction 2; threonine, proline, methionine, leucine, alloisoleucine, isoleucine, phenylalanine, β-aminobutyric acid, tyrosine, lysine, and histidine in fraction 3; and one unidentified component in fraction 4. Prior to hydrolysis, the absence of tryptophan was proved. Proline was the main component of fraction 1 and 3.

From the aqueous extracts of white ginseng, urasil was isolated, and from its lateral roots, guanine and adenine were obtained.

By HPLC analysis using a TSK gel LS 160 column with water and G3000W and G-2000SW columns with H_2O/AcOH/Et_3N (100 : 0.3 : 0.3) as a solvent system, the presence of uridine, guanidine, and adenine in white ginseng and of uridine, uracil, and adenosine in the lateral roots was demonstrated. The contents of these bases and nucleosides were determined using a dual-wavelength chromatogram zig-zag scanner (Hiyama *et al.*, 1978).

IX. PHARMACOLOGY AND BIOCHEMISTRY OF GINSENG

As mentioned previously, ginseng has been known among the people in East Asia for thousand years as an elixir, while Western people have not taken any special interest in it until recently, or have believed its effects to be entirely psychosomatic.

In orthodox traditional medicine in China, ginseng has been neither panacea nor tonic, in a strict meaning. Ginseng is combined in the prescriptions as a drug restoring the Yang or against the void condition to recover balanced homoeostasis. Some pharmacological studies on ginseng were reported earlier, but the newer approach to the biological activities of Ginseng has been stimulated by the pioneering works reported by Brekhman and Dardymov (1969a,b) in the U.S.S.R. and by Petkov and Staneva-Stoicheva (1963) in Bulgaria, whose studies were designed on the basis of Pavlov's behavioural pharmacology using ginseng extracts.

Brekhman's work started on people before using mice. Soviet soldiers given an extract of ginseng ran faster in a 3-km race than those given a placebo of a similar flavour. Radio operators tested under the administration of ginseng extract transmitted text significantly faster and with fewer mistakes than those given placebo. These results and others performed by other European research groups suggested that ginseng results in improvement in stamina.

Brekhman tried to confirm his results in humans by animal experiments using mice. Tested with complete exhaustion after a course of swimming, a group of mice administered ginseng extract showed much longer duration on the second swimming test than did the others. In these experiments ginseng has shown an antifatigue action.

Petkov published several papers (1959, 1961) on the pharmacological actions of ginseng extracts. He suggested from his experimental results a stimulative effect of ginseng upon the central nervous system (CNS), hypotensive effect, stimulation of respiration, lowering of blood sugar, potentiation of insulin action, and increase of erythrocyte count and haemoglobin. Petkov also demonstrated that ginseng increases the reactivity of cerebrocortical cells.

In this stage of investigation, several papers on the pharmacology of ginseng have been published to estimate its activities. However, some

TABLE XIII

ESTIMATED ACTIONS OF GINSENG BY BLIND SCREENING

Estimated actions	References
Slight CNS-stimulant action	Petkov (1968); Brekhman and Dardymov (1969a,b); Kim *et al.* (1971)
CNS-depressant action	Kim *et al.* (1971)
Tranquillizing action	Kim *et al.* (1971)
Cholinergic action	Petkov (1968); Wood *et al.* (1964)
Histaminelike action	Lee *et al.* (1960)
Blood-pressure fall	Wood *et al.* (1964); Kitagawa *et al.* (1963)
Blood-pressure elevation	Sakai (1915)
Papaverinelike action	Petkov (1959)
Serotoninlike action	Hwang and Park (1960)
Ganglion stimulant action	Takagi (1975)
Analgestic antipyretic, and antiinflammatory actions	Takagi (1975)
No antihistaminelike action	Petkov (1959)
Antihistaminelike action	Kitagawa *et al.* (1963)

inconsistent results have been reported, due mostly to the different procedures in the preparation of extracts, which were not chemically pure (Table XIII).

Takagi and his collaborators studied systematically the pharmacological actions of ginseng principles on every step of fractionation (Fig. 12). The aqueous extracts of ginseng showed a stimulative action, whereas the methanolic extracts revealed mainly a CNS-sedative action.

Takagi *et al.* (1972, 1974) and Saito *et al.* (1972, 1973a, 1974b), the first pharmacologists who employed pure ginseng saponins for their studies, demonstrated CNS-sedative, tranquillizing, and hypotensive actions in ginsenoside R_{b-1}, and CNS-stimulating, hypertensive and antifatigue actions in ginsenoside R_{g-1}. The results are summarized in Table XIV. Ginsenoside R_{g-1} shows a slight CNS-stimulating activity in the lower dose, but the results of higher-dose administration of R_{g-1} showing sedative actions seemed to be toxic effects.

Ginsenoside R_{g-1} at 10 mg/kg (i.p.) in rats stimulates discrimination between two sounds similar to the sound connected with the electric shock in the pole climbing test, while it does not show any increase of spontaneous motor activity. With ginsenoside R_{g-1}, Saito *et al.* (1977a) also demonstrated successfully the stimulation of exploratory behaviour of rats. When the rats were put into a Y-maze with a cheese at one end, they got the cheese in a shorter time and with very few mistakes after

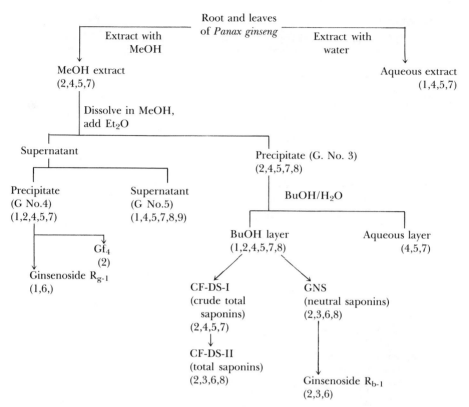

FIG. 12   Pharmacological activities of extracts of *Panax ginseng*. Type of activity: 1, weak CNS-stimulating; 2, CNS-sedative; 3, tranquillizing; 4, cholinergic; 5, histaminelike; 6, hypotensive; 7, hypertensive; 8, papaverinelike; 9, antifatigue. CF-DS-II is total saponins of *Panax ginseng* leaves.

learning. Running time and response latency were shortened by the administration of ginsenoside R_{g-1}. The known psychostimulants, such as caffeine and methamphetamine (MA), increased spontaneous motor activity, but showed elongation in running time and response latency with many failures in reversal learning (Fig. 13).

Then the cheese was switched to the other end (reversal learning experiment), and the time and the failure in learning the new location were measured. The administration of ginsenoside R_{g-1} in rats remarkably increased the correct response of rats (Fig. 14).

Saito and Lee (1978) studied the effect of ginsenoside R_{b-1} on the acquisition of conditioned avoidance response (pole-climbing test) in the

TABLE XIV

PHARMACOLOGICAL PROPERTIES OF GINSENOSIDE R_{b-1} AND GINSENOSIDE R_{g-1}

Ginsenoside R_{b-1}	Ginsenoside R_{g-1}

CNS-depressant action
  Anticonvulsant
  Analgesic action
	A weak CNS-stimulant action
	  A slight increase of motor activity
	  Potentiation of DB performace (Y-
	    maze test)

Antipyretic action Antipsychotic action	Behavioural effects
Inhibition of conditioned avoidance response (pole-climbing and shuttle-box tests) Protection of stress ulcer (antistress action)	Acceleration of acquisition of conditioned avoidance response and discrimination behavior (pole-climbing and Y-maze tests), reversal learning (Y-maze test), and one-trial passive avoidance learning (step-down method)
Increase of gastrointestinal motility A weak antiinflammatory action Potentiation of the NGF-mediated fibre production in chicken embryonic dorsal root ganglia and sympathetic ganglia. Antihaemolytic action Acceleration of glycolysis, cholesterol synthesis (serum and liver), and nuclear RNA synthesis Acceleration of serum protein synthesis	Antifatigue action Aggravation of stress ulcer

chronic intraventricularily canulated rats (Fig. 15). The rats administered 10 and 50 μg/kg of ginsenoside R_{g-1} did not show facilitation of the first success in the learning trial, but they showed significantly lowered mean numbers of trials to achieving 10 continuous success responses. They also showed a tendency toward facilitation of the first success in the testing trials. This result indicated the acceleration of acquisition of learning.

Hong *et al.* (1974, 1975) in Korea reported behavioural pharmacological experiments with rats and mice, concluding that ginseng saponins show stimulative activities with lower doses (2.5–5.0 mg/kg, i.p.) and sedative actions with higher doses (50 mg/kg or more, i.p.).

The pharmacological actions of Japanese Chikusetsu ginseng were studied by Saito *et al.* (1977b) in comparison with those of ginseng.

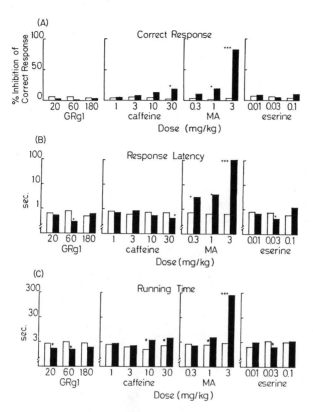

FIG. 13 Effect of caffeine, eserine, methamphetamine (MA), and ginsenoside R_{g-1} (GRg1) on rat learning in Y-maze test. White bars, control group; Black bars, drug-treated group. (A) Correct response means that the response latency and the running time are within 2 sec and the rats run straight to the object. (B) Response latency is the duration from opening of the gate to start of the rat. (C) Running time is the duration from start of the rat to opening the chamber where a piece of cheese was placed. Significant difference between control and the drug-treated rat (Student's test) $*p < .05$; $***p < .01$.

The blind screening consisted of five tests, and the tests on pharmacological actions induced from the clinical experiences in traditional medicine were performed using every fraction of Chikusetsu ginseng extracts. The results are shown in the Fig. 16. The extracts of Chikusetsu ginseng did not show any stimulating and antifatigue actions, but revealed sedative, tranquillizing, antipyretic, anti-stress-ulceric, antitussive, and expectorant actions. These results suggested that Japanese Chikusetsu ginseng could not be used as a mere substitute of Korean ginseng but would have its own therapeutic character.

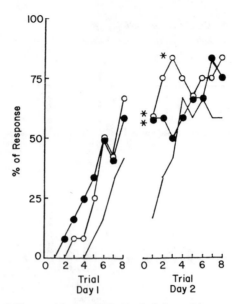

FIG. 14 Effect of Ginsenoside R_{g-1} (GRg1) administered i.p. on reversal learning behaviour in Y-maze test. ——, Control; —O—, 10 mg/kg; —●—, 50 mg/kg.

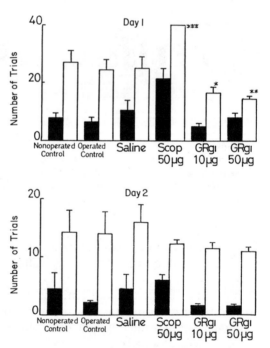

FIG. 15 Effect of ginsenoside R_{g-1} (GRg1) on the acquisition of learning in chronic intraventricularly canulated rats, compared with saline and scopolamine (Scop). Black bars

Rhizome of *Panax pseudoginseng* subsp. *japonicus*
|
MeOH extract
(1,3,4,5,6,7,8,11,12,13,15,16,17,19-ii,19-iii)

BuOH layer
(1,3,4,5,6,7,8,12,15,16,17,18,19-i,19-iii,20)

Aqueous layer
(3,9,13,14,19-iii)

Chikusetsu saponin III
(1,2,4,6,7,8,9,10,12,15,
16,17,18,19-i,19-iii,20)

Chikusetsu saponin IV
(1,6,9,10,12,18,
19-i,19-iii,20)

Chikusetsu saponin V
(1,9,10,11,20)

FIG. 16 Pharmacological actions of extracts of Japanese Chiketsu ginseng. Activity: 1, CNS-depressive; 2, tranquillizing; 3, cholinergic; 4, anticholinergic; 5, histaminelike; 6, antihistamine; 7, papaverinelike; 8, antinicotinic; 9, elevation of blood pressure; 10, fall of blood pressure; 11, antiwrithing (analgesic and antiinflammatory); 12, local irritation; 13, anticholinesterase; 14, diuretic; 15, sedative; 16, antipyretic; 17, antitussive; 18, expectorat; 19, on gastroenteric disorders (i, prevention of stress ulcer; ii, depression of gastric secretion; iii, acceleration of intestinal motility); 20, antiinflammatory. Actions 15–20 were predicted from clinical experiences in traditional medicine.

X. POTENTIATION OF NERVE GROWTH FACTOR WITH GINSENG SAPONINS

Saito *et al.* (1980, Saito, 1980) found that ginsenosides R_{b-1} and R_d potentiated nerve growth factor (NGF) or submandibular extracts to promote nerve-fibre production in organ cultures of chick embryonic dorsal root and sympathetic ganglia. Potentiation of NGF occured in the presence of 3 μM of R_{b-1} and reached a plateau at the presence of 300 μM R_{b-1}. They are all (20S)-protopanaxadiol oligosides having one or two glucosyl moieties in their side chains. The removal of glucosyl group or hydroxylation at the side chain attached to C-17 were formed to

---

show the mean number of first successes in the pole-climbing test. White bars show the mean number of trials taken from the first success to ten continuous successes. Significant differences at $*p = .05$, $**p = .02$, $***p = .01$. Mean + standard deviation (vertical lines) shown. One week after the operation the canulated rats were used in the pole-climbing test. The nontrained canulated rat was placed in the apparatus, then 500-Hz sounds that occurred intermittently once a second were delivered as a conditioned stimulus for 20 sec: sounds only for the first 10 sec, and sounds with electric shock for the other 10 sec. Shocks were delivered from the grid at the same time. When the rat climbed the pole, stimulations were immediately terminated. Rats were exposed to the conditioned stimulations until they had 10 continuous successes of conditioned avoidance response. Drugs were given 5 min before the beginning of learning.

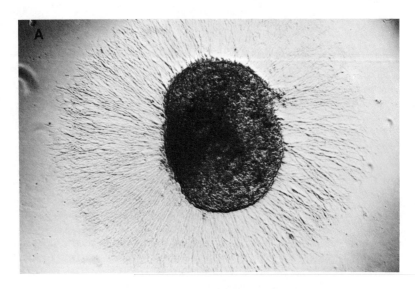

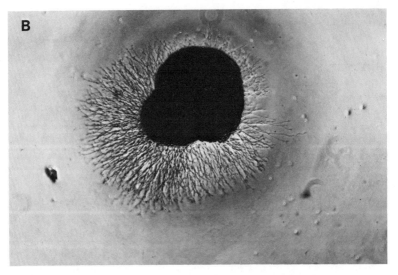

FIG. 17 (A) Effect of nerve growth factor (NGF) on nerve fibre production in organ culture of chick embryonic dorsal root ganglia (DRG), index 4 NGF at 18 ng/ml in culture medium. (B) Effect of ginsenoside R_{b-1} (GRb1) on NGF-mediated nerve-fibre production. Nerve fibre production (index 5) was obtained with 4.5 ng/ml of NGF in presence of 30 μM *of GRb1.*

reduce the activity. Ginsenoside R_{g-1} showed no remarkable activity in NGF potentiation.

An optimal neurite outgrowth (index 4) at the cultured chick embryonic dorsal root ganglia was observed at the presence of 18 ng NGF/ml in the medium. Almost the same grade of outgrowth (index 5) was observed with 4.5 ngNGF/ml in the presence of 30 μM of ginsenoside R_{b-1} (Fig. 17).

The mechanism of NGF potentiation with ginsenoside R_{b-1} has not been clarified, but it would have some important effect in the process of nerve-fibre production.

Effects of ginsenoside R_{b-1} on the activity of tyrosine hydroxylase (TH) and NGF content in superior cervical ganglia (SCG) and submandibular gland (SubG) of irradiated mice were also studied. As shown in Table XV, on administration of ginsenoside R_{b-1} once a day for 1 week after a 600-R irradiation, TH activity in SubG in the irradiated mice remained unchanged, whereas NGF content decreased. Exposure to irradiation might cause a gradual destruction of the sympathetic nerve terminals in SubG, and ginsenoside R_{b-1} would prevent this effect.

TABLE XV

NERVE GROWTH FACTOR (NGF) AND TYROSINE HYDROXYLASE (TH) ACTIVITIES IN SUPERIOR CERVICAL GLAND (SCG) AND SUBMANDIBULAR GLAND (SUBG) OF IRRADIATED MICE

Drugs treated[a] (dose mg/kg)	NGF (B.U. $\times 10^3$/SubG)	Total activity Th (nmol DOPA formed/hr·pairs of organs)	
		SCG	SubG
Control	8.1 ± 1.5	1.85 ± 0.20	0.346 ± 0.017
Irradiated mice with saline	2.1 ± 0.6^c	1.82 ± 0.18	0.239 ± 0.035^d
Irradiated mice with GRb-1[a] (100)	2.3 ± 0.6^c	1.93 ± 0.10	0.366 ± 0.041^e
Irradiated mice with GRb-1[a] (300)	2.7 ± 0.6^c	1.55 ± 0.04	0.401 ± 0.053^e

[a] Drugs were given s.c., once a day for a week after a 600-R irradiation, and the biological activity of NGF in SubG was determined in an organ culture assay using chick embryonic dorsal root ganglia, and TH activity by a radioassay using ^{14}C-labeled L-tyrosine as a substrate.
[b] GRb-1: Ginsenoside R_{b-1}, orally administered.
[c] Student's t-test, $p < .01$.
[d] Student's t-test, $p < .05$.
[e] Significant difference between irradiated control and the drug-tested mice, $p < .05$.

XI. BIOCHEMICAL STUDIES ON GINSENG

A. EFFECTS OF GINSENG SAPONINS ON LIVER RNA BIOSYNTHESIS AND SERUM PROTEIN METABOLISM

Oura *et al.* (1971a) screened the effects of Chinese drugs on nuclear RNA biosynthesis in rat liver under intraperitorial administration of drug extracts (corresponding to 0.15 g of drug) using radioactive [^{14}C]orotic acid (2.5–7.0 μci/rat) as a tracer. Among 18 kinds of crude drugs that have been used in Chinese medicine as restorative, ginseng shows the most remarkable effects on incorporation (50–60%) of [^{14}C]orotic acid in the rat liver RNA.

On fractionation of these active principles of ginseng, Oura *et al.* (1972a,b) obtained F_3 and F_4 fractions, which contain 50 and 92% saponins, respectively. The activation of nuclear RNA biosynthesis by F_3 and F_4 fractions has been shown to be followed by the promotion of biosyntheses of ribosome RNA and messenger RNA, activation of RNA polymerase, increase of polysome, and activation of biosynthesis of protein, revealing activation of every step of biochemical reactions in the protein biosynthesis by ginseng saponins (Oura *et al.*, 1971a,b, 1972a,b, Nagasawa *et al.*, 1977).

Using pure ginsenosides, Oura *et al.* (1975) proved that incorporation of [^{3}H]leucine into serum protein was most remarkably promoted by ginsenoside R_d, while other members of ginsenosides appeared to be more or less effective (Table XVI).

TABLE XVI

EFFECT OF GINSENOSIDE ON THE INCORPORATION^a OF [^{3}H]LEUCINE INTO SERUM PROTEIN

Ginsenoside R_X	Dose (mg)	Number of Mice	[^{3}H]Leucine (cpm/mg protein)	Incorporation (%)
b-1	2	9	1.142 ± 42	106
b-2	2	10	1.320 ± 58	122
c	2	8	1.310 ± 36	121
d	0.1	5	1.410 ± 119	130
d	2	9	1.539 ± 85	147
e	2	10	1.430 ± 80	132
g-1	2	10	1.294 ± 79	120
Control (saline)	—	20	1.082 ± 38	100

^{a}Measured 6 hr after administration (i.p.) of ginsenosides; [^{3}H]leucine (5 μCi per mouse) as 1-hr pulse.

On oral administration of ginseng extracts to rats for 4 weeks, the amount of rough endoplasmic reticulum of hepatocyte was increased, as observed by electron microscope. This result was confirmed by a colorimetric determination of the membrane-bound ribosomes separated from the liver of rats fed with Ginseng extracts.

Iijima *et al.* (1979) and Iijima and Higashi (1979) determined the incorporation of [^{3}H]orotic acid into the liver nuclear RNA in the rat 4 hr after the i.p. administration of ginsenosides and found that R_{b-1} increased the incorporation 18%, whereas R_c decreased it 14%.

The incorporation of [^{3}H]cytidine triphosphate into liver nuclear RNA in rats was also promoted by ginsenoside R_{b-1} (+33%), while it was decreased by ginsenoside R_c (−16%).

The effects of ginsenoside on serum protein biosynthesis in rats were also studied by Y. Shibata *et al.* (1976) via incorporation of [^{14}C]leucine (10 μCi/100 g i.p.) to find that R_{b-1} promoted the incorporation level to 2.2 times and R_c to 1.9 times that of the control, while R_{g-1} gave no remarkable result.

B. EFFECTS OF GINSENG SAPONINS ON CHOLESTEROL METABOLISM

Yamamoto *et al.* (1977) reported that the cholesterol and triglycerides levels in blood plasma in male Sprague–Dawley rats fed a high-cholesterol diet containing 1% cholesterol and 0.5% cholic acid were significantly decreased by the intramuscular administration of ginseng saponins (fraction 4: 1 mg/100 g·day for 14 days). On the other hand, the biosynthesis of cholesterol was increased by ginseng saponins, while the turnover of cholesterol into bile acids and the excretion of cholesterol were also promoted to keep the lower level of cholesterol in blood.

Yokozawa *et al.* (1982) found that ginsenoside R_{b-2} remarkably decreased the total cholesterol and low-density lipoprotein (LDL) cholesterol in the blood of rats fed a high-cholesterol (1%) diet, while it significantly increased high-density lipoprotein (HDL) cholesterol.

Sakakibara *et al.* (1975) reported a similar line of work on the effects of pure ginsenosides in cholesterol metabolism, using [^{14}C]acetate incorporation, to reveal that ginsenoside R_{b-1} promoted significantly the biosynthesis of cholesterol in the liver and serum of rats.

Using liver slices of rats sacrificed 2.5 hr after feeding of 5 mg of ginsenosides, the incorporation of [^{14}C]acetate into cholesterol was determined. Gommori *et al.* (1976) demonstrated in this experiment a remarkable enhancement of cholesterol biosynthesis (4.15 times) by ginsenoside R_{b-1}. The incubation of normal liver slice of rat with ginsenoside R_{b-1} *in vitro* gave no effect on [^{14}C]acetate incorporation into cholesterol. Therefore, there is some intermediate metabolic mechanism for inducing the action of ginsenosides in the biosynthesis of cholesterol.

Joo (1980) also demonstrated that ginsenosides promoted the metabolism of cholesterol in rabbits fed a high-cholesterol diet, and lowered blood cholesterol level to prevent aortic atheroma formation.

C. EFFECTS OF GINSENG SAPONINS ON SUGAR AND LIPID METABOLISM

On administration of crude saponins of ginseng (fraction 3) to the adrenalectomized rats, Yokozawa *et al.* (1975) observed acute hypoglycemia.

In the normal rats, ginseng saponins gave no such a sudden reduction of blood sugar level, but showed a remarkable increase of lipid accumulation in liver and epididymal adipose tissue and a decrease of liver glycogen. However, the decrease in liver glycogen caused by ginseng saponin administration varied under different nutritional conditions. It suggested that hormonal control could act to mediate the action of ginseng saponins.

D. ENDOCRINOLOGICAL EFFECTS OF GINSENG SAPONINS

The biochemical studies on ginseng extracts and saponins suggested that their indirect actions might be mediated by some hormonal systems. On the basis of his behavioural pharmacological studies on ginseng, Brekhman proposed calling ginseng an "adaptogen," since it potentiates nonspecific resistance against stress (Brekhman and Dardvmov, 1969a,b). Some other naturally occurring drugs that might be expressed in traditional Chinese medicine as restoring Yang or harmony, such as wujiapi (the root bark of *Eleutheroccus senticosus*) and tu-chung (the bark of *Eucommia ulmoides*) might be included in this group.

As the characteristic properties of an "adaptogen," Brekhman referred to (1) antifatigue action against physical and mental loads, (2) nonspecific resistance against stress, and (3) normalization from an abnormal state, which might be caused by some excess or deficient physiological factors.

These properties, especially (2), would suggest corticosteroidal action. According to Fulder (1978), the concept of "adaptogen" leaves behavioural and antifatigue effects of ginseng unexplained, and he claimed that ginseng is a harmony drug that restores the body from the invasion of various stresses by stimulating the hypothalamic–pituitary–adrenal axis.

The behavioural and antifatigue effects of ginseng might relate to this system. Hiai *et al.* (1979a,b) showed that ginseng saponins (fraction 5) increased adrenal cyclic adenosine monophosphate (cAMP) in intact rats, but not in hypophysectomized rats. This result revealed that ginseng saponins react directly with the hypothalamus or hypophysis (pituitary) to secrete adrenocorticotrophic hormone (ACTH), which stimulates the adrenal cortex. The plasma ACTH of male Wistar rats administered ginseng saponins i.p. (fraction 5; ginsenosides R_{b-1}, R_{b-2}, R_c, R_d, R_e, and R_{g-1}) was determined by radioimmunoassay. Plasma ACTH was significantly increased by feeding ginsenosides, while the plasma corticosterones were also increased in a parallel kinetic pattern. Since this increase of ACTH and corticosterones was blocked by dexamethazone, which acts at the pituitary and hypothalamus, ginsenosides react primarily on the hypothalamus or pituitary to secrete ACTH, which first stimulates cAMP in the adrenal, and then promotes corticosteroid synthesis. Bombardelli *et al.* (1980) pursued experiments to study the acute and chronic effects of a pure ginseng saponin mixture, whose ginsenosides composition was acurately determined, on the adrenal functions of rats. Ginseng saponins counteracted significantly the body temperature decline that would be caused by the stress of exposure to cold, without affecting blood glucose and corticosterone levels. In the adrenalectomized rats, ginseng saponins had no significant effects, but the rats' ability to maintain their temperature when exposed to cold was revived by the exogenous administration of hydrocortisone. The sign of hyperfunction was observed histologically in the supraoptic and paraventicular nuclei of hypothalamus of the rats fed ginseng saponins. A remarkable increase of haematoxylin-stainable, basophilic cells in the adrenohypophysis, which produce ACTH, was also found. In the adrenal of the rats treated with ginseng saponins, hyperplasia of the zona fasciculata was observed, indicating that the hyperfunction of adrenal was promoted by the ginseng saponins. These results also revealed the stimulating action of ginseng saponins on the function of pituitary–adrenal axis.

Some pharmacological and biochemical activities of ginseng and ginseng saponins, such as antifatigue action and promotion of selectivity and discrimination, might be ascribed to the effect of ACTH and cortical steroids whose secretions are induced by them.

In comparison with the effects of usual stimulants, the antifatigue action of ginseng shows an essential difference. The stimulants give effects under most situations, whereas ginseng reveals its action only under the challenge of stresses.

From these experimental results, in physiological actions the cortical steroids could be replaced by ginseng and ginseng saponins. However, contrary to exogenous administration of corticoids, the feeding of ginseng does not cause any shrinkage of the adrenal, but potentiates its function (Tanizawa *et al.*, 1981).

It has been known that corticosteroids and ACTH bind directly to brain tissue to increase mental activities during stress (De Wied *et al.*, 1976). Fulder (1981) demonstrated that corticosterones labeled with tritium were bound to nuclei in different brain regions—septum, hippocampus, hypothalamus, amygdala, and pituitary—of the adrenalectomized and ovarectomized rats severalfold more in the group of animals administered ginseng saponin mixture than in controls. Fulder suggested, as one of the reasons, that the ginseng saponins would give a specific effect on cell membrane of hypothalmus leading to a more rapid passive diffusion of corticosterones across the membrane. Such an effect would occur also at the membrane of adrenal gland independently from the function of ACTH.

The endocrinological concepts of ginseng and ginseng saponins suggest that ginseng is not a remedy for curing any certain disease, but rather that it acts through the hypothalamic–pituitary–adrenal axis to adjust metabolic and functional systems governing hormonal control of homeostasis against the challenge of stresses. This might agree with the essential principle of traditional Chinese medicine of restoration of Yang or recovery of physical and mental balance.

E. CLINICAL TRIALS OF GINSENG

Some clinical trials of red ginseng powder were performed by Yamamoto and Uemura (1980) in eight normal volunteers and three diabetic patients to investigate its effects on plasma levels of various hormones and lipids. Single oral administrations of 4.5 to 6.0 g of ginseng powder reduced plasma glucose, while plasma-immunoreactive insulin (IRI) and glucagon were remained unchanged. Plasma cortisol was increased in the normals subjects, but decreased in the diabetics, while ACTH in the plasma was increased in both cases. Effects on plasma adrenaline showed either increase or decrease, while noradrenaline was slightly decreased. Gastrin was especially increased in the diabetic patients.

One week administration of ginseng powder (4.5 g/day for 7 days, p.o.) in five volunteers and six hyperlipemic patients gave the following effects: significant decline of glucose level in plasma; no change in the levels of IRI and cortisol; increase of plasma gastrin, HDL cholesterol, and triglyceride; and decline of atherogenic index.

Chang *et al.* (1978) reported a clinical trial of ginsenoside R_{g-1} on the postoperative gynecological patients. Ginsenoside R_{g-1} (0.23 g) administered daily to 60 patients for 3 weeks after surgery, and placebo was given to 60 patients as control. The results were as follows: no side effects were observed; hemoglobin and hematocrit were increased more in the treated group than in the controls; serum total protein was more increased in treated group than in control; serum glucose level was significantly decreased in both groups; body weight was significantly increased in treated group.

This would have been the first clinical trial using a pure ginseng saponin. Apart from classical medical use of ginseng, most modern clinical studies on ginseng have been made so far, by using crude extracts or a mixture of saponins, which might sometimes yield ambiguous results.

A report presented by Siegel (1979) on the ginseng abuse syndrome, for example, was made using various kinds of ginseng preparations of different origins, including not only Korean ginseng but also American ginseng and an *Eleutherococcus* preparation ambiguously named Siberian ginseng. Among the subjects who showed some side effects even in taking the real ginseng there would be those who are Yang-dominant. This is obviously abuse of ginseng from the viewpoint of traditional Chinese medicine.

On the basis of fundamental scientific works on ginseng and its principles described in this chapter, further clinical trials will be developed in the future using well-defined pure ginsenosides or other constituents of ginseng to evaluate their efficacies more precisely.

A number of papers have been published since 1960 on the chemistry, biochemistry, pharmacology, endocrinology, and some clinical studies on ginseng and its chemical principles, other than those described in this chapter. In the following four publications, there are almost complete compilations of references of scientific works. Those who may have further interest in ginseng studies could find reliable informations in Bae (1978), Research Institute Office of Monopoly Korea (1975), Fulder (1980), and Oura *et al.* (1981).

REFERENCES

Ando, T., Tanaka, O., and Shibata, S. (1971). *Shoyakugaku Zasshi* **25,** 28.
Aoyama, S. (1929). *Yakugaku Zasshi* **49,** 678.

Aoyama, S. (1930). *Yakugaku Zasshi* **50**, 1076, 1163.

Asahina, Y., and Taguchi, B. (1906). *Yakugaku Zasshi* **26**, 549.

Bae, H. -W. (1978). "Korean Ginseng." Korean Ginseng Res. Inst., Seoul, Korea.

Biellmann, J. F. (1966). *Tetrahedron Lett.* p. 4803.

Besso, H., Saruwatari, Y., Futamura, K., Kunihiro, K., Fuwa, T., and Tanaka, O. (1979). *Planta Med.* **37**, 226.

Besso, H., Saruwatari, Y., Fuwa, T., Kasai, R., and Tanaka, O. (1981). *Abstr. Pap., 28th Annu. Meet. Jpn. Soc. Pharmacogn.* p. 53.

Besso, H., Saruwatari, Y., Fuwa, T., Kasai, R., Taniyasu, S., and Tanaka, O. (1982a). *Abstr. Pap., 102nd Annu. Meet. Pharm. Soc. Jpn.* p. 531.

Besso, H., Kasai, R., Wei, J.-X., Wang, J.-F., Saruwatari, Y., Fuwa, T., and Tanaka, O. (1982b). *Chem. Pharm. Bull.* **30**, 4534.

Bombardelli, E., Cristomi, A., and Lietti, A. (1980). *Proc. Int. Ginseng Symp., 3rd, 1980* p. 9.

Brekhman, I. I. (1957). "Zen-shen." State Publ. House Med. Lit., Leningrad.

Brekhman, I. I., and Dardymov, I. V. (1969a). *Lloydia* **32**, 46.

Brekhman, I. I., and Dardymov, I. V. (1969b). *Annu. Rev. Pharmacol.* **9**, 419.

Chang, Y.-S., Lee, J.-Y., and Kim, Ch.-W. (1978). *Proc. Int. Ginseng Symp., 2nd, 1988* p. 79.

Chen, S. E., and Staba, E. J. (1978). *Lloydia* **41**, 361.

Chen, S. E., Staba, E. J., Taniyasu, S., Kasai, R., and Tanaka, O. (1981). *Planta Med.* **42**, 406.

Dabrowski, Z., Wrobel, J. T., and Wojtasiewicz, K. (1980). *Phytochemistry* **19**, 2464.

Davydow, L. (1889). *Pharm. Z. Russland* **29**, 97.

De Wied, D., Bolms, B., Gispen, W. H., Urban, I., and Greidanus, Tj-B. R.W. (1976). *In* "Hormones Behaviour and Psychopathology" (E. J. Sachen, ed.), pp. 1–14. Raven Press, New York.

Elyakov, G. B., Striena, L. T., Churlin, A. Y., and Kochetkov, N. K. (1962). *Izv. Akad. Nauk SSSR* p. 1125.

Elyakov, G. B., Stringa, L. I., Uvarova, N. I., Vaskovsky, V. E., Dzizenko, A. K., and Kochetkov, N. K. (1964). *Tetrahedron Lett.* p. 3591.

Elyakov, G. B., Strigna, L. I., and Kochetkov, N. K. (1965). *Khim. Prir. Soedin.* p. 149; *Chem. Abstr.* **63**, 16444 (1965).

Elyakov, G. B., Stringna, L. I., Shapkina, E. V., Aladyina, N. T., and Kornilova, S. A. (1968). *Tetrahedron* **24**, 5483.

Fischer, F. G., and Seiler, N. (1959). *Justus Liebigs Ann. Chem.* **626**, 185.

Fischer, F. G., and Seiler, N. (1961a). *Justus Liebigs Ann. Chem.* **644**, 146.

Fischer, F. G., and Seiler, N. (1961b). *Justus Liebigs Ann. Chem.* **644**, 162.

Fujita, M., Itokawa, S., and Shibata, S. (1962). *Yakugaku Zasshi* **82**, 1634, 1638.

Fulder, S. J. (1978). *Proc. Int. Ginseng Symp., 2nd, 1978* p. 25.

Fulder, S. J. (1980). "The Root of Being. Ginseng and the Pharmacology of Harmony." Hutchinson, London.

Fulder, S. J. (1981). *Am. J. Clin. Med.* **9**, 112.

Furuya, T., and Ishii, T. (1973). *Japanese Patent* **48**, 13917.

Furuya, T., and Ishii, T. (1981). *Japanese Patent* **56**, 12119.

Furuya, T., Kojima, H., Syono, K., and Ishii, T. (1970). *Chem. Pharm. Bull.* **18**, 2371.

Furuya, T., Kojima, H., Syono, K., Shii, T., Uotani, K., and Nishio, M. (1973). *Chem. Pharm. Bull.* **21**, 98.

Garriques, S. (1854). *Ann. Chem. Pharm.* **90**, 231.

Gommori, K., Miyamoto, F., Shibata, Y., Higashi, T., Sanada, S., and Shoji, J. (1976). *Chem. Pharm. Bull.* **24**, 2985.

Gstirner, F., and Vogt, H. J. (1966). *Arch. Pharm. (Weinheim. Ger.)* **299**, 936.

Hiai, S., Oura, H., Tsukada, K., and Hirai, Y. (1971). *Chem. Pharm. Bull.* **19**, 1656.
Hiai, S., Yokozawa, H., Oura, H. and Yano, S. (1979a). *Endocrinol. Jpn.* **26**, 661.
Hiai, S., Sasaki, S., and Oura, H. (1979b). *Planta Med.* **37**, 15.
Hiyama, C., Miyai, S., Yoshida, H., Yamazaki, K., and Tanaka, O. (1978). *Yakugaku Zasshi* **98**, 1132.
Hong, S. A., Park, C. W., Kim, J. H., Chang, H. K., Hong, S. K., and Kim, M. S. (1974). *Korean J. Pharmacol.* **10**, 1.
Hong, S. A., Park, C. W., Kim, J. H., Chang, H. K., Hong, S. K., and Kim, M. S. (1975). *Proc. Int. Ginseng Symp., 1st, 1975* p. 33.
Hörhammer, L., Wagner, H., and Lay, B. (1961). *Pharm. Ztg.* **106**, 1307.
Hwang, W. T. (1960). *Bull. Chonnam Univ.* **5**, 425.
Iida, Y., Tanaka, O., and Shibata, S. (1968). *Tetrahedron Lett.* p. 5449.
Iijima, M., and Higashi, T. (1979). *Chem. Pharm. Bull.* **27**, 2130.
Iijima, M., Higashi, T., Sanada, S., and Shoji, J. (1976). *Chem. Pharm. Bull.* **24**, 2400.
Iljin, S. G., Dzizenko, A. K., and Elyakov, G. B. (1978). *Tetrahedron Lett.* p. 593.
Joo, C. N. (1980). *Proc. Int. Ginseng Symp., 3rd, 1980* p. 27.
Kaizuka, H., and Takahashi, K. (1983). *J. Chromatogr.* **258**, 135.
Kaku, T., and Kawashima, Y. (1977). *Rep. Yamanouchi Cent. Res. Lab.* No. 3, p. 45.
Kasai, R., Matsuura, K., Tanaka, O., Sanada, S., and Shoji, J. (1977a). *Chem. Pharm. Bull.* **25**, 3277.
Kasai, R., Asakawa, J., Yamasaki, K., and Tanaka, O. (1977b). *Tetrahedron* **33**, 1935.
Kasai, R., Suzuo, M., Asakawa, J., and Tanaka, O. (1977c). *Tetrahedron Lett.* p. 175.
Kasai, R., Okihara, M., Asakawa, J., Mizutani, K., and Tanaka, O. (1979). *Tetrahedron* **35**, 1427.
Kawashima, K. (1981). *J. Pharm. Dyn.* **4**, 534.
Kim, J. Y., and Staba, E. J. (1973). *Korean J. Pharmacogn.* **4**, 193.
Kim, J. Y., and Staba, E. J. (1975). *Proc. Int. Ginseng Symp., 1st, 1975* p. 77.
Kitagawa, H., and Iwaki, R. (1963). *Folia Pharmacol. Jpn.* **59**, 348.
Kitagawa, O., Yoshikawa, M., Hayashi, T., Taniyama, T., and Arichi, S. (1981). *Abstr. Pap., 28th Annu. Meet. Jpn. Soc. Pharmacogn.* p. 52.
Kitasato, Z., and Sone, C. (1932). *Acta Phytochim.* **6**, 218.
Klyne, W. (1950). *Biochem. J.* **47**, xli.
Komatsu, M., and Tomimori, T. (1966). *Shoyakugaku Zasshi* **20**, 21.
Komatsu, M., Tomimori, T., and Makiguchi, Y. (1966). *Yakugaku Zasshi* **89**, 122.
Komori, T., Tanaka, O., and Nagai, Y. (1974). *Org. Mass Spectrom.* **9**, 744.
Komori, T., Kawamura, M., Miyahara, K., Kawasaki, T., Tanaka, O., and Yahara, S. (1979). *Z. Naturforsch., C. Biosci.* **34C**, 1094.
Kondo, H., and Amano, U. (1920). *Yakugaku Zasshi* **40**, 1027.
Kondo, H., and Tanaka, O. (1915). *Yakugaku Zasshi* **35**, 779.
Kondo, H., and Yamaguchi, S. (1918). *Yakugaku Zasshi* **38**, 747.
Kondo, N., and Shoji, J. (1962). *Yakugaku Zasshi* **88**, 325.
Kondo, N., and Shoji, J. (1975). *Chem. Pharm. Bull.* **23**, 3282.
Kondo, N., Aoki, K., Ogawa, H., Kasai, R., and Shoji, J. (1970). *Chem. Pharm. Bull.* **18**, 1558.
Kondo, N., Shoji, J., and Tanaka, O. (1973). *Chem. Pharm. Bull.* **21**, 2702.
Kondo, N., Shoji, J., Nagumo, N., and Komatsu, N. (1969). *Yakugaku Zasshi* **89**, 846.
Kotake, M. (1930). *Nippon Kagaku Kaishi* **51**, 357.
Kotake, M., and Kimoto, Y. (1932). *Sci. Pap. Inst. Phys. Chem. Res. (Jpn.)* **18**, 83.
Kotake, M., Matsubara, T., and Kimoto, Y. (1930). *Nippon Kagaku Kaishi* **51**, 396.
Lee, T. M., and Marderosian, A. D. (1981). *J. Pharm. Sci.* **70**, 89.

Lee, W. C., Chang, W. S., and Lee, S. K. (1960). *Ch'oesin Uihak* **3,** 37.

Lin, Y.-T. (1961). *J. Chin. Chem. Soc. (Taipei)* [2] **8,** 109, *Chem. Abstr.* **58,** 4374 (1963).

Lui, J. H., and Staba, E. (1980). *Lloydia* **43,** 340.

Matsuura, H., Kasai, R., Tanaka, O., Saruwatari, Y., Fuwa, T., and Zhou, J. (1983). *Chem. Pharm. Bull.* **31,** 2281.

Mills, J. S. (1956a). *J. Chem. Soc.* p. 2196.

Mills, J. S., and Werner, A. E. A. (1955). *J. Chem. Soc.* p. 3132.

Mizutani, K., Kasai, R., and Tanaka, O. (1980). *Carbohydr. Res.* **87,** 19.

Morita, T., Kasai, R., Tanaka, O., Zhou, J., Yang, T.-R., and Shoji, J. (1982). *Chem. Pharm. Bull.* **30,** 4341.

Morita, T., Kasai, R., Kohda, H., Tanaka, O., Zhou, J., and Yang, T. (1983). *Chem. Pharm. Bull.* **31,** 3205.

Murayama, Y., and Itagaki, T. (1923). *Yakugaku Zasshi* **43,** 783.

Murayama, Y., and Tanaka, S. (1927). *Yakugaku Zasshi* **47,** 526.

Nagai, M., Tanaka, O., and Shibata, S. (1966). *Tetrahedron Lett.* p. 4797.

Nagai, M., Ando, T,, Tanaka, O., and Shibata, S. (1967). *Tetrahedron Lett.* p. 3579.

Nagai, M., Tanaka, O., and Shibata, S. (1971). *Chem. Pharm. Bull.* **19,** 2349.

Nagai, M., Ando, T., Tanaka, O., and Shibata, S. (1972). *Chem. Pharm. Bull.* **20,** 1212.

Nagai, Y., Tanaka, O., and Shibata, S. (1971). *Tetrahedron* **27,** 881.

Nagasawa, T., Oura, H., Hiai, S., and Nishinaga, K. (1977). *Chem. Pharm. Bull.* **25,** 1665.

Nagasawa, T., Yokozawa, T., Nishino, Y., and Oura, H. (1980a). *Chem. Pharm. Bull.* **28,** 2059.

Nagasawa, T., Choi, J. H., Nishio, Y., and Oura, H. (1980b). *Chem. Pharm. Bull.* **28,** 3701.

Ohsawa, T., Tanaka, N., Tanaka, O., and Shibata, S. (1972). *Chem. Pharm. Bull.* **20,** 1890.

Otsuka, H., Morita, Y., Ogihara, Y., and Shibata, S. (1977). *Planta Med.* **32,** 9.

Oura, H., Hiai, S., Nakashima, S., and Tsukada, K. (1971a). *Chem. Pharm. Bull.* **19,** 453.

Oura, H., Hiai, S., and Seno, H. (1971b). *Chem. Pharm. Bull.* **19,** 1598.

Oura, H., Tsukada, K., and Nakagawa, H. (1972a). *Chem. Pharm. Bull.* **20,** 219.

Oura, H., Nakashima, S., Tsukada, K., and Ohta, Y. (1972b). *Chem. Pharm. Bull.* **20,** 980.

Oura, H., Hiai, S., Odaka, Y., and Yokozawa, T. (1975). *J. Biochem. (Tokyo)* **77,** 1057.

Oura, H., Kumagai, A., Shibata, S., and Takagi, K., eds. (1981). "Yakuyoninjin (Recent Studies on Ginseng)." Kyoritsu Publ. Co., Tokyo (in Japanese).

Petkov, W. (1959). *Arzneim.-Forsch.* **9,** 305.

Petkov, W. (1961). *Arzneim.-Forsch.* **11,** 288, 418.

Petkov, W. (1968). *Pharm. Ztg.* **113,** 1281.

Petkov, W., and Staneva-Stoicheva, D. (1963). *Proc. Int. Pharmacol. Congr., 2nd, 1963* pp. 39–45.

Poplawski, J., Wrober, J. T., and Glinka, T. (1980). *Phytochemistry* **19,** 1539.

Research Institute Office of Monopoly Korea (1975). "Abstracts of Korean Ginseng Studies (1687–1975)." Res. Inst. OSP. Monopoly Korea, Seoul, Korea.

Saito, H. (1980). *Proc. Int. Ginseng Symp., 3rd, 1980* p. 181.

Saito, H., and Lee, Y. M. (1978). *Proc. Int. Ginseng Symp., 2nd, 1978* p. 109.

Saito, H., and Nabata, H. (1972). *Jpn. J. Pharmacol.* **22,** 245.

Saito, H., Morita, M., and Takagi, K. (1973a). *Jpn. J. Pharmacol.* **23,** 43.

Saito, H., Tsuchiya, M., and Takagi, K. (1973b). *Jpn. J. Pharmacol.* **23,** 49.

Saito, H., Tsuchiya, M., and Takagi, K. (1974a). *Jpn. J. Pharmacol.* **24,** 119.

Saito, H., Yoshida, Y., and Takagi, K. (1974b). *Jpn. J. Pharmacol.* **24,** 119.

Saito, H., Tsuchiya, M., Naka, S., and Takagi, K. (1977a). *Jpn. J. Pharmacol.* **27,** 509.

Saito, H., Lee, Y.-M., Takagi, K., Shibata, S., Shoji, J., and Kondo, N. (1977b). *Chem. Pharm. Bull.* **25,** 1017.

Saito, H., Lee, Y.-M., and Handa, S. (1980). *In* "Recent Advances in Catecholamines and

Stress" (E. Usdin, R. Kvetnanski, and K. P. Kopin, eds.), p. 371. Elesvier/North-Holland, Amsterdam.

Sakai, K. (1915). *Toikaishi* **30**, 935.

Sakakibara, K., Shibata, Y., Higashi, T., Sanada, S., and Shoji, J. (1975). *Chem. Pharm. Bull.* **23**, 1009.

Sakamoto, M., Morimoto, K., and Tanaka, O. (1975). *Yakugaku Zasshi* **95**, 1456.

Sanda, S., and Shoji, J. (1978a). *Chem. Pharm. Bull.* **26**, 1694.

Sanada, S., and Shoji, J. (1978b). *Shoyakugaku Zasshi* **32**, 96.

Sanada, S., Kondo, N., Shoji, J., Tanaka, O., and Shibata, S. (1974a). *Chem. Pharm. Bull.* **22**, 421.

Sanada, S., Kondo, N., Shoji, J., Tanaka, O., and Shibata, S. (1974b). *Chem. Pharm. Bull.* **22**, 2407.

Sanada, S., Shoji, J., and Shibata, S. (1978). *Yakugaku Zasshi* **90**, 1048.

Sankawa, U., Sung, C.-K., Han, B.-H., Akiyama, T., and Kawashima, K. (1982). *Chem. Pharm. Bull.* **30**, 1907.

Shibata, S., Fujita, M., Itokawa, H., Tanaka, O., Itokawa, H., Tanaka, O., and Ishii, T. (1962a). *Tetrahedron Lett.* p. 419.

Shibata, S., Tanaka, O., Nagai, M., and Ishii, T. (1962). *Tetrahedron Lett.* p. 1239.

Shibata, S., Fujita, M., Itokawa, H., Tanaka, O., and Ishii, T. (1963a). *Chem. Pharm. Bull.* **11**, 759.

Shibata, S., Fujita, M., Itokawa, H., Tanaka, O., Itokawa, H., Tanaka, O., and Ishii, T. (1963b). *Chem. Pharm. Bull.* **11**, 759.

Shibata, S., Tanaka, O., Nagai, M., and Ishii, T. (1963c). *Chem. Pharm. Bull.* **11**, 762.

Shibata, S., Tanaka, O., Sado, M., and Tsushima, S. (1963d). *Tetrahedron Lett.* p. 795.

Shibata, S., Tanaka, O., Ando, T., Sado, M., Tsushima, S., and Ohsawa, T. (1966a). *Chem. Pharm. Bull.* **14**, 595.

Shibata, S., Ando, T., and Tanaka, O. (1966b). *Chem. Pharm. Bull.* **14**, 1157.

Shibata, S., Ando, T., Tanaka, O., Meguro, Y., Soma, K,, and Iida, Y. (1971). *Shoyakugaku Zasshi* **25**, 28.

Shibata, Y., Nozaki, T., Higashi, T., Sanada, S., and Shoji, J. (1976). *Chem. Pharm. Bull.* **24**, 2818.

Siegel, R. K. (1979). *JAMA, J. Am. Med. Assoc.* **241**, 1614.

Sticher, O., and Soldati, F. (1979). *Planta Med.* **36**, 30.

Sticher, O., and Soldati, F. (1980). *Planta Med.* **39**, 348.

Takagi, K. (1975). *Proc. Int. Ginseng Symp., 1st, 1974,* p. 119.

Takagi, K., Saito, H., and Tsuchiya, M. (1972). *Jpn. J. Pharmacol.* **22**, 339.

Takagi, K., Saito, H., and Tsuchiya, M. (1974). *Jpn. J. Pharmacol.* **24**, 41.

Takahashi, M., and Yoshikura, M. (1964). *Yakugaku Zasshi* **84**, 757.

Takahashi, M., and Yoshikura, M. (1966). *Yakugaku Zasshi* **86**, 1051, 1053.

Takahashi, M., Isoi, K., Yoshikura, M., and Osugi, T. (1961). *Yakugaku Zasshi* **81**, 771.

Takahashi, M., Isoi, K., Kumura, Y., and Yoshioka, M. (1962). *Yakugaku Zasshi* **82**, 752.

Takemoto, T., Arihara, S., Nakajima, T., and Okuhira, M. (1983). *Yakugaku Zasshi* **103**, 173.

Tanaka, O., and Yahara, S. (1978). *Phytochemistry* **17**, 1353.

Tanaka, O., Nagai, M., and Shibata, S. (1964). *Tetrahedron Lett.* p. 2291.

Tanaka, O., Nagai, M., and Shibata, S. (1966). *Chem. Pharm. Bull.* **14**, 1150.

Tanaka, O., Nagai, M., Ohsawa, T., Tanaka, N., and Shibata, S. (1967). *Tetrahedron Lett.* p. 391.

Taniyasu, S., Tanaka, O., Yang, T. R., and Zhou, J. (1982). *Planta Med.* **44**, 124.

Tanizawa, H., Numano, H., Kotani, D., Takino, Y., Hayashi, T., and Arichi, S. (1981). *Yakugaku Zasshi* **101**, 169.

Wagner-Jauregg, Th., and Roth, M. (1962). *Pharm. Acta Helv.* **37,** 352.

Woo, L. K., Han, B. H., Baik, D. W., and Park, D. (1973). *Yakhak Hoe Chi* **17,** 129.

Wood, W. B., Roh, B. L., and White, R. P. (1964). *Jpn. J. Pharmacol.* **14,** 284.

Yahara, S., Tanaka, O., and Komori, T. (1976a). *Chem. Pharm. Bull.* **24,** 2204.

Yahara, S., Matsuura, K., Kasai, R., and Tanaka, O. (1976b). *Chem. Pharm. Bull.* **24,** 3212

Yahara, S., Kasai, R., and Tanaka, O. (1977). *Chem. Pharm. Bull.* **25,** 2041.

Yahara, S., Tanaka, O., and Nishioka, I. (1978). *Chem. Pharm. Bull.* **26,** 3010.

Yahara, S., Kaji, K., and Tanaka, O. (1979). *Chem. Pharm. Bull.* **27,** 88.

Yamamoto, M., and Uemura, T. (1980). *Proc. Int. Ginseng Symp., 3rd 1980* p. 115.

Yamamoto, M., Hayashi, Y., and Kumagai, A. (1977). *Proc. Symp. Wakan-yaku* **10,** 90.

Yang, T. R., Kasai, J., Zhou, J., and Tanaka, O. (1983). *Phytochem.* **22,** 1473.

Yokozawa, T., Seno, H., and Oura, H. (1975). *Chem. Pharm. Bull.* **23,** 3095.

Yokozawa, T., Izawa, K., Kobayashi, T., Oura, H., and Yamamoto, M. (1982). *Abstr. Pap.,*
    102nd Annu. Meet. Pharm. Soc. Jpn. p. 193.

Zhou, J., Wu, M.-Z., Tanyasu, S., Besso, H., Tanaka, O., Saruwatari, Y., and Fuwa, T.
    (1981). *Chem. Pharm. Bull.* **29,** 2844.

Index